Sustainable Development Goals Series

The **Sustainable Development Goals Series** is Springer Nature's inaugural cross-imprint book series that addresses and supports the United Nations' seventeen Sustainable Development Goals. The series fosters comprehensive research focused on these global targets and endeavours to address some of society's greatest grand challenges. The SDGs are inherently multidisciplinary, and they bring people working across different fields together and working towards a common goal. In this spirit, the Sustainable Development Goals series is the first at Springer Nature to publish books under both the Springer and Palgrave Macmillan imprints, bringing the strengths of our imprints together.

The Sustainable Development Goals Series is organized into eighteen subseries: one subseries based around each of the seventeen respective Sustainable Development Goals, and an eighteenth subseries, "Connecting the Goals," which serves as a home for volumes addressing multiple goals or studying the SDGs as a whole. Each subseries is guided by an expert Subseries Advisor with years or decades of experience studying and addressing core components of their respective Goal.

The SDG Series has a remit as broad as the SDGs themselves, and contributions are welcome from scientists, academics, policymakers, and researchers working in fields related to any of the seventeen goals. If you are interested in contributing a monograph or curated volume to the series, please contact the Publishers: Zachary Romano [Springer; zachary.romano@springer.com] and Rachael Ballard [Palgrave Macmillan; rachael.ballard@palgrave.com].

Güliz Karaarslan-Semiz

Editor

Education for Sustainable Development in Primary and Secondary Schools

Pedagogical and Practical Approaches for Teachers

 Springer

Editor
Güliz Karaarslan-Semiz
Department of Mathematics and Science Education
Ağrı İbrahim Çeçen University
Ağrı, Türkiye

The content of this publication has not been approved by the United Nations and does not reflect the views of the United Nations or its officials or Member States.

ISSN 2523-3084   ISSN 2523-3092   (electronic)
Sustainable Development Goals Series
ISBN 978-3-031-09114-8   ISBN 978-3-031-09112-4   (eBook)
https://doi.org/10.1007/978-3-031-09112-4

This Springer imprint is published by the registered company Springer Nature Switzerland AG
The registered company address is: Gewerbestrasse 11, 6330 Cham, Switzerland

To all teachers working for a sustainable world

As if Everything Depends on It: Educating for Life

This book was written as the Covid-19 pandemic was causing death and economic havoc throughout the Earth. The planet itself is becoming hotter, more capricious, and biologically more impoverished. Governments of, by, and for the people everywhere are under assault by autocrats, oligarchs, and anti-democratic forces. On the brink of irrevocable catastrophe, violence and economic injustice seem immovable. What appear to be separate crises, however, are parts of a single global crisis with various manifestations, which is to say, the working out of the rules of a system built on exploitation of people and nature alike maintained by military power. Their goal is economic growth come hell or high water, both quite likely; the drivers are the overabundance of greed and ignorance. The consequences are beyond measure.

An antidote, and I think the most important, is education, but only the kind of education that informs and enlarges the mind and engages the heart and the hands in the process of learning. The purpose is to cultivate the complex traits we call wisdom, compassion, and the kind of humble intelligence that grows from the awareness of our implicatedness in the larger enterprise of life. The aim of life-centered education, in other words, is to foster competent, thoughtful people with good hearts and disciplined but imaginative minds capable of learning over a lifetime. There is nothing fast about a good education, however. Educing, drawing forth, from the young entails the patient investment in talents not yet evident, character not yet fully formed, courage not yet summoned, and a life path not yet charted. We, their teachers, are mid-wives to the process of emergence. At our best we help to enlarge their vision, validate their higher self, hone their skills, and encourage them to serve worthy causes larger than themselves. Across different cultures and circumstances, there is no one formula for what constitutes good education, but several principles standout.

The first is that our various problems are rooted in how we think and what we think about and so a problem for people and institutions presuming to improve how we think. It is the central problem for all schools, colleges, and universities and the larger society. For a durable culture, education is the priority.

Second, all education, everywhere, and by whatever means is environmental education. By what we include or exclude from courses and curricula we teach the young that they are part of or apart from the natural world. Environmental education should not be a separate subject; rather it should pervade the entire learning experience connecting sciences, social sciences, and humanities.

This leads to a third and related point: the most radical word in any language is "system" because it implies relatedness and interdependence. We are made of stuff that exploded out of stars billions of years ago. We are part of all that was, is, and ever will be. In turn, that implies obligations owed to others in the past, present, and future. Obligations, further, put limits on what we can do to others now and in the future including the places in which they live or will live. We live in a world of reciprocity, much of it hidden from our eyes and comprehension. Seen or unseen, we live by the work of myriads of creatures that are part of the surrounding ecological community. The same is true of our reliance on others in which "no [one] is an island entire unto him/herself." We are, in short, beneficiaries of the love and care of mothers, fathers, brothers, sisters, extended families, teachers, artists, writers, workers, officials, and even the kindness of strangers. Our existence extends out into ever widening circles of humanity where borders and barriers disappear and even time dissolves between those generations long departed or yet to be born.

Fourth, we are then members of two communities, one of our making and one that exists independent of us. In both cases, we are citizens—dual citizens—of a polity and an ecological system, and these systems are themselves intimately linked. A political system is the framework in which decisions are made about "who gets what, when, and how." The "what" in this process includes air, water, land, resources, forests, atmosphere, and animals. The upshot is that environmental education is fundamentally political but also a call to "a politics which is far-sighted and capable of a new, integral and interdisciplinary approach (Francis, 2015)."

Fifth, environmental education is unavoidably radical, in the original sense of the word: getting to the root of the matter. The fact is that humans have made the Earth into "a pile of filth" in Pope Francis' words. But we know better than we act and that is grounds for hope that we can learn to do better and cause for overthrowing the powers that diminish our better possibilities. Those economic and political forces that prey on our fears and weaknesses have rendered us dumb and vulnerable to exploitation and lies in the cause of division. At its best, environmental education liberates the student from the many illusions conjured in a commercialized, unequal, and violent society. It also connects students to the practical causes of repairing, restoring, and regenerating damaged lands, forests, waters, and ecologies.

Sixth, as architect Lance Hosey writes, "70 percent of the human body's sense receptors cluster in the eyes. Vision dominates (Hosey, 2012)." We are creatures of sight, and so the influence of what we see is a powerful agent in education for good or bad. Most schools, however, are designed by architects oblivious to the fact that architecture is a kind of crystallized pedagogy. Instead, they should be designed to mirror the sustainable world we wish to create, giving hope a form and substance at a comprehensible scale. Schools should be places of light, powered by solar systems, using natural materials, purifying wastewater in biological water treatment systems, and surrounded by gardens and forests. They should be systems that instruct by the way they function as buildings and landscapes and by the way they operate with full student participation in the making and management of the places we call schools (Orr, 2006, 2018, 2021). It is an idea going back at least to the writings of John Dewey, Mahatma Gandhi, Rabindranath Tagore, Maria Montessori, Rudolf Steiner, and Paulo Freire.

Finally, a word about this extraordinary collection of scholarship and thought assembled so well by Dr. Karaarslan-Semiz. The contributors are a remarkable group of teachers and educational philosophers. The range of work and depth of the ideas presented here is considerable: democracy, justice, and education (Chaps. 3 and 6); systems thinking in the curriculum (Chaps. 4, 11, and 12); the proper role of experience and experiential learning; (Chaps. 4, 5, 7, 8, 9 and 10) rigorous assessment of the results (Chaps. 13, 14, 15, 16 and 17); and more.

I hope that this book is widely read and deeply influential in shaping the directions of education and that it inspires other teachers, educational administrators, and students to respond fully and creatively to the peril and the promise of our situation.

Paul Sears Distinguished Professor Emeritus, Oberlin College; Professor of Practice, Arizona State University, author of *Dangerous Years* (Yale University Press, 2017) and co-editor of *Democracy Unchained* (New Press, 2020).

Oberlin College   David W. Orr
Oberlin, OH, USA

References

Francis, P. (2015). *Encyclical on climate change & inequality* (p. 120). Melville House.

Hosey, L. (2012). *The shape of green* (p. 77). Island Press.

Orr, D. W. (2006). *Design on the edge*. MIT Press.

Orr, D. W. (2018). The political economy of design in a hotter time. In R. B. Egenhoefer (Ed.), *Routledge handbook of sustainable design*. Routledge.

Orr, D. W. (2021, February 21). Building as pedagogy: Oberlin's Adam Joseph Lewis Center. *Buildings and Cities*. www.buildingandcities.org/insights/commentaries/building-pedagogy.html

Editor's Preface

When we started to compile the content of this book, the world stood at the beginning of the pandemic; now we are completing the work in the shadow of terrible forest fires and severe environmental events all around the world. We live in an age of crisis because we are losing the life support systems of Earth, and we are on the brink of environmental breakdown. The latest IPCC report warns of climate catastrophe unless we act quickly. We are already experiencing the irreversible impacts of climate change. Disasters such as heatwaves and droughts, wildfires, and flooding have increased on a global scale since the 1950s because of human actions (IPCC, 2021). We have no time for climate delay. We need to act to limit the impacts of the anthropogenic climate crisis before it is too late.

In these unpredictable times, education has a key role to play. How we achieve the sustainability of our planet is closely related to how well we can put education for sustainable development (ESD) into effect at the classroom level. Over the years, ESD has been considered challenging to implement in schools due to a lack of conceptual understanding of sustainability and to the challenges encountered in bringing ESD into the school curriculum. As some policy documents related to ESD are abstract, teachers struggle to integrate ESD into the school curriculum and achieve sustainability-related learning objectives. Teachers need more support to develop holistic and transformative approaches to implement ESD.

The overarching goal of this volume is to provide teachers with pedagogical approaches and practical applications for the implementation of ESD, as well as assessment strategies to evaluate learning outcomes in primary and secondary education. In addition to the appropriate pedagogical tools for ESD, the book also presents some case studies that teachers can use as a guide in their classes. These examples span a variety of pedagogical approaches such as outdoor learning, nature-inspired learning, environmental justice, real-world learning, and the whole-school approach.

The pedagogical methods associated with ESD aim not only to facilitate knowledge of sustainability but also to promote attitudes, new perspectives, values, and skills related to sustainability. This book provides some empirical examples of measuring sustainability competencies in primary and secondary education in different contexts. It is essential to develop a framework and assessment strategies for evaluating the intended outcomes of ESD and monitoring the learning process in schools. Teachers and educators can determine the most appropriate tools for evaluating their ESD practices. This volume presents good practices in ESD implementation at primary and secondary levels through the close and active collaboration of 22 authors from Germany, Italy, Slovenia, Sweden, Türkiye, and the UK. The authors who contributed to this edited collection are at the same time active researchers in the field and teachers and educators with extensive experience of teaching sustainability issues at schools or education centers. All the authors have put in extraordinary efforts to share their knowledge, wisdom, and experience, and we believe that the pedagogical tools, research examples, and theoretical underpinnings presented here will guide teachers in developing their own sustainability-oriented programs.

For this book, we focused on primary and secondary education as young people worldwide have started to question the purpose of education and declared their desire for sustainability issues, especially climate change, to be embedded in school curricula. As teachers and educators, we need to respond to this call of the youth and develop our skills and capacity to integrate ESD topics into curricula and transform our schools and institutions into evolving models of sustainability. With respect to terminology, we have used the term education for sustainable development (ESD) in the title of this book, as this is the term most commonly used in academic and policy documents internationally. Most of the authors have used the term ESD in the text, but some have preferred to use learning for sustainability or sustainability education. Although we are aware of the differences between these terms, we do not discuss them here, as it is beyond the purpose of the book.

The book consists of three main sections: Introduction to ESD, Implementation of ESD: Integrating Theory and Practice, and Assessment in ESD: Measuring ESD Learning Outcomes. In the first section of the book, we provide an introduction to ESD, covering the historical background, critical trends, pedagogical approaches, and links with the SDGs and Agenda 2030. The second section contains ten chapters on theory and practice in ESD which offer learner-centered, action-oriented, experiential, and holistic pedagogical approaches for embedding sustainability into the curriculum. Here, the authors present examples based on their experiences, teaching practices, and research studies. The last section of the book contains five chapters that focus on approaches to assessment and measurement tools for the effective evaluation of ESD learning process and outcomes. We hope this section will guide teachers and educators who want to develop effective assessment strategies while formulating their ESD practices.

In these complicated times, teaching and learning for a sustainable future is fundamental. We have come to understand better that our education, culture, environment, economy, and society are all interdependent. A change in any one of these systems could have a domino effect on the others. One of the goals of SDGs is to develop the quality of education and ensure that all individuals are equipped with the necessary knowledge, values, and skills to deal with the complex problems of the world and create a sustainable future together. If we are to empower our students, we need supportive teachers and educators equipped with the necessary knowledge and skills for teaching ESD successfully.

We hope that this book will contribute to resolving teachers' confusion and uncertainties about the implementation of ESD in schools, will increase their confidence to integrate sustainability issues into their teaching, and will initiate a transformation towards sustainability in their schools. We need to reconsider the role of education today and redesign our school curricula and teaching and learning approaches in such ways as to develop individuals' key competences and achieve a sustainable future. We know that it is not an easy path to teach complex, controversial sustainability issues in class. However, to reach the goal of a sustainable transformation in society, we must promote ESD in school education through action-oriented and transformative pedagogical approaches.

I would like to end by expressing my deepest gratitude to the authors of the chapters of the book for contributing examples of their research and projects, their theoretical perspectives and ideas, and numerous case studies of the implementation of ESD in primary and secondary schools. I greatly appreciate their valuable contributions to this volume.

Ağrı, Türkiye   Güliz Karaarslan-Semiz
2021

Reference

The Intergovernmental Panel on Climate Change (IPCC). (2021). Sixth assessment report. https://www.ipcc.ch/report/ar6/wg1/#FullReport. Accessed 10 September 2021.

Acknowledgement

I would first and foremost like to thank all authors for their tremendous efforts and valuable contributions that made this book possible.

Many thanks to all teachers and students who were involved in the empirical studies at some time and allowed to share information from their side.

I wish to thank the series editors and production team at Springer for their patience and all the work they have done after the writing was completed.

Lastly, I would like to thank to my dear family and my friends for their endless support and encouragement at all times.

Contents

Part I Introduction to ESD

1 Conceptualisation of ESD: Theoretical and Pedagogical
Considerations...................................... 3
Güliz Karaarslan-Semiz

2 Prospects for a Better Future: The Significance
of ESD in Primary and Secondary Education 13
Gaye Teksöz

Part II Implementation of ESD: Integrating Theory and Practice

3 Curriculum Change and Selective Teaching
Traditions: Consequences for Democracy and
the Role of Education 25
Per Sund

4 Outdoor Education for Sustainability
with Systems Thinking Perspective...................... 39
Güliz Karaarslan-Semiz

5 Real-World Learning as a Frame for Sustainability
in Education... 55
Daniela Conti and Richard Dawson

6 The Environment: A Question of Justice? 73
Ben Ballin

7 You Are Part of the Sustainability Picture: Ideas
for Implementation of ESD 93
Armağan Ateşkan and Jennie Farber Lane

8 Earth as Self: Healing Our Connection with
the Earth Through Education 107
Deniz Dinçel and Birgül Çakır-Yıldırım

9 Nature-Inspired Learning: How Nature Can Teach
Us to Be Sustainable? 119
Richard Dawson

10 Inspirational Outdoor Education: Learning
    for Sustainability with the Educational Polygon
    for Self-Sufficiency in Dole............................. 137
    Ana Vovk, Janja Lužnik, and Danijel Davidović

11 Implementation of Education for Sustainable Development
    Through a Whole School Approach 153
    Niklas Gericke

12 Building Primary Schools as a Model of
    Sustainable Communities: Hints for Teachers.............. 167
    Elvan Şahin

Part III   Assessment in ESD: Measuring ESD Learning Outcomes

13 Learning Our Way Forward and How
    We Might Assess That.................................. 181
    Paul Vare

14 Developing and Assessing Sustainability
    Competences in the Context of Education
    for Sustainable Development 191
    Marco Rieckmann

15 Application-Oriented Development of Outcome
    Indicators for Measuring Students' Sustainability
    Competencies: Turning from Input Focus to
    Outcome Orientation 205
    Eva-Maria Waltner, Anne Overbeck, and Werner Rieß

16 Assessing Learning Outcomes for Sustainability
    in Primary and Secondary Schools in the UK 221
    Vasiliki Kioupi and Nikolaos Voulvoulis

17 The Body in Mind: Ideas for Assessing
    Sustainability Literacy 247
    Jennie Farber Lane and Armağan Ateşkan

Index 261

Contributors

Armağan Ateşkan Department of Educational Sciences, Bilkent University, Ankara, Türkiye

Ben Ballin Geography and Sustainability Education Consultant, Birmingham, UK

Daniela Conti Centre for Environmental Research, Documentation and Education CREDA, Monza, Italy

Danijel Davidović Faculty of Arts, Department for Geography, International Center for Ecoremediations, University of Maribor, Maribor, Slovenia

Richard Dawson Wild Awake, Shrewsbury, UK

Deniz Dinçel Department of General Education, Istanbul Bilgi University, Istanbul, Türkiye

Niklas Gericke Department of Environmental and Life Sciences, Karlstad University, Karlstad, Sweden

Vasiliki Kioupi Centre for Environmental Policy, Imperial College London, London, UK

Jennie Farber Lane Department of Educational Sciences, Bilkent University, Ankara, Türkiye

Janja Lužnik Project Office, Faculty of Arts, University of Maribor, Maribor, Slovenia

Anne Overbeck Personality and Social Psychology Division, Otto-von-Guericke University Magdeburg, Magdeburg, Germany

Marco Rieckmann University of Vechta, Department of Education, Vechta, Germany

Werner Rieß Department of Biology and Pedagogy of Biology, University of Education Freiburg, Freiburg, Germany

Elvan Şahin Department of Mathematics and Science Education, Middle East Technical University, Ankara, Türkiye

Güliz Karaarslan-Semiz Department of Mathematics and Science Education, Ağrı İbrahim Çeçen University, Ağrı, Türkiye

Per Sund Department of Teaching and Learning, Stockholm University, Stockholm, Sweden

Environmental and Life Sciences, Karlstad University, Karlstad, Sweden

Gaye Teksöz Department of Mathematics and Science Education, Middle East Technical University, Ankara, Türkiye

Paul Vare School of Education and Humanities, University of Gloucestershire, Cheltenham, UK

Nikolaos Voulvoulis Centre for Environmental Policy, Imperial College London, London, UK

Ana Vovk Faculty of Arts, Department for Geography, International Center for Ecoremediations, University of Maribor, Maribor, Slovenia

Eva-Maria Waltner Department of Biology and Pedagogy of Biology, University of Education Freiburg, Freiburg, Germany

Birgül Çakır-Yıldırım Department of Primary Education, Ağrı İbrahim Çeçen University, Ağrı, Türkiye

Abbreviations

DEFRA	Department for Environment, Food & Rural Affairs
EE	Environmental Education
EfS	Education for Sustainability
ESD	Education for Sustainable Development
FEE	Foundation for Environmental Education
GAP	Global Action Programme
GEB	General Ecological Behavior
IFPRI	International Food Policy Research Institute
LoS	Learning Outcomes
PISA	Programme for International Student Assessment
RSP	A Rounder Sense of Purpose
RWL	Real World Learning
SC	Sustainability Competences
SD	Sustainable Development
SDG	Sustainable Development Goals
SSI	Socio-Scientific Issues
UN	United Nations
UNCED	UN Conference on the Environment and Development
UNDESD	UN Decade of Education for Sustainable Development
UNECE	United Nations Economic Commission for Europe
UNESCO	United Nations Educational, Scientific and Cultural Organization
VBN	Value Belief Norm
WB	World Bank
WWF	World Wildlife Fund

Part I

Introduction to ESD

Conceptualisation of ESD: Theoretical and Pedagogical Considerations

1

Güliz Karaarslan-Semiz

Abstract

This chapter introduces Education for Sustainable Development (ESD) by focusing on the historical development of ESD and some theoretical and pedagogical approaches in the context of primary and secondary school education. I also briefly describe the outline of the chapters in this book, which represent examples of research implemented in ESD, case studies and practical examples of teaching sustainability issues. The role of education for building a sustainable future is then discussed, and some critical ideas are presented to support teachers who are the actual implementers of ESD.

Keywords

ESD (Education for Sustainable Development) · sustainability · teacher education

G. Karaarslan-Semiz (✉)
Department of Mathematics and Science Education, Ağrı İbrahim Çeçen University, Ağrı, Türkiye
e-mail: gkaraarslan@agri.edu.tr

1.1 Introduction

Today, there is unequivocal evidence that humanity is threatening life on Earth. Dramatic changes are occurring in Earth's biological systems, and reversing this situation becomes more difficult with each passing day. Termed the "Great Acceleration", this process began more than 50 years ago, and we have begun to witness its negative impacts on Earth (Rockström, 2015). As described by Rockström (2015), some of the several environmental events that constitute signs of the Great Acceleration are the collapse of marine life, the rapid melting of glaciers, the warming of ocean water, extreme droughts and the collapse of coral reef ecosystems. Rockström et al. (2009) defined a set of nine "planetary boundaries for estimating a safe operating space for humanity with respect to the functioning of the Earth System" (p.3). This group of scientists underlined the importance of these boundaries for the continuity and resilience of the Earth system and indicated that humanity has already transgressed three of them – namely, climate change, the loss of biodiversity and the changes in the nitrogen cycle (Rockström et al., 2009). Recent studies also show that chemical pollution levels in water and soil stemming from the use of pesticides and heavy metals are bringing greater and greater risks to human health and biodiversity and approaching the planetary boundary (Stockholm Resilience Center, 2017).

One of the most notable of the reactions of humanity to all these rapid changes on Earth has been the recent global increase in youth activism. An international movement started by students has spread around the world that aims to present their demands and invite governments to take action over climate change. All around the world, young people are particularly worried about the impacts of climate change and their future (see, e.g. Clayton, 2020; Hickman et al., 2021). Recent studies show that climate change and ecological crisis negatively affect the psychological well-being of both young people and adults. In an international study on climate anxiety, more than half of the participants reported feeling "very" or "extremely" worried about climate change, and nearly half of the respondents mentioned that their feelings about climate change negatively affected their daily lives (Hickman et al., 2021). All in all, environmental and sustainability crises are critically affecting the well-being of both humans and non-humans. So, what are we educators to do now? How can education respond to these critical changes in the world? How can we heal our relationship with the planet and put sustainability at the centre of our lives? In these uncertain and complicated times, Education for Sustainable Development (ESD) should be a major focus area for developing our understanding of complex sustainability issues. In the rest of this chapter, I will try to elaborate on ESD by briefly recounting its history and describing some theoretical and pedagogical approaches in the context of primary and secondary school education.

1.2 Education for Sustainable Development: Brief History and Theory

The roots of environmental and sustainability education has been appeared on the historical documents in the 1970s. The UN Stockholm conference on the Human and Environment was the first conference which addressed environmental issues globally (UN, 1972). After that,

Belgrade Charter (UNESCO, 1976) and Tbilisi Declaration (UNESCO-UNEP, 1977) stated the impact of humans on the natural environment and clarified the crucial role of environmental education. A common concern for the environment and sustainability was expressed through these earliest international documents. 20 years later, education was first mentioned as a keystone of the transition to sustainable development at the UN Conference on the Environment and Development (UNCED) in 1992. The seeds of ESD were thus planted in the UN document *Agenda 21*, which was the landmark publication of UNCED (1992) Chapter 36 identified four major goals of ESD; 1) improving the quality and access to basic education, 2) reorienting existing education 3) increasing public understanding and awareness of sustainable development and 4) developing training for all sectors (UNCED, 1992, p.32). It is generally emphasized that ESD should include a vision that integrates environment, economy and society (McKeown & Hopkins, 2003). Now, we better understand that all disciplines have a responsibility to transform education for a sustainable society.

Over the years that followed, several international conferences were held around this theme. The UN initiated its Decade of Education for Sustainable Development (UNDESD) programme (2005–2015) in order to reorient education at all levels and in all regions towards learning how to live and work sustainably (UNESCO, 2014). This decade of work to integrate sustainable development into education systems concluded with a final monitoring report prepared by UNESCO (2014). Shortly afterwards, in 2015, the UN launched its 17 Sustainable Development Goals (SDGs) programme as the new agenda for promoting sustainable development in all sectors. The SDGs bring together the social, environmental and economic aspects of sustainable development and reflect an integrated perspective (Sterling, 2016). However, the academic literature has emphasised that achieving the SDGs will not be sufficient without changes to societies' dominant values, beliefs and consciousness (see, e.g.

Sterling, 2016). One of the global goals addressed in SDG 4 is the need to provide quality education for all individuals and to develop their knowledge and skills to promote sustainability. Education is also seen as a catalyst to reach all of the global goals, and teachers throughout the world need support to teach sustainability (Van Poeck et al., 2019).

Since education plays a critical role in achieving the SDGs, this book aims to support and inspire primary and secondary school teachers by presenting them with some practical examples, pedagogical approaches and assessment strategies related to ESD. For this purpose, the book brings together the various perspectives of researchers and educators in the field and encompasses a diversity of topics related to sustainability. This volume acquaints teachers with a number of different points of view and will enable them to reflect on their own worldviews. It will therefore assist them in clarifying which ESD perspectives they wish to focus on as primary or secondary school teachers.

ESD is a broad, dynamic and complex concept. It is open to interpretations, and approaches for teaching ESD vary. The common understanding is that ESD aims to integrate sustainability into our learning, teaching and school practices in a holistic and integrative way. In general, ESD has been described as "an education that allows every human being to acquire knowledge, skills, attitudes and values necessary to shape a sustainable future" (Leicht et al., 2018). ESD aims to empower all people at all ages to take responsibility for creating a sustainable future together (UNESCO, 2014). ESD does not belong to one discipline; rather, it requires an interdisciplinary, participatory and transformative approach. Every discipline from art and history to science and mathematics can bring different perspectives, knowledge, skills and values to the teaching of sustainability, and an interdisciplinary connection must be created among the different subjects (McKeown & Hopkins, 2003). In the following section, I will elaborate the content and pedagogy of ESD and look at critical themes, pedagogical approaches and effective assessment strategies related to ESD.

1.3 Critical Themes and Pedagogical Approaches in ESD

As ESD is a comprehensive term, teachers may find it challenging to decide which topics related to sustainability need to be included and how to teach them. The two main components of the term ESD are "education" and "sustainable development". That is to say, content and pedagogy are both important when implementing ESD. We may focus on three questions – *What*, *How* and *Why*. These have been described in the literature as the three dimensions of teaching (e.g. Hofman, 2015; Sund & Gericke, 2020).

The *What* dimension refers to the content of ESD. The key themes of ESD listed in UNESCO documents are climate change, biodiversity, sustainable consumption and production and ending poverty (e.g. Rieckmann, 2018). These critical themes indicate the multidisciplinary nature of ESD as teachers and educators need to highlight interrelationships between these disciplines. The 17 SDGs also include multiple aspects of sustainability like environmental challenges (water, land, oceans, etc.) and fundamental issues related to the development of societies (equity, gender issues and justice) (UNESCO, 2020). However, ESD is generally criticised for its vagueness and the lack of concrete examples of how to integrate sustainability into school curricula. The chapters in this book cover a variety of topics such as democracy in education, elements of nature, biomimicry, food and consumption, rivers and urbanisation and environmental justice. They will therefore assist teachers in connecting ESD with a variety of subjects and provide them with examples that align with their subject areas.

The *How* dimension refers to the teaching methods and strategies associated with ESD. It concerns the pedagogy that teachers can adopt in order to deliver this range of diverse and evolving issues. UNESCO documents state that ESD pedagogy should be based on learner-centred, action-oriented, exploratory and transformative approaches (UNESCO, 2014). In the literature, it is noted that ESD should promote sustainability competences (Wiek et al., 2011) and prepare individuals as active sustainable citizens who can take responsibility for shaping a sustainable future (Rieckmann, 2018). In line with this literature, ESD pedagogy should be linked to participatory, transformative, holistic and competence-based approaches.

Here, the importance of pluralistic teaching in ESD becomes apparent. The pluralistic approach to teaching is usually put forward as a way of developing students' competences to act (Hofman, 2015). Pluralistic teaching aims to develop students' competences to discuss different perspectives critically and analyse them with respect to sustainability issues, to create a democratic environment and to strengthen the students' autonomy (Öhman & Östman, 2019). The democratic perspective plays a vital role in pluralistic teaching, as students are able to encounter different viewpoints and values in classroom discussions (Hofman, 2015). While implementing a pluralistic tradition, teachers can provide students with opportunities to investigate different points of view related to sustainability issues. Panel debates and role-playing activities can be designed so that students can participate in classroom discussions related to real conflicts in society (Öhman & Östman, 2019). Adopting a pluralistic approach to teaching ESD and designing lessons in accordance to this approach might prove time-consuming for teachers, and a crowded curriculum might inhibit them from integrating a democratic perspective on education into their classes. The third chapter in this book focuses on science education and ESD and describes how introducing a new school curriculum can influence the teaching approaches adopted by science teachers while teaching Environmental Education and ESD. The author argues that opportunities for science teachers to engage in democratic and pluralistic ESD teaching decreased after the introduction of a new curriculum. As emphasised in the literature, multiple perspectives and conflicting issues related to sustainability should be seen as a starting point, rather than an obstacle, for the teaching of ESD based on the pluralistic approach to teaching (Tryggvason & Öhman, 2019).

As teachers and educators determine their teaching strategies, it is important to consider the goal of the teaching in terms of the *Why* dimension of teaching (Sund, 2008). In this context, it is essential to discuss whether the teaching focuses on the outcome or the learning process. Specifically, we can focus on two key approaches to education for sustainability: the *intrinsic view of education* and the *instrumental view of education*. Sterling (2011) discusses these two views in relation to education for sustainability. The instrumental view of education, also called the transmissive approach to education, is based on a simple and linear process of learning that sees increasing knowledge and awareness as leading to behavioural change (Sterling, 2011). This view is reflected in international policy documents such as UNESCO (2005), which speaks of encouraging individuals to acquire values and behaviour for building a sustainable future. Examples of activities based on the *instrumental view of education* are school programmes on recycling, waste reduction and energy conservation or the preparation of environmental campaigns to increase people's awareness (Wals & Benavot, 2017). In short, the instrumental approach promotes specific pro-environmental forms of behaviour. On the other hand, the intrinsic view of education seeks to support individuals' capacity and skills to make responsible choices for sustainability. Sterling (2011) points out that the intrinsic view is based on an idealistic position and social constructivist view of education which emphasises the development of learners' capacities to think critically, systemically and reflexively. This is also called the emancipatory approach. The emancipatory approach to ESD prompts learners to reflect on their values and behaviour and to develop their abilities to find their own solutions for complex sustainability problems (Wals & Benavot, 2017).

These two views (intrinsic and instrumental) are aligned with the ESD-1 and ESD-2 approaches of Vare and Scott (2007). Vare and Scott note that focusing on ESD-1 (the instrumental view of education) alone is not a good way of promoting sustainable development. Instead, ESD-1 should be complemented by ESD-2, in which learning strategies are designed which focus on both the learning outcome and the learning process. Many researchers and educators have noted that a paradigm shift is needed in our way of thinking in favour of more holistic, integrative and systems thinking modes in order to develop a transformative approach to ESD (e.g. Capra, 1996; Sterling, 2003; Sterling, 2011). Systems thinking is an important tool for understanding and analysing the complex structure of systems and resolving problems within a systemic approach, and it is closely linked to ESD as one of the core sustainability competences (Wiek et al., 2011). The fourth chapter of this book focuses on systems thinking in the ESD context. The author suggests outdoor learning for sustainability as a pedagogical approach to the development of students' systems thinking skills and presents a case study from an urban environment that can help students see connections between the social, environmental and economic aspects of sustainability as well as develop their systems thinking abilities.

While learner-centred, action-oriented, experiential and transformative pedagogical approaches constitute the general teaching principles and learning processes to be adopted in ESD, the implementation of ESD also requires some specific teaching strategies (Rieckmann, 2018). The chapters in the part of this book that focuses on implementation emphasise experiential, inquiry-based, place-based learning strategies and include case studies of the development of ESD practices both inside and outside the classroom. In Chap. 5, for example, the authors concentrate on Real World Learning as a learning model that teachers can use to shift mindsets and induce actions for sustainability. They suggest solid planning between activities in and out of the classroom and the treatment of the school environment as open learning spaces for students to experience sustainability. In Chap. 6, the author discusses environmental, social and economic justice within the framework of two case studies and invites us to consider how to explore environmental justice issues with children and young people and make them aware of their right to know the impact of these issues on their lives. In Chap. 7, the authors propose several ideas for

teachers to implement ESD in the classroom to raise awareness related to the SDGs. Primarily, they propose that teachers use more problem-based, place-based and discussion-based teaching strategies to achieve SDG learning objectives. In Chaps. 9 and 10, the authors draw attention to learning in/from nature. A nature-inspired learning strategy focuses on how to learn from nature's mentoring to achieve a sustainable future. The authors describe *biomimicry* as a pedagogical tool to demonstrate to young people how nature can be a model for creating sustainable systems. The key take is that nature is the best teacher to show us how to live sustainably on Earth. Designing outdoor learning environments while teaching sustainability contributes to improvements in the development of the well-being of both people and the planet and improves our connection with Earth, as elaborated in Chaps. 8 and 10. The implementation section presents teachers with some case studies and practical examples of nature- and place-based learning strategies which are still infrequently used in ESD teaching.

Implementing ESD is not only about what we teach and how we teach; it is also related to reorienting school practices for sustainability and internalising the *walk your talk* principle. Known as the whole-school approach, this requires the participation of the entire school and its community in sustainability activities (Davis & Cooke, 2007). The whole-school approach to ESD aims to embed sustainability in all aspects of education, including the curriculum, community partnership, student engagement, school leadership and school organisational and institutional practices (Henderson & Tilbury, 2004; Shallcross & Robinson, 2008; UNESCO, 2014). Schools have a great potential to disseminate sustainability thinking and practices for realising the transition to a sustainable society. Green schools such as eco-schools and ESD schools draw on various resources and certification systems to promote the whole-school approach to sustainability. Two of the chapters in this volume highlight the idea of whole-school approaches in ESD. The author of Chap. 11 presents different frameworks for the whole-school approach and shares a validated theoretical model with four dimensions: *holistic idea, routines and structures, professional knowledge creation* and *teaching and learning*. At the end of the chapter, the author addresses the need for schools to focus on both internal and external characteristics and structures which can contribute to a transition in the education system in favour of a successful whole-school approach in ESD. In Chap. 12, the author examines some practical examples related to sustainability from primary green schools and discusses their practices within the whole-school approach lens. The main conclusion of the chapter is that the schools mainly seek to increase knowledge of environmental issues, but do not foster non-cognitive skills such as values, beliefs, attitudes or transformative actions. The author emphasises that non-cognitive constructs also need to be taken into consideration and evaluated in order to create a whole-school approach and to observe long-term impacts as a result of implementing ESD.

All in all, these learner-centred, action-oriented and experiential learning strategies, together with the whole-school approach to ESD, create more possibilities for developing students' competence for action for sustainable development. Mogensen and Schnack (2010) note that action competence development is linked to democratic, participatory and action-oriented teaching and learning, which allow students to develop their ability and desire to participate in finding democratic solutions to sustainability problems. The importance of pluralistic and democratic teaching approaches is observed to be growing in ESD, as these approaches prompt students to consider different views and perspectives, develop their critical thinking skills and their willingness to take action for sustainability and enable them to learn how to be active citizens in a democratic society (Rudsberg & Öhman, 2010; Sass et al., 2020).

The chapters in the last section of this book focus on assessment frameworks and present empirical examples of assessment methods in ESD. The assessment of learning outcomes and learning processes is an important element of ESD teaching. Assessment has been found to be a challenging aspect of ESD, and more effort is therefore needed to make use of systemic assessment strategies in schools (UNESCO, 2014).

One of the key goals of ESD is to develop students' sustainability competences so that they can contribute to the transformation of society for sustainability (Rieckmann, 2018; UNESCO, 2017). Previous studies have pointed to a lack of tools and guidance to enable teachers and educators to assess their students' sustainability competences (Redman et al., 2021). More empirical data is needed to establish whether our pedagogical approaches and teaching strategies are effective in achieving our ESD goals. UNESCO (2017) notes that assessment in ESD should go beyond assessment *of* learning and argues that there should be a form of assessment *for* learning and assessment *as* learning. Teachers should use not only summative assessment strategies but also formative assessment strategies such as self-assessment of individual progress or performance-based strategies (UNESCO, 2017). Teachers may select different assessment tools depending on their goals. Self-assessment, peer feedback and feedback from educators could help learners to monitor their own learning processes and reveal ways in which the learning process could be improved.

In Chap. 13, the author discusses ways of assessing learning outcomes in the ESD context and presents a competency-based assessment framework called *A Rounder Sense of Purpose*, which holistically covers 3 learning outcomes and 12 competences. The chapter questions the purpose of assessment strategies and rethinks *assessment as learning* in a form better linked to the idea of transformative learning for sustainability. At the end of the chapter, the author emphasises the need for a different kind of assessment for achieving a different kind of education in ESD. In Chap. 14, the author focuses on assessing competencies in ESD from a general perspective and discusses the ongoing need to develop competence models and measurement tools in ESD in both the cognitive and non-cognitive dimensions of competencies. How to assess sustainability competencies and develop competence models is a current debate in the literature. In the rest of the section on assessment, the authors provide some assessment ideas and measurement tools for evaluating students' sustainability competences based on research results.

For instance, the authors of Chap. 15 present a research-based example of the measurement of students' sustainability competences. They assess secondary school students' sustainability competences in three dimensions – knowledge, attitude and behaviour – by developing a reliable and valid quantitative measurement tool to assess the impacts of ESD implementations. They conclude that higher levels of sustainability attitudes and knowledge among students could indicate more impact-relevant sustainability behaviour. Chap. 16 discusses another research study concerned with measuring the sustainability competences of primary and secondary school students. As in Chap. 15, this chapter focuses on measuring the cognitive (knowledge and skill), affective (attitude) and behavioural (action) dimensions of learning. In Chap. 17, the authors discuss different ideas for measuring students' sustainability literacy based on the whole-body learning approach. They suggest both formative and summative assessment strategies based on parts of the body: the *head*, *heart*, *gut*, *arms* and *legs*. The authors provide some examples of measuring different aspects of sustainability literacy through various measurement tools such as interviews, questionnaires, journals and self-assessment tools.

This book once again underlines the importance of assessment in ESD by discussing the evaluation and assessment aspects of the topic, which are often neglected in both teaching practices and academic literature. There is a need to develop specific assessment tools that align with our ESD aims, teaching content and teaching methods. We can only determine the effectiveness of our pedagogical approaches and explore ways of revising and improving them through a systematised assessment strategy.

1.4   Conclusion

In this chapter, I have attempted to provide an introduction to the conceptualisation of ESD by discussing both the academic literature and policy documents. I have described ESD from various perspectives, considering its content, teaching approaches and methods, and strategies for assessment. Today, we are in a better position to understand that while teaching sustainability issues, we need to focus not only on what we teach but also on our teaching process and our pedagogical approach (Hofman, 2015).

There are currently almost 85 million teachers worldwide, and 64 million of these teachers work in primary and secondary schools (WB, 2021). Compulsory education is a vitally important field for promoting ESD, and teachers are critical agents as they shape children's future, affecting their lives both academically and socially. The role of teachers in achieving sustainability-oriented programmes in schools should not be underestimated (Gough, 2005; Timm & Barth, 2021). Teachers can foster learners' knowledge, understanding, skills, values and dispositions and in this way respond to the current problems facing the world, such as climate change, land degradation and population growth (Evans, 2020). However, there still remains a gap between national and international policy documents and ESD teaching practices. The goals, proposals and recommendations contained in ESD policy documents do not easily extend to teachers and educators around the world (Waltner et al., 2020). Teachers still experience uncertainties about teaching ESD and achieving the learning outcomes (Andersson, 2017). The International Network of Teacher Education Institutions notes that teachers see ESD as an unclear and unrecognised concept. There is a lack of awareness and conceptual clarity about teaching sustainability in many teacher training organisations (Hopkins & Kohl, 2019). Teachers need more tangible, practical examples to formulate and implement ESD teaching. This book aims to provide teachers with some concrete examples of the implementation of ESD and the assessment of ESD learning outcomes and processes at the primary and secondary school levels.

In addition to teachers implementing ESD, schools should be places to practise what we teach. Schools should be organised in a way that supports and facilitates sustainability thinking and sustainability-oriented activities. In this way, students can see the implications of what they learn in the curriculum related to ESD. Through ESD, we need to promote more ethical values related not only to human beings but also to non-human beings, care for future generations, critical thinking, systems thinking and action competence. All these competences can be achieved through a holistic understanding and transformation of our way of learning and teaching.

In conclusion, we may list some of the critical ideas emphasised in the chapters of this book as follows:

- Firstly, we need to have an applicable, accessible educational policy for implementing ESD at the national level. Some countries have already put new strategies in place.
- To integrate ESD into school programmes, we need to listen together to teachers from all disciplines as well as to educators working on promoting ESD and to the opinions of students about creating sustainability-integrated programmes.
- To create a sustainable school plan connected with the broader community, all the school's key actors (teachers, students, school administrators and other school staff) need to be involved.
- Priority should be attached to learning approaches for ESD which are more nature-based, outdoor and place-based, inquiry-based, real, active and transformative, along with practical examples of implementing these strategies, and these should be included in teacher training programmes.
- We need valid and reliable assessment frameworks and tools for evaluating learning processes and outcomes related to ESD.
- As educators, we should promote perspectives which are more Earth-centred, nature- and environmental justice-based and systemic and formulate our education programmes accordingly.

References

Andersson, K. (2017). Starting the pluralistic tradition of teaching? Effects of education for sustainable development (ESD) on pre-service teachers' views on teaching about sustainable development. *Environmental Education Research, 23*(3), 436–449.

Capra, F. (1996). *The web of life.* Harper and Collins.

Clayton, S. (2020). Climate anxiety: Psychological responses to climate change. *Journal of Anxiety Disorders, 74*, 102263. https://doi.org/10.1016/j.janxdis.2020.102263

Davis, J., & Cooke, S. (2007). Educating for a healthy, sustainable world: An argument for integrating health promoting schools and sustainable schools. *Health Promotion International, 22*(1), 346–353. https://doi.org/10.1093/heapro/dam030

Evans, N. (2020). What ought to be done to promote education for sustainability in teacher education? *Journal of Philosophy of Education, 54*(4). https://doi.org/10.1111/1467-9752.12482

Gough, A. (2005). Sustainable schools: Renovating educational processes. *Applied Environmental Education, 4*(4), 339–351. https://doi.org/10.1080/15330150500302205

Henderson, K., & Tilbury, D. (2004). Whole-school approaches to sustainability: An international review of sustainable school programs. Report Prepared by the Australian Research Institute in Education for Sustainability. (ARIES) for The Department of the Environment and Heritage, Australian Government. ISBN, 1(86408), 979.

Hickman, C., Marks, E., Pihkala, P., Clayton, S., Lewandowski, R. E., Mayall, E. E., Wray, B., Mellor, C., & van Susteren, L. (2021). Climate anxiety in children and young people and their beliefs about government responses to climate change: A global survey. *Lancet Planet Health, 5*(12), e863–e873. https://doi.org/10.1016/S2542-5196(21)00278-3

Hofman, M. (2015). What is an education for sustainable development supposed to achieve-A question of what, how and why. *Journal of Education for Sustainable Development, 9*(2), 213–228. https://doi.org/10.1177/0973408215588255

Hopkins, C., & Kohl, K. (2019). Teacher education around the world: ESD at the heart of education - responsibilities and opportunities towards a sustainable future for all. In D. Karrow & M. DiGiuseppe (Eds.), *Environmental and sustainability education: Canadian perspectives.* Springer Nature AG.

Leicht, A., Combes, B., Byun, A. J., & Agbedahin, A. V. (2018). From agenda 21 to target 4.7: The development of ESD. In A. Leicht, J. Heiss, & W. J. Byun (Eds.), *Education on the move. Issues and trends in education for sustainable development* (pp. 25–38). United Nations Educational, Scientific and Cultural Organization.

McKeown, R., & Hopkins, C. (2003). EE ≠ ESD: Diffusing the worry. *Environmental Education Research, 9*(1), 117–128. https://doi.org/10.1080/1350462032000034395

Mogensen, F., & Schnack, K. (2010). The action competence approach and the 'new' discourses of education for sustainable development, competence and quality criteria. *Environmental Education Research, 16*(1), 59. https://doi.org/10.1080/13504620903504032

Öhman, J., & Östman, L. (2019). Different teaching traditions in environmental and sustainability education.

In K. Van Poeck, L. Östman, & J. Öhman (Eds.), *Sustainable development teaching. Ethical and political challenges* (Routledge studies in sustainability) (pp. 70–82). Routledge.

Redman, A., Wiek, A., & Barth, M. (2021). Current practice of assessing students' sustainability competencies: A review of tools. *Sustainability Science, 16*, 117–135. https://doi.org/10.1007/s11625-020-00855-1

Rieckmann, M. (2018). Learning to transform the world: Key competencies in ESD. In A. Leicht, J. Heiss, & W. J. Byun (Eds.), *Education on the move. Issues and trends in education for sustainable development* (pp. 39–59). United Nations Educational, Scientific and Cultural Organization.

Rockström, J. W. (2015). Bounding the planetary future: Why we need a great transition. Great Transition Initiative. https://www.tellus.org/pub/Rockstrom-Bounding_the_Planetary_Future.pdf. Accessed 12 Sept 2021.

Rockström, J., Steffen, W., Noone, K., Persson, Å., Chapin, F. S., Lambin, E., Lenton, T. M., Scheffer, M., Folke, C., Schellnhuber, H., Nykvist, B., de Wit, C. A., Hughes, T., van der Leeuw, S., Rodhe, H., Sörlin, S., Snyder, P. K., Costanza, R., Svedin, U., ... Foley, J. (2009). Planetary boundaries: Exploring the safe operating space for humanity. *Ecology and Society, 14*(2), 32–65. https://www.ecologyandsociety.org/vol14/iss2/art32/. Accessed 12 Sept 2021.

Rudsberg, K., & Öhman, J. (2010). Pluralism in practice – Experiences from Swedish evaluation, school development and research. *Environmental Education Research, 16*(1), 95–111. https://doi.org/10.1080/13504620903504073

Sass, W., Boeve-de Pauw, J., Olsson, D., Gericke, N., De Maeyer, S., & Van Petegem, P. (2020). Redefining action competence: The case of sustainable development. *The Journal of Environmental Education, 51*(4), 292–305. https://doi.org/10.1080/00958964.2020.1765132

Shallcross, T., & Robinson, J. (2008). Sustainability education, whole school approaches, and communities of action. In B. J. J. N. Alan Reid & V. S. Bjarne (Eds.), *Participation and learning - Perspectives on education and the environment, health and sustainability* (pp. 299–320). Springer.

Sterling, S. (2003). *Whole systems thinking as a basis for paradigm change in education: Explorations in the context for sustainability* (Unpublished doctoral dissertation). University of Bath, UK.

Sterling, S. (2011). Transformative learning and sustainability: Sketching the conceptual ground. *Learning and Teaching in Higher Education., 5*(11), 17–33.

Sterling, S. (2016). A commentary on education and sustainable development goals. *Journal of Education for Sustainable Development., 10*(2), 208–213. https://doi.org/10.1177/0973408216661886

Stockholm Resilience Center. (2017). *Chemical pollution.* https://www.stockholmresilience.org/research/research-news/2017-06-09-how-far-are-we-pushing-chemical-boundaries.html. Accessed 10 Feb 2022.

Sund, P. (2008). Discerning the extras in ESD teaching: A democratic issue. In J. Öhman (Ed.), *Values and democracy in education for sustainable development-contributions from Swedish research* (pp. 57–74). Liber.

Sund, P., & Gericke, N. (2020). Teaching contributions from secondary school subject areas to education for sustainable development – A comparative study of science, social science and language teachers. *Environmental Education Research., 26*(6), 772–794. https://doi.org/10.1080/13504622.2020.1754341

Timm, J. M., & Barth, M. (2021). Making education for sustainable development happen in elementary schools: The role of teachers. *Environmental Education Research, 27*(1), 50–66. https://doi.org/10.1080/13504622.2020.1813256

Tryggvason, A., & Öhman, J. (2019). Deliberation and agonism: Two different approaches to the political dimension of environmental and sustainability education. In K. Van Poeck, L. Östman, & J. Öhman (Eds.), *Sustainable development teaching. Ethical and political challenges* (Routledge studies in sustainability) (pp. 115–124). Routledge.

UN (United Nations) (1972). *Report of the United Nations conference on the human environment.* Stockholm: UN.

UNESCO (1976). The Belgrade charter. A global framework for environmental education. https://naaee.org/sites/default/files/153391eb.pdf Accessed 5 May 2022.

UNCED. (1992). *United Nations Conference on environment and development.* UN.

UNESCO. (2005). *United Nations decade of education for sustainable development (2005–2014): International implementation scheme.* UNESCO.

UNESCO. (2014). *Shaping the future we want. UN decade education for sustainable development (2005–2014) Final Report.* UNESCO.

UNESCO. (2017). *Education for sustainable development goals. Learning objectives.* UNESCO. http://unesdoc.unesco.org/images/0024/002474/247444e.pdf. Accessed 10 Aug 2021.

UNESCO. (2020). *Education for sustainable development. A roadmap. ESD for 2030.* UNESCO. https://unesdoc.unesco.org/ark:/48223/pf0000374802. Accessed 15 Sept 2021.

UNESCO-UNEP (1977). Tbilisi declaration. http://www.gdrc.org/uem/ee/tbilisi.html. Accessed 5 May 2022.

Van Poeck, K., Östman, L., & Öhman, J. (2019). *Sustainable development teaching* (Routledge studies in sustainability). Routledge.

Vare, P., & Scott, W. (2007). Learning for a change: Exploring the relationship between education and sustainable development. *Journal of Education for Sustainable Development, 1*(2), 191–198. https://doi.org/10.1177/097340820700100209

Wals, A. E. J., & Benavot, A. (2017). Can we meet the sustainability challenges? The role of education and lifelong learning. *European Journal of Education., 52*(4), 404–413. https://doi.org/10.1111/ejed.12250

Waltner, E. M., Scharenberg, K., Hörsc, C., & Rieß, W. (2020). What teachers think and know about education for sustainable development and how they implement it in class. *Sustainability, 12*(4), 1690. https://doi.org/10.3390/su12041690

Wiek, A., Withycombe, L., & Redman, C. L. (2011). Key competencies in sustainability: A reference framework for academic program development. *Sustainability Science, 6*(2), 203–218. https://doi.org/10.1007/s11625-011-0132-6

World Bank [WB]. (2021). *Teachers.* https://www.worldbank.org/en/topic/teachers#1. Accessed 2 Aug 2021.

Güliz Karaarslan-Semiz is an assistant professor in the Department of Mathematics and Science Education at Ağrı İbrahim Çeçen University in Türkiye. She studied science education (BS) and then received her master degree (focus on environmental education). She completed her PhD at Middle East Technical University in Ankara (focus on ESD). Her study was awarded as the best PhD thesis of the year at Middle East Technical University in 2016. Her research interests are education for sustainable development, systems thinking, and recently whole school approach to sustainability. She has worked on national and international projects related to ESD and science education. She is a member of the Environmental and Sustainability Education Research (ESER) network in Europe and recently she has worked as a visiting researcher at Stockholm University in Sweden.

Prospects for a Better Future: The Significance of ESD in Primary and Secondary Education

Gaye Teksöz

Abstract

Nowadays, it is a widely accepted discussion among academic circles that ESD is a key enabler for UN Sustainable Development Goals, but it would also be a key for the world to prepare for potential crises such as COVID-19. Contrary to this conviction, what comes to a critical mind would be questions which strive for an accurate answer revolving around what urged humanity to launch Sustainable Development, then to set SDGs, and eventually to redefine education as ESD. This chapter will attempt to answer these questions both by touching upon writer's personal experiences and by putting for the history about human awakening on the necessity for a Sustainable Development which would accentuate the roles of future generations as change-makers and since they stand as hope for the future.

Keywords

Education for Sustainable Development · Sustainable Development · SDGs · Young people

G. Teksöz (✉)
Middle East Technical University, Department of Mathematics and Science Education, Ankara, Türkiye
e-mail: gtuncer@metu.edu.tr

2.1 My Personal Trip to ESD

There was a strange stillness. The birds, for example – where had they gone? Many people spoke of them, puzzled and disturbed. The feeding stations in the backyards were deserted. The few birds seen anywhere were moribund; they trembled violently and could not fly. It was a spring without voices…
Rachel Carson – Silent Spring

Nowadays, it is reported frequently that "ESD is a key enabler for Sustainable Development Goals, but it also may be key to preparing the world for future crises like COVID-19" (UNESCO). This sentence actually is an answer for the question asked years ago in *Earth in Mind* by David Orr (2004)9: "what is education for?"

The Sustainable Development concept was launched in 1987 through *Our Common Future*, Education for Sustainable Development has been implemented since 1992, and Sustainable Development Goals were set in 2015. To create a clear vision, it is suggested to look back to the past and remember what made humanity launch Sustainable Development, to set SDGs, and to redefine education as ESD.

In my personal history, the first time I realized the human impact on the environment literally dates back to times when I worked on my master's thesis (Teksöz, 1989). I have a B.S. degree on environmental engineering. During this study, I took courses on how to treat wastewaters after domestic, industrial, or agricultural use; in other words, my education taught me how to use technology in order to decrease the human impact on the natural resources or, to put it in a nutshell, to increase replenishment rate of natural resources. Nevertheless, I was not aware of the "cradle-to-cradle approach" when I pursued my undergraduate degree. At later stages of my history, I was lucky enough to take sediment samples from one of the most re-known spots on earth when I was working on my master's thesis: from the Golden Horn of Bosphorus (Istanbul/Türkiye). The Golden Horn, a unique inland water channel of Istanbul, gradually has turned into a hub of discharges of domestic wastewaters which resulted from a rapidly increasing population in the city as well as from the industrial discharges coming from hundreds of small metal works situated just at the shores of the Golden Horn during the early 1950s.

Coming back to my research as briefly mentioned above, the sediment samples, which had been taken from the Golden Horn to trace the pollution's chronology, were made use of so that the sources of the pollution in the area and the metals would be evidenced. All the work and the study related to substantiating my thesis work were very stimulating, as particularly there were traces in the sediments collected from the Golden Horn which proved the impacts of increasing population and increasing number of industries on the pollution on the mentioned spot. When it comes to the details of my research, a 3-m-long undisturbed core sample was collected from the Golden Horn while on board with RV Knorr, the research vessel, which coincided with the third phase of Turkish-American Black Sea joint expedition of 1989. The core was sliced, and its date was defined by making use of the 210Pb isotope technique. The result was that the bottom of the core corresponded to 1912. Then, each slice was analyzed for major, minor, and trace elements by inductively coupled plasma emission spectrometry (ICP). The masses of the measured elements would account for approximately the half of the collected sediment mass. The lithophilic elements Li, K, Rb, Mg, Ca, Ba, Al, La, Ti, V, Mn, Fe, Co, and Ni accounted for more than 90% of the elemental mass and did not indicate any change in their concentrations for the years

between 1912 and 1987. Although anthropogenic elements Mo, Zn, Cr, Cu, Ag, and Cd accounted for a minor fraction of the elemental mass, it was observed that their concentrations increased along the core, which signified the human impact on the chemical composition of the Golden Horn sediments. Moreover, it was found out that lead was enriched at the bottom of the core which suggests pollution of Golden Horn sediments by the element lead even at the beginning of the century, but observed concentrations of the remaining anthropogenic elements, at the bottom of the core, would be explained by sedimentary material. My results indicate on the one hand that concentrations of pollution-derived elements did not change significantly between from 1912 to 1950; however, their concentrations increased sharply in the second half of the century. In order to create an explanation, a factor analysis was applied to the data set; the end result has shown that the inorganic fraction of the Golden Horn sediments included crustal, marine, and two anthropogenic components. One of the anthropogenic components has been attributed to the discharges from an iron and steel plant. Another anthropogenic component, which accounted for a larger fraction of system variance, was due to discharges from industrial installations along the Golden Horn, particularly metalwork plants.

Metal fluxes observed in the Golden Horn had significantly different patterns in industries discharges than the domestic. Discharges of Cr, Cu, Zn, and Ni into the Golden Horn increased gradually between from 1940 to 1980 as a result of a large number of small-scale production units which were accommodated on the spot every year. Likewise, commissioning a number of large-capacity plants has resulted in dramatic increases in metal discharges into the Golden Horn. Two of these were metalwork plants. One of them started its operations in 1948; the other one was commissioned in 1959. Although metal discharges into the Golden Horn increased gradually, it started to decrease only after 1960 since large industrial installations were not allowed to operate along the Golden Horn, whereas the domestic discharges of the same elements between from the 1950s to 1985 increased logarithmically. Concerning the domestic discharges of elements, its development had similarity with the population increase in the region where it had been ceaseless. Such pace of increase in the volume of population did not increase abruptly unlike industrial discharges and did not level off after 1960 or any other time.

Based on the mentioned results, it would be concluded that the key factor which represented the major anthropogenic input into the Golden Horn was domination of industrial discharges, particularly driven from the metal producing plants in the vicinity of the Golden Horn. However, the proportion of metals resulting from the domestic discharges was negligible and too small to be detected by the factor analysis. Furthermore, our calculations demonstrated that 24,000 tons of Cr, 300 tons of Cu, 130 tons of Ni, and 7500 tons of Zn were discharged by the plants in 1980, while, in the same year, domestic discharges of Cr, Cu, Ni, and Zn were 50, 90, 13, and 120 tons, respectively. The substantive difference between the industrial and domestic discharges of elements, except for Cu, supported our conclusion which asserted that the elemental profiles were determined by industrial rather than domestic discharges.

The consequences of my MS research (Teksöz, 1989) had triggering effects which contributed my future philosophy and understanding on the relations between the nature and the human. Interestingly, the point I have reached had more research agenda which posed one of the grand challenges related to "how human impact would be decreased on the natural resources even before designing plants which are planned to treat the already polluted sources, and how could education help?". The reason for these questions in my mind was because of the results I had obtained from the Golden Horn sediments, which indicate that the humans were one of the constant/vital factors in relation to the decreasing replenishment rate of nature. Since my realization, I have a strong driving force to inspire people and to infuse particularly the young with the knowledge about how natural resources support our lives and how we respond to the cause and effect relationship. Regrettably to say, the world's response to

environmental degradation, however, was far earlier than mine.

2.2 The World's Response

If we make nights bright, we cannot see the stars.
(from Dreams – Akira Kurosawa)

The awakening began with the sentimental impacts of the Green Revolution. Therefore, I chose to report about the Green Revolution as one of the milestones of environmental awareness:

> The revolution was green at the beginning, because, from the 1960s through the 1990s, yields of rice and wheat in Asia doubled. Even as the continent's population increased by 60 percent, grain prices fell, the average Asian consumed nearly a third more calories, and the poverty rate was cut in half. When Borlaug won the Nobel Peace Prize in 1970, the citation read, "More than any other person of this age, he helped provide bread for a hungry world" (Folger, 2013)

But in fact, there are two sides of the same coin: The Green Revolution itself has also been extensively criticized for inducing environmental damage. The criticism has its roots in excessive and inappropriate use of fertilizers and pesticides which has not only polluted water sources but also poisoned agricultural workers as well as killed insects favorable to agriculture and damaged other wildlife. Additionally, irrigation practices have led to salt build-up and eventual abandonment of some of the most fertile farming lands. Over and above, the groundwater levels have been retreating in the areas where excessive water was pumped for irrigation rather than it would be replenished by the rains; and what was more, burdensome dependence only on few major cereals has led to loss of biodiversity on farms. It has been understood that some of these outcomes were inevitable as unconscious farmers who own large territory of lands started to make use of modern input for the first time; nevertheless, inadequate knowledge and training, the absence of an effective regulation related to water quality, as well as input pricing and subsidy policies which made modern inputs too cheap and, thus, encouraged excessive use also created negative environmental impacts. These issues have gradually been rectified without yield loss and sometimes even with yield increases, thanks to policy reforms, improved technologies, and enhanced management practices such as but not limited to pest-resistant varieties, biological pest control, precision farming, and crop diversification. However, there is a significant point which has been often ignored: positive impact of higher yields on saving large forested areas and other environmentally fragile lands, which would have otherwise been allocated for farming. This positive trend has been observed in some regions in the world. Particularly, in Asia, cereal production doubled between 1970 and 1975, yet the total land area cultivated with cereals increased by only 4% (International Food Policy Research Institute [IFPRI], 2002.

"However, under today's global threats, climate change and population growth, between now and 2050, we'll need another green revolution to provide bread for a hungry World" as Tim Folger wrote in *National Geographic* magazine. While imagining emergence of another green revolution, it should also be pinpointed that "despite the presence of over-organized and over mechanized age, individual initiatives and courage still do count: change can be brought about, not through incitement to war or violent revolution, but rather by altering the direction of our thinking about the world we live in," as did Rachel Carson's *Silent Spring* in 1962.

Carson's groundbreaking book has shaken our rigid worldviews not just in the United States but around the globe (Culver et al., 2012). The book's message about the threat of pesticide abuse reached a wide audience; there has been an evidence that the so-called ecological revolution was caused in no small part by the 1962 publication of Carson's book (Carson, 1962). *Silent Spring* became an immediate bestseller and remained on *The New York Times* list for 31 years. Predicting the threat to humanity through overpopulation and resouce exploitation several years before Paul Ehrlich (Ehrlich, 1971) and Barry Commoner (Commoner, 1971), *Silent Spring* led to new environmental awareness and a vision that translated into tangible political action.

More immediately, *Silent Spring* had an influence on the government policy. Each one of the toxic chemicals named in the book was either banned or severely restricted in the United States by 1975. Farm chemicals, pest control chemicals, and household chemicals undergo much greater scrutiny, regulation, and control than before Rachel Carson published the book, and the chemicals allowed were transformed into less deadly and used in smaller amounts. As well, the book has been regarded one of the iconic works/studies which would have made people conscious about the environment. 2007 Nobel Peace Prize laureate Al Gore once stated that:

> For me, personally, Silent Spring had a profound impact. It was one of the books we read at home at my mother's insistence and then discussed around the dinner table…Rachel Carson was one of the reasons why I became so conscious of the environment and so involved with environmental issues. Her example inspired me to write Earth in the Balance…Her picture hangs on my office wall among those of political leaders…Carson has had as much or more effect on me than any of them, and perhaps than all of them together. (Vice President Al Gore, "Introduction," *Silent Spring* (1994 ed.)

London fog in 1952, Bhopal disaster in 1984, Chernobyl Nuclear Disaster in 1986, Exxon Valdez oil spill in 1989, and many others (disasters) have been the events which have also stimulated people, with extraordinary personal ordeals, to comprehend the expense of wasting natural resources unconsciously.

There existed many valuable books, articles, and researches which made us realize our limits to growth (Meadows et al., 1972), population as a bomb (Ehrlich, 1971) or that we have only one Earth (Ward & Dubos, 1972). Despite the abundance of the sources arousing environmental awareness, two pieces of artistic works have a unique value in my personal history since they triggered my awakening on the sustainable use of the resources: the 2009 documentary "HOME" as well as Akira Kurosawa's "Dreams" (especially water mill village). As first, the documentary "HOME" (Yann, 2009) depicts a story telling about how Earth's problems are interlinked, whereas in Kurosawa's "Dreams," a 101-year-old man teaches the audience with his striking words "it is not right to cut trees, the ones go down should be enough for us." Nevertheless, I must confess that these do not belong only to my personal awakening history but also among the most effective stories that my students eagerly select for their academic and personal development.

2.3 My ESD Experience

> The history of life on earth has been a history of interaction between living things and their surroundings. To a large extent, the physical form and the habits of the earth's vegetation and its animal life have been molded by the environment. …. Only within the moment of time represented by the present century has one species – man – acquired significant power to alter the nature of this world…
> Rachel Carson – Silent Spring

Once I used to include not only stories revolving around the preservation of nature in my ESD plans by which would support pedagogical strategies of the related lecture similar to the ones reported by Tilbury (2011) but also stimulus activities, critical incidents, case studies, and reflexive accounts. Particularly, the stories I utilized for the lectures have been selected from the real cases along with documentaries or PowerPoint presentations. Among the stories which have been regarded the most striking, thought provoking, and awakening by my students have been a PPT presentation accompanied by some short videos of "Cape Town – Day Zero" and "The Burden of Thirst (by Tina Rosenberg)." In order to analyze how the stories influenced my students, I designed a set of questions. And below (Table 2.1) are the responses of my 231 students who were enrolled in "Education and Awareness for Sustainable Development" course during 2019–2020 Academic Year.

> Most important, freedom from water slavery means girls can go to school and choose a better life. The need to fetch water for the family, or to take care of younger siblings while their mother goes, is the main reason very few women in Konso have attended school. Binayo is one of only a handful of women I met who even know how old they are. Tina Rosenberg

Table 2.1 Responses of the students: "Which of the stories told during the course impressed you most?"

Title of the story	% agreement
Cape Town – Day Zero (ppt presentation – real story)	36
Climate Issue (documentary)	36
The Burden of Thirst by Tina Rosenberg (article – real story)	33
Kızılırmak Catchment Area and Water Buffalos (ppt presentation – real story)	33
Little Prince – Antoine de Saint-Exupéry (book)	32
Before the Flood – Leonardo DiCaprio (documentary)	31
HOME (documentary)	22
Story of Palm Oil (ppt presentation – real story)	18
Dreams – The No Name Village – Akira Kurosawa (movie)	14
Small Islands and Climate Change Kiribati-Tuvalu – Seychelles (ppt presentation and video – real story)	12
Bhutan – National Happiness Index (documentary – real story)	10
Disappearance of Aral Sea (documentary – real story)	10
Elephant Poachers (ppt presentation – real story)	7
At the Frontier Young People and Climate Change – Marjorie's Story (ppt presentation – real story (by Martin Caparros, Dr. Laura Laski, Victor Bernhardtz))	3
Story of Ganges River (ppt presentation – real story)	3

Based on my personal experience, I would confidentially state that the stories inspired from the real lives as well as the recent ones such as the first 4 in the above list are considered as the most impressive when sustainability involves though it would have been hard to categorize the stories in the mentioned list. What is more, these stories do not possess a common characteristic and do not represent local stories all the time unlike very few examples which would be possibly found in the related literature. Quite the contrary to this representation, these stories do embrace and combine more one issue from politics, human life, society, economy, and natural sources.

As to my motivation for telling stories during my sustainability awareness course, it has not derived only from my personal experience of awakening (related to pollution chronology of the Golden Horn) but also emerged out of the literature and the historical background of ESD.

The mentioned background began with the environmental education: in the 1970s, the educators specialized in the Environmental Education in the United States requested support from the United Nations Educational, Scientific and Cultural Organization (UNESCO) and the UN Environment Programme for the development of the field. Following the positive decision on such an assistance, the UN sponsored EE conferences in Belgrade, Yugoslavia, in 1975 and in Tbilisi, the Soviet Republic of Georgia, in 1977. The Belgrade Charter of 1975 and Tbilisi Declaration of 1977, both similar in nature, emphasized necessary action; nevertheless, none were sufficiently clear. Therefore, in order to clarify this ambiguity, Professor Harold Hungerford and his colleagues published "Goals for Curriculum Development in Environmental Education" in the Journal of Environmental Education in 1980. Their explanations eliminated this ambiguity and helped prioritizing the action:

> ...to aid citizens in becoming environmentally knowledgeable and, above all, skilled and dedicated citizens who are willing to work, individually and collectively, toward achieving and/or maintaining a dynamic equilibrium between quality of life and quality of the environment. (Hungerford et al., 1980, p.44)

2.4 Brief History of Human Awakening on the Need for Sustainable Development: Human Environment (Stockholm) – Environment and Development (Rio) and Sustainable Development (Johannesburg)

The Green Revolution and *Silent Spring* gave way to increased discussions on the environmental problems, while international response was established through conferences. On the other hand, education became a useful means for the solution of environmental problems during the Stockholm Conference of 1972. This conference is regarded the first world conference on the environment issues. The conference resulted with a declaration and a subsequent Action Plan for the Human Environment. Following the 20th anniversary of the Stockholm Conference, the United Nations Conference on Environment and Development (UNCED) was held in Rio de Janeiro of Brazil. The Rio differed much from the Stockholm since it brought together political leaders, diplomats, scientists, and representatives from the media as well as from the non-governmental organizations (NGOs). As the title of the conference implied (Environment and Development), delegates from 179 countries convened to concentrate on the deteriorative effects of the socioeconomic activities on the environment. Within 20 years of time between 1972 and 1992, human understanding and conceptualization of the issue have evolved from "human environment" to "the world's future in relation to the environment and socioeconomic development." Accordingly, one of the primary outcomes of the Rio Conference also referred to as Earth Summit was that it produced a wider agenda for international action on environment and development issues of the twenty-first century. This vision would also be observed in one of the UN reports stating that "the Earth Summit concluded that the concepts of sustainable development were an attainable goal for all the people of the World, regardless of whether they were at the local, national, regional or international level. It also recognized that integrating and balancing economic, social and environmental concerns in meeting our needs is vital for sustaining human life on the planet" (UN, 1992). Apart from what has been cited in the report, Rio had another significant conclusion asserting that integrating the three dimensions requires new understanding our production, consumption, and working methods as well as our lifestyle and decision making habits. At this point, being one of the key results of the Earth Summit, Agenda 21, among many others, excels as an interesting the program for the action of new strategies including education.

Nevertheless, the origin of Agenda 21 dates back to the Brundtland Report of 1987 when sustainable development was first endorsed at the UN General Assembly. In parallel to the endorsement, this was the time when the concept of education as a support to the sustainable development was also explored. From 1987 to 1992, the concept of sustainable development progressed/enhanced, while the committees discussed, negotiated, and eventually formed the 40 chapters of Agenda 21. Initial ideas concerning ESD were touched upon in the Chapter 36 titled as "Promoting Education, Public Awareness, and Training" (McKeown, 2002). Therefore, it is called the re-orientation of environmental education toward sustainability.

As Tilbury (1995) reported, environmental education literature has been influenced by the outputs of Rio Conference and re-directed its focus toward Environmental Education for Sustainability (Huckle, 1990; Orr, 1992; Sterling, 1992). Therefore, the key principle of ESD was set in the mid-1990s outlining the relevance as:

> Environmental work will need to be relevant to the student, through increasing their understanding of themselves and the world around them. It must encourage pupils to explore links between their personal lives and wider environmental and development concerns, by dealing with issues like consumerism and how the practices of business and industry influence their lives. In doing so, ESD prepares students for contemporary reality. (Tilbury, 1995, p.6)

Parallel to the relevance, holistic outlook and values were highlighted for their significance in ESD. By the help of this new understanding, we should recognize that the decision to participate in the environmental improvement is not only stimulated by the cognitive realm but it is also dependent on personal motivation as well as on awareness tempting a sort of responsibility, all of which are the outcomes of the development of a personal environmental ethic (UNESCO, 1986; UN, 1992).

As the leading UN agency on ESD, UNESCO sets the targets for ESD as to "empower people to change the way they think and work towards a sustainable future." Since it requires, in order to attain this objective, for the society an adoption to a greener economy as well as an enhanced political society through ESD, the next international conference, also known as Rio+20, was held in 2002 in Johannesburg, South Africa. Titled as "World Conference on Sustainable Development," the conference implied the evolution of the human's mind: within 30 years from 1972 to 1992 and then to 2002, this mentality shifted from "human environment" to "the world's future in relation to the environment and socioeconomic development" and eventually to 'Sustainable Development'." Participating states and other key actors agreed in Johannesburg that more ESD progress was needed and that the UN Decade of Education for Sustainable Development (UNDESD) was endorsed.

In her proposal which evaluates the progress during UNDESD, Tilbury (2011) reported the educational shifts proposed by ESD (Table 2.2). As Tilbury reported, as a mid-term conference for the DESD, the Bonn Declaration (2009) called governments to develop ESD policies and frameworks which would ensure the quality education for all and raise awareness about sustainability issues. In the report, it is also emphasized that "learning" in ESD would happen not only in

Table 2.2 Educational shifts proposed by ESD (Tilbury, 2011, p.25)

From	To
Passing on knowledge	Understanding and getting to the root of issues
Teaching attitudes and values	Encouraging values clarification
Seeing people as the problem	Seeing people as facilitators of change
Sending messages	Dialogue, negotiation, and action
Behaving as expert – formal and authoritarian	Acting as a partner – informal and egalitarian
Raising awareness	Changing the mental models which influence decisions and actions
Changing behavior	More focus on structural and institutional change

formal education system but also in daily and professional life.

During the Rio Conference, participating states and other key actors adopted "The Future We Want," a document outlining the conference outcomes and setting a set of goals, which was regarded the first step toward the sustainable development (SDGs). In the latter one, Rio +20 also contained measures for the implementation of sustainable development. Its results included mandates related to the works of future programs particularly for the matters in financial dimension of development and situation of developing small island states and more. Interestingly, in 2015, negotiation process on the post-2015 development agenda ended up with 17 SGDs. The purpose was proclaimed as transformation of our world and has been launched as 2030 Agenda for Sustainable Development (UNESCO, 2021).

Among these goals, SDG4 stands for "Quality Education" which aims inclusive and equitable quality education and lifelong learning opportunities for all. In the same year with the proclamation of 17 SDGs, the World Education Forum adopted a global education strategy in order to implement SDG4 by merging SDG4 and ESD. In other words, the new overarching vision of ESD is accurately inserted in the 2030 Agenda due to its crucial importance. Nowadays, ESD is at the core of all 17 SDGs, which aims to create a sustainable future for our planet and for all a critical point which deserves the attention of both formal and non-formal educators. Thanks to this agenda formed in 2015, a new global framework on ESD was renamed in 2019 as "Education for Sustainable Development: Towards achieving the SDGs or ESD for 2030" (UNESCO, 2020a).

2.5   ESD for 2030

It's been more than 30 years since I first recognized the significance of humans in finding solutions to the environmental problems; and this was when I looked at the results of history of metal pollution in the Golden Horn. It's been more than 20 years since I first began to offer courses on environmental awareness and sustainability education. Today, the world needs education for a sustainable future more than ever. The climate change and the pandemic are the grand challenges we face since the beginning.

In March 2020 when the COVID-19 pandemic dominated the headlines, Xu et al. (2020) found out that in 50 years' time since now, one billion people have managed to live under intolerable heat. The areas which were home to a third of the world's population have been hotter than the Sahara, while hundreds of millions more would have to abandon their homes to rising sea levels (Watts, 2019).

As Giannini (2020) reported, school closures have laid bare inequalities in education. This would be a once-in-a-generation opportunity to improve education to fight global climate change. We should turn ESD into a key to prepare the world for future crises. The global challenges of today have taught us what individuals and societies should have responded to the environmental challenges. This tells us that we, as humans, need the implementation of ESD as the core principle more than before. Moreover, it urges us to embrace the relevance that was set in the 1990s together with the ability to think systematically; to understand complexity and the relations; to anticipate different scenarios; to negotiate trade-offs; to react promptly with the limited information; and to collaborate in finding the best solutions.

2.6   Young Change-Makers and Hope for Future

The young in the society shape our future as they represent our hope. For the last part of this chapter, allow me to introduce young people with the most creative and ingenious solutions to tackle sustainability challenges (UNESCO, 2020b).

In November 2020, UNESCO organized an online workshop with the participation of three inspiring young change-makers to discuss the following questions:

- What are the enabling conditions for young people to become change agents for sustainability?
- How can Education for Sustainable Development (ESD) help young people stay resilient in challenging situations?

The response to the question "what do you need in order to stay resilient?" displayed the in a mind map: Support, hope, love, education, to be heard, family, motivation, a network of support...

The young panelists highlighted several factors such as the necessity of being connected to their local context, a sustainable observation of their own communities, and identification of social and environmental issues which they would address in their societies with their own skills and creativity. Coming to education, the young panelists highlighted that entrepreneurial skills, social and emotional competencies, and supportive learning environments would be necessary to be able to contribute positively to their communities and to become resilient actors of change for sustainable development. The young also agreed that schools and education systems should help young people to nourish their aspirations so that they would build confidence in themselves.

Before closing up with the hope on young people, one last significant work would be once more the one related to the young. Dabija et al. (2019) carried out a study to highlight the sustainable behavior of young consumers (members of Generation Z) and their preferences and attitudes and the rationale behind why they are more willing to choose particular retailers which would be eager to implement green and sustainable offers and to adapt their products accordingly. As a result of a systematic literature review, it was concluded that unlike their parents and older brothers, members of Generation Z, belonging to Generation X and/or Millennials, have indicated a totally different behavior pattern such as being greener, sustainability-oriented, and tech savvy, taking into consideration mainly those companies and especially brands with which they would able to connect and enhance their experiences and feelings. The outcome of this study reveals that when designing their offers and promoting their brands, retailers have to be more aware of young consumers' preferences and expectations. Nowadays, selling products to young consumers might not be even possible without relying on green strategies, either in the production processes or in marketing stages which would promote sustainable principles.

Hopefully, 2030 Agenda implementing SDG4 by merging SDG4 and ESD will help to develop this trend so that a shift to a more sustainable future would be possible. The answer for the question "what urged humanity to launch sustainable development, then to set SDGs, and eventually to redefine education as ESD" comes through the increasing discrepancy between human consumption and the nature's capacity to renew itself. However, the rate for decreasing the discrepancy will be achieved by bringing young change-makers and ESD together.

References

Carson, R. (1962). *Silent spring*. Fawcett Publications.

Carson, R. (Ed.). (1994). *Silent spring with an introduction by vice president Al Gore*. Houghton Mifflin Company.

Commoner, B. (1971). *The closing circle. Nature, man and technology*. Dover Publications, Inc.

Culver, L., Mauch, C., & Ritson, K. (2012). *Rachel Carson's silent spring encounters and legacies*. http://www.environmentandsociety.org/sites/default/files/rcc_issue7_web-3.pdf. Accessed 10 Feb 2021.

Dabija, D. C., Bejan, B. M., & Dinu, V. (2019). How sustainability oriented is generation Z in retail? A literature review. *Transformations in Business & Economics, 18*(2), 140–155.

Ehrlich, Paul R. (1971). *The Poulation Bomb*, Ballantine Books.

Folger, T. (2013). *The next green revolution*. https://www.nationalgeographic.com/foodfeatures/green-revolution. Accessed 15 Feb 2021.

Giannini, S. (2020). *Build back better: Education must change after Covid-19 to meet the climate crisis.* https://en.unesco.org/news/build-back-better-education-must-change-after-covid-19-meet-climate-crisis. Accessed 20 Feb 2021.

Huckle, J. (1990). Environmental education: Teaching for a sustainable future. In B. Dufour (Ed.), *The new social curriculum* (pp. 150–166). Cambridge University Press.

Hungerford, H., Peyton, R. B., & Wilke, R. J. (1980). Goals for curriculum development in environmental education. *The Journal of Environmental Education, 11*(3), 42–47. https://doi.org/10.1080/00958964.1980.9941381

International Food Policy Research Institute [IFPRI]. (2002). *Green revolution: Curse or blessing.* http://ebrary.ifpri.org/utils/getfile/collection/p15738coll2/id/64639/filename/64640.pdf. Accessed 10 Feb 2021.

McKeown, R. (2002). *Education for sustainable development toolkit.* Center for Geography and Environmental Education. University of Tennessee USA.

Meadows, D. H., Meadows, D. L., Randers, J., & Behrens, W. W. (1972). *The limits to growth.* Universe Books.

Orr, D. W. (1992). *Ecological literacy: Education and the transition to a postmodern world.* State University of New York Press.

Orr, D. W. (2004). *Earth in mind: On education, environment, and the human prospect.* Earth Island Press.

Sterling, S. (1992). Review of the year. *Annual Review of Environmental Education, 1992*(5), 7–8.

Teksöz, G. (1989). *Pollution history of the Golden Horn.* [Master's thesis, Middle East Technical University], METU Library Archive.

Tilbury, D. (1995). Environmental education for sustainability: Defining the new focus of environmental education in the 1990s. *Environmental Education Research, 1*(2), 195–212. https://doi.org/10.1080/1350462950010206

Tilbury, D. (2011). *Education for sustainable development: An expert review of processes and learning.* UNESCO. Available in Spanish, French and English. ED-2010/WS/46.

UN. (1992). *UN conference on environment and development: Agenda 21.* https://sustainabledevelopment.un.org/content/documents/Agenda21.pdf. Accessed 11 February 2021.

UNESCO. (1986). Educating in environmental values. *Connect*, Vol. XI, No. 3 (Paris, UNESCO). https://unesdoc.unesco.org/ark:/48223/pf0000156775. Accessed 11 Feb 2021.

UNESCO. (2009). Bonn Declaration https://unesdoc.unesco.org/ark:/48223/pf0000188799/PDF/188799eng.pdf.multi. Accessed 11 February 2021.

UNESCO. (2020a). *ESD for 2030. What is next for education for sustainable development?* https://en.unesco.org/news/esd-2030-whats-next-education-sustainable-development. Accessed 12 Feb 2021.

UNESCO. (2020b). *Meet three inspiring young change makers in Education for Sustainable Development.* https://en.unesco.org/news/meet-three-inspiring-young-change-makers-education-sustainable-development. Accessed 12 Feb 2021.

UNESCO (2021). *Education for sustainable development.* https://en.unesco.org/themes/education-sustainable-development. Accessed 15 Feb 2021.

Ward, B., & Dubos, R. (1972). *Only one earth.* W.W. Norton and Company Inc.

Watts, J. (2019). *Rising sea levels.* https://www.theguardian.com/environment/2019/oct/29/rising-sea-levels-pose-threat-to-homes-of-300m-people-study. Accessed 12 Feb 2021.

Xu, C., Kohler, A. T., Lenton, T. M., Svenning, J. C., & Scheffer, M. (2020). Future of the human climate niche. *Proceedings of National Academy of Sciences of the United States of America, 117*(21), 11350–11355. https://doi.org/10.1073/pnas.1910114117

Yann A. B. (Director) (2009). *HOME (Documentary).* Europa Corp.

Gaye Teksöz is a professor in Science Education at Middle East Technical University in Türkiye. She got her BS, MS and PhD in Environmental Engineering andshe teaches Education and Awareness for Sustainable Development, Environmental Science and Climate Change Education for Sustainability courses. Her research area is mainly on the implementation of Education for Sustainable Development (ESD); especially on the impact of real stories on higher education as well as middle school students' awareness on sustainability and climate change. Nowadays she is working on systems thinking skills as a requirement of ESD.

Part II

Implementation of ESD: Integrating Theory and Practice

Per Sund

Abstract

The purpose of this chapter is twofold: it provides an introduction to research and discussions about selective teaching traditions and discusses the possible consequences for education for sustainable development at lower secondary school level following the implementation of a new Swedish national curriculum in 2011. By comparing a former national report (2002) and a small-scale empirical study (2020), this research shows that the distribution of *teaching traditions* amongst science teachers teaching environmental education/education for sustainable development changed with the introduction of the new curriculum in 2011. The possible consequences of this shift towards a more fact-based teaching are discussed in the light of Dewey's view of *democracy* and Biesta's *functions of education*. The core content that was introduced for science subjects in the new curriculum made science teachers emphasise factual content knowledge and reduce the time for group work. Opportunities for teachers to teach in accordance with a *pluralistic teaching* tradition have decreased as a result of the new curriculum, where the overall role of education is now less focused on enhancing pupils' *emancipation* by an education aimed at *subjectification*. These two consequences of the curriculum change risk limiting science teachers' *democratic* teaching of education for sustainable development, thereby reducing their possibilities to support pupils' development of an *action competence*.

Keywords

Teaching traditions · Selective traditions · Democracy in education · Functions of education · Action competence

3.1 Introduction

The Swedish national curriculum was changed in 2011, primarily to solve problems with lower secondary school pupils' low levels of scientific knowledge in international PISA study comparisons (Wahlström & Sundberg, 2015). After the curriculum change, science teachers experienced a more crowded curriculum that challenged them in their everyday classroom activities. This affected their possibilities to conduct education for sustainable development (ESD), which supports the enhancement of pupils' *action competence*.

P. Sund (✉)
Department of Teaching and Learning, Stockholm University, Stockholm, Sweden

Environmental and Life Sciences, Karlstad University, Karlstad, Sweden
e-mail: per.sund@su.se

It is important for teachers to have support in the national curriculum for different school subjects (content and concepts), for example, maths and science. The teaching of the complex concepts of sustainable development at the lower secondary school level (pupils aged between 13 and 16 years) is supported in the Swedish curriculum (Education, 2011). Sustainable development is mentioned in the curriculum as a whole and also in the objectives of many subject syllabuses, such as those for science, social science and home economics. There is a long and strong support for ESD teaching and the development of pupils' abilities to enhance action competence in the Swedish curriculum. However, this support for ESD was in danger of being reduced when the PISA results in science for pupils aged between 15 and 16 years in Sweden declined dramatically from 2000 to 2012 (Education, 2019). This decline led to concern amongst politicians, and, as a result, a new national curriculum was launched in 2011. The goal-oriented curriculum for the 9-year compulsory school that had been introduced in 1994 (Education, 1994) focused on the development of pupils' abilities and enabled teachers and pupils to work their way through the subject content, methods and timetable towards the knowledge goals in the ways they found best. In contrast, the new goal-oriented curriculum that was introduced in 2011 (Education, 2011) specified a considerable amount of core content in all school subjects. This became especially obvious in the science subjects of biology, chemistry and physics. This introduction of core content was seen as a way of supporting and guiding teachers and pupils towards the knowledge goals and, ultimately, increasing the results in these subjects in forthcoming international PISA studies (Wahlström & Sundberg, 2015). The Swedish PISA results from 2015 and 2018 show a slight increase in the results in mathematics, reading and science, which could be related to the introduction of core content in the 2011 curriculum (Education, 2019).

However, the increase in science core content in the new curriculum has reduced the teaching time for pupil participation in ESD (Sund & Gericke, 2020). As science teachers now have a lot of detailed core content to convey to pupils, there is little time left for group discussions, classroom debates and group work with authentic issues. The empirical comparison in this chapter shows that due to the increase in core content, there is a risk that science teaching will become more fact-oriented and for the dominating teaching methods to be lectures and tests. Earlier ESD research has shown that pupils' active participation is crucial for ESD teaching (Jensen, 2000; Lundegård & Wickman, 2007; Scott & Gough, 2003). Being able to participate in classroom activities is a *democratic issue*. In the everyday classroom setting, pupils should be included as the co-creators of education (Öhman, 2008). In the end, it becomes a democratic issue for the entire education system, which begs the question: What is education for? (Biesta, 2009).

In relation to science teachers and ESD, this question is approached using research on science teachers' selective teaching traditions both before and after the curriculum change in 2011 (Education, 2011). The consequences of the changes in the curriculum are examined using John Dewey's (1916/1966) discussions about the purposes of and *democratic* approaches in education. Biesta's (2009) discussions about the role of education can be approached using the three functions of *qualification*, *socialisation* and *subjectification*. Biesta's educational philosophy is heavily grounded in Pragmatist theory and the work of John Dewey. The increase of science knowledge in a PISA study can generate a decrease in *democracy* in education and, subsequently, limit pupils' possibilities to develop action competence (Jensen & Schnack, 1997) and participate in the development of a more sustainable society in the future.

3.2 Selective Teaching Traditions and Democracy

3.2.1 Tacit Frameworks

Teachers make different choices in their teaching, and these often become visible in the form of different selective teaching traditions (Williams,

1973). One consequence of teachers' habitual choices is that important value-laden content can be hidden in the teaching content, which in turn can contribute to the development of teachers' often tacit selective teaching traditions (Östman, 1995). Knowledge about selective teaching traditions helps teachers to be clearer about what they need to emphasise most in their teaching, e.g. facts, attitudes, the development of competences, etc.

Selective traditions in environmental education (EE) can be understood as collective habits. This way of approaching the selective traditions is based on pragmatist philosophy. According to Dewey (1922), an analysis of such habits means looking at more complex activities and their meaning through their consequences in practice. In this sense, a habit cannot always be explicitly articulated by the teacher, although it can be discerned through reflection, by either an interviewer or interviewee, in terms of the patterns of one's own actions, e.g. choices of teaching content and methods. However, before they can be reflected on, these habits need to be acknowledged before they can be adjusted or changed (Wickman, 2004, 2012; Wickman et al., 2018). These habits should not be regarded as simple, repetitive actions, but as habits that are acquired. A habit is something that is continuously developed as a result of encounters with earlier and current teaching experiences. If many teachers act in similar ways in similar situations, they can be described as having a collective habit, in that they are all participants in a similar selective teaching tradition. Individual teachers develop their personal habits on the basis of the contextual situations that have been created by earlier generations of teachers and educators. An individual does not live in a vacuum (Dewey, 1938/1997), but interacts with and encounters artefacts and physical and social environments.

The reason for studying selective traditions through habits is to find functional ways of reflecting on actions in order to change them, so that they better fit a teacher's teaching purposes. Individual habits are changed and developed by collective habits, while at the same time the individuals involved affect the collective habits. In addition, teachers design their teaching based on the traditions they themselves have been exposed to at school or in higher education. Dewey's (1922) discussion of individual habits and their interplay at an individual and collective level seems to be an accurate description of how selective teaching traditions in EE evolve and are consolidated in the formal education system.

3.2.2 The Origins of Research in Science Education

The purpose of science education in formal schooling has been discussed by researchers for many years (Goodson, 1987). Fensham (1988) claimed that there are two main reasons for teaching science, which he called *induction into science* and *learning from science. Induction into science* aims to make pupils become scientists and focus on learning core science, such as basic scientific concepts and knowledge about science. *Learning from science* aims to teach scientific knowledge for everyday use or to become an active citizen in societal development.

Roberts (1982) developed the concept of *curriculum emphasis* when discerning the purpose of science education in written North American curriculum material. The same scientific content can be taught with two different emphases. The seven different emphases should not be regarded as clearly limited categories, but as capturing 'the essence of very broadly different orientations which science can assume' (Roberts, 1982, p. 246.). Roberts developed these discussions further and described two different visions of how to make pupils more scientifically literate (Roberts, 2007). Vision I is product-oriented towards science knowledge, while Vision II concerns the application of scientific knowledge and stresses that teaching in school also has to focus on the skills that are needed for the application of knowledge. Researchers focusing on socioscientific issues (SSI), such as Zeidler (2003), also claim that pupils need to learn how to use knowledge in practical and societal contexts. In Vision II, it is pointed out that in addition to subject knowledge, education must also include knowledge and skills that enable pupils to apply scientific knowledge in everyday, existential, moral and political contexts (Roth & Désautels, 2002).

When science teaching becomes politicised, it can prescribe solutions and become more normative. This is where science education related to SSI and EE/ESD overlaps with subject/thematic areas in classroom teaching (Sund, 2016). Normativity can also be an obstacle, in that there are many possible fruitful ways towards societal development. Patterns for future actions should not be built into the teaching aims from the start, but need to be scrutinised in democratic discussions (Öhman, 2004). There are some similarities in the discussions about the characteristics of the selective traditions and Visions I and II. There are also differences in the research starting points. Earlier research on the two Visions departs from written curriculum material and theoretical discussions, whereas ESD research applies a methodology to empirical data that is close to classroom practice and includes enquiries and teacher interviews.

3.2.3 Selective Teaching Traditions in Environmental Education

Research on the history of school subjects has shown that they all have different ways, or traditions, of selecting educational content and methods. These traditions can thus be termed selective traditions (Williams, 1973). Since the 1960s, three selective traditions have evolved in Swedish EE, with reference to roots in educational philosophy and how environmental and developmental problems are perceived by teachers. Sandell et al. (2005) described three educational philosophies connected to the selective teaching traditions: *essentialism, progressivism* and *reconstructivism. Essentialism* means that the content of the education ought to be based on science, that the actual subject has priority and that the teaching in school is conducted using adapted scientific terminology and models. Here, the role of a teacher is to be an expert in the subject and convey knowledge and facts to their pupils.

Progressivism puts the pupils in a central position, where the teaching is organised in accordance with the needs and interests of the group of pupils being taught. The choice of teaching methods takes priority, and the emphasis is on collaborations and discussions as important aspects of the learning process. In addition, pupils' knowledge is developed through first-hand authentic experiences in nature and society. *Reconstructivism* emphasises the role of the school in the democratic development of a future sustainable society. Values are in focus here, and the aim is that pupils learn how to critically evaluate possible alternative viewpoints. The solution to environmental problems can be understood in three different ways: as a lack of relevant scientific knowledge, as weakly developed attitudes and un-reflected lifestyles and as informed attempts to solve conflicting human interests. The selective teaching traditions are fact-based environmental education, normative environmental education and pluralistic environmental education, the latter of which is also called ESD by various authors (Lundegård & Wickman, 2007; Sandell et al., 2005). It is important to point out that the descriptions of these traditions (outlined below) have been summarised in order to make them more succinct and easier for the reader to grasp. They were studied for the first time in a large empirical study of teachers (*N* = 568) in the Swedish school system by the Swedish National Agency for Education (2002). Initially, the traditions were categories that resulted from an analysis of a written enquiry. The categories were later made clearer by follow-up interviews with selected teachers. The descriptions outlined below follow the original descriptions (Education, 2002) closely.

The *fact-based tradition* was formed during the development of EE in the 1960s. Environmental issues are regarded as ecological issues. In this tradition, environmental problems are based on a lack of knowledge and can often be solved by science. There is an assumption that if teachers teach scientific knowledge to everyone in schools, then environmental problems caused by human activities will disappear more or less automatically. From an environmental ethics perspective, this tradition is situated within modern anthropocentrism. The natural world is considered to be separate from humanity. In terms of educational philosophy, this tradition is closest to essentialism, where the teaching focuses on the subject knowledge that is needed to solve the current problems. The pedagogic task is to teach pupils the right knowledge and proper knowledge. This knowledge is then assumed to enable pupils to take good decisions and carry out fruitful actions. But, pupils tend not to have opportunities to practise such actions during the actual teaching, because teachers often take it for granted that active competence follows automatically from knowing science. The teaching style is mainly through lectures with very little group discussion or activities in which the learned knowledge can be authentically used. Pupils do not participate in lesson planning either, which implies that teachers take their earlier observations of pupils' attitudes and opinions into account in the planning of future lessons (Sandell et al., 2005).

The *normative tradition* emerged during the societal debate about environmental issues in the 1980s, e.g. the nuclear power referendum in Sweden. Environmental issues are primarily a question of values, where people's lifestyles and their consequences become the main threats to the natural world. Scientific knowledge can offer hints about the best ways of living and be prescriptive in decision-making. The development of an environmentally friendly society is obvious and unambiguous. According to the teachers of this tradition, right knowledge is assumed to automatically lead to better values that make people want to behave more ecologically correctly. From an ethical point of view, humans are regarded as an indispensable part of nature and should therefore adapt to its conditions. The teaching content is partly organised in a thematic way and requires content from disciplines other than science, e.g. social science. In order to ensure that the lessons achieve their intended objectives, close attention is paid to the use of pupils' everyday experiences and attitudes when creating teaching examples and tasks (Sandell et al., 2005). Solving problems in groups in com-

bination with scientific factual information makes this tradition appear as a combination of essentialism and progressivism.

The *pluralistic tradition* developed during discussions in the 1990s in connection with the Rio de Janeiro Earth Summit in June 1992. An increasing uncertainty about environmental issues and the number of different standpoints in environmental debates are important points of departure for this tradition. Environmental issues are viewed as both moral and political problems, and environmental problems are regarded as conflicts between different human interests. Science does not provide guidance on how to act when it comes to environmental issues. In this tradition, EE includes the entire spectrum of social and economic development and is replaced with the concept of ESD (Öhman, 2004). The conflict-based perspective of ESD highlights the democratic processes of classroom activities, where everyone's view is regarded as being equally relevant when deciding on courses of action in environmental and developmental issues. Pluralism is an important starting point for the conduct of teaching in ESD. In terms of environmental ethics, humans are back in focus, in that ESD is anthropocentric. As pupils develop their abilities to engage in democratic discussions about the development of a sustainable society or a more sustainable world, it suggests that the lessons are reconstructivist in character. The various problems that are encountered in the lessons indicate that the teaching methods and approaches also vary from an individual search for more scientific facts to writing articles or formulating collective arguments that can be used and published in newspapers. Democracy in education is a special character of ESD.

3.2.4 Democracy and Action Competence

At present, there are several theoretical publications on the selective traditions in EE (Callahan & Dopico, 2016; Öhman, 2004; Sund, 2008; van Poeck et al., 2019), and some empirical studies also theoretically underpin the traditions

(Östman, 1995; Sund et al., 2020; Sund & Wickman, 2011). The theoretical discussions inspired by John Dewey concern important aspects of teaching, such as *democracy* (Öhman, 2004). Is democracy something that pupils in compulsory schools have to learn about in order to become well-informed democratic citizens, or can they claim to have been born into a democracy with democratic rights and can use them *in* the ongoing educational situation? Another similar discussion about democracy at a public level is 'citizenship as achievement versus citizenship as participation' (van Poeck & Vandenabeele, 2012). Other publications suggest that the democratic approach in the pluralistic tradition can be perceived to be important for developing EE into ESD (Lundegård & Wickman, 2007). This development requires much more than simply adding subject matter content from social science and economics to ecology and science. Here, the overall teaching approach is changed from one that is often normative character to an inclusive participatory democratic approach (Sund & Wickman, 2011).

The democratic participatory approach is a prerequisite for developing pupils' *action competences* (Jensen & Schnack, 1997). In teaching practice and research and at the policy level for global development, the learning outcomes of EE/ESD/Environmental and Sustainability Education (ESE) have increasingly been translated into a number of competences for sustainable development, e.g. critical thinking, collaborative decision-making, future scenario skills and action competence (Leicht et al., 2018). The underlying educational idea is to empower young people by developing key competences. Key competencies are something to achieve, whereas action competence is an ongoing teaching approach that encourages pupils to use the knowledge and abilities they learn at school to guide their actions. Action competence is an educational ideal (Jensen & Schnack, 1997). A developed action competence teaching enables pupils to handle the often-complex societal challenges of sustainable development. The long-term ambition of the development of pupils' action competences is to support young people's

capabilities for reflection and analysis and to support their willingness and possibilities to act based on informed views, decisions and knowledge about how to work for change towards sustainability in democratic societies.

3.2.5 Education and Democracy

The selective teaching traditions in this study are approached by discussing how values can be incorporated into the teaching of sustainability issues and that democracy is not solely structural or organisational, but an approach *in* teaching (Öhman, 2004). The discussion about values focuses on the fact that sustainability issues, such as climate change, do not only include measurable variables like temperature and carbon dioxide emissions. Sustainable development is also a political concept, in that its use is tied to specific interests and priorities. While science often discerns and warns about issues that affect humanity, for example, climate change and the ozone layer, the solutions need to be arrived at internationally. There is an interplay between facts and warnings and the democratic processes for solutions.

According to the Swedish curriculum material from 1994, it is not enough to convey knowledge about basic democratic values. Rather, the teaching should be conducted in a democratic way and prepare pupils for an active participation in societal development (Education, 1994). Sustainable development and democracy are strongly promoted in the new curriculum from 2011 (Education, 2011), which makes the relation between them important to study. The relationship between democracy and education is somewhat problematic though. There is a paradox between socialising pupils into informed and democratic members of society and at the same time supporting young people to become free-thinking, independent, autonomous intellectuals who may even want to challenge the societal system that educated them. This double pedagogical mission is important for ESD teaching. How do you get young people involved and engaged in solving specific sustainability issues and, at the same time, leave the door open for different opinions about possible solutions? The pluralistic teaching tradition is one way of meeting this educational challenge.

In educational philosophy, the balance between an expert-driven and socialising education is discussed in relation to an inclusive and democratic participation-driven education. Evidence-based teaching, or 'what works', is often put forward as a prerequisite for good teaching (Biesta, 2007). Biesta (2007) argues that this type of expert-driven teaching sets the democratic influence of the participants aside. There is more than one way towards a more sustainable future. This type of steering and normativity towards an unknown future is important in the discussion about using the preposition *for* in ESD. The preposition indicates that we know what knowledge and skills will be needed in the future (Jickling, 1992). The basic aim of education is to educate pupils – not for any specific future reasons, but to support their emancipation and empowerment (Jickling & Wals, 2008). Dewey (1929/1958) reminds us to handle educational situations with the 'ends-in-view' perspective and that it is important not to set long-term purposes for the use of the knowledge and skills. The purpose of education is to work and understand sustainability issues in the actual educational situation and to discuss the consequences of our actions as far as we can foresee them with some degree of accuracy.

This discussion about the double mission of education highlights the question posed earlier in the chapter: What is education for? Biesta (2009) describes the purpose of education in terms of three functions. First, that education has a role in pupils' *socialisation* into society by conveying social, political and cultural values and behaviour that aim to preserve a specific democratic society. Historically, socialisation has been a school's important assignment in fostering citizens. Second, education contributes to pupils' *qualifications*, thereby advancing their knowledge, skills and competences for their lives in various arenas, such as the labour market (different professions), further studies and as citizens. This mission is perhaps now the most important

assignment in schools due to the neo-liberal market view of 'value for taxpayers' money'. Third, education has a role to play in pupils' *subjectification*. This is about the emancipation of pupils as humans and providing them with agency as citizens. However, the function of subjectification does not imply pure individualism or an independent individual subject. The term subjectification is chosen with the intention that individuals will be interrelated-to (or subjected-to) in the surrounding world at large and with other people.

The functions are separate, yet overlap in different kinds of education and generally aim in some way at societal participation. Tensions arise in the intersections between the functions, and teachers are faced with several educational challenges. An example is the earlier mentioned double pedagogical mission, i.e. the meeting between legitimising common perspectives and individual critical thinking. Teachers are expected to uphold democratic values and trust political institutions and to allow pupils to critically review them. Thus, there is a tension in allowing pupils to 'be acting citizens' and seeing them as 'becoming citizens'.

Biesta (2009) claims that subjectification is the most important aspect to stress in contemporary schooling. It is a function that school systems in general have not emphasised enough. Education should take pupils' thoughts and lives seriously, allow their values to matter in school discussions and not always decide in advance what the answers should be. It is about democracy *in* education (Dewey, 1916/1966). This does not imply that pupils should only express personal opinions about societal development and sustainability issues but also encounter other pupils' opinions in school. Pupils should have opportunities to experience difference, plurality and resistance to their own worldviews. Becoming an emancipated individual is not just an individual process, but is about being 'subjected-to others'. If education only conveys facts and figures, prescribes specific norms and does not allow pupils to independently express themselves, the teaching will solely focus on the qualifying and socialising functions that promote the existing societal order. However, this does not mean that qualifications and socialisation are not important (Biesta, 2012). On the contrary, the knowledge and abilities that according to the national curriculum schools are to stress are significant contributions and vital prerequisites for pupils' citizenship education.

Dewey (1916/1966) concluded that practising competences without a meaningful content is non-sense. ESE teaching focuses on the development of competences. The learning of content is a simultaneous process, in which content is suggested to be learned in action. To Dewey, democracy is more than a way of ruling or guiding societal development. It is also more than electing a government. Rather, it is:

> a mode of associated living, of conjoint communicated experience. The extension in space of the number of individuals who participate in an interest so that each has to refer his [or her] own action to that of others, and to consider the actions of others to give point and direction to his [or her] own, is equivalent to the breaking down of those barriers of class, race, and national territory which kept [people] from perceiving the full import of their activity. (Dewey, 1916/1966: 87)

According to Dewey, democracy does not mean a struggle where people relate to each other by expressing predetermined standpoints, but is rather a life form, where people with different experiences together create new possibilities to influence each other. This change in the view of democracy from a form of ruling to a life form gives democracy in teaching a deeper meaning than everybody's right to express themselves and claim other formal rights. It is about how we as humans encounter each other in everyday situations. It is not just about what we do to others but also how we listen and adjust to other people in group situations, how we learn to cherish other people's differences and thereby how we learn to grow as individuals. This is described as the resistance, or the subjected-to, that Biesta (2009) points out as important in education aiming at subjectification. Dewey (1916/1966) stressed that formal schooling was one of the most important arenas for democratic practice and communication in our modern society. The main reason for this view is that school is a place in which

people with different experiences and backgrounds meet in terms of social class, race, gender, religion and ethnicity. Pupils therefore have the possibility to *live through* democracy. When learning *about* democracy, they can practise *in* and *through* democracy as an integrated educational approach. A shift towards a more fact-based or normative teaching, due to a curriculum change, can thus be problematic for achieving a more sustainable society. In this way, attention to the development of a more pluralistic education for sustainable development in compulsory school becomes even more important. If not, schools will risk losing their roles as democratic arenas for individual emancipation and societal development.

3.3 Distribution of Selective Traditions: Before and After the Curriculum Change in 2011

In order to illustrate the shift towards the fact-based teaching tradition for science teachers in the curriculum changes in 2011, this empirical part of the chapter makes use of three different sources: a Swedish national report from 2002 (Education, 2002), results from a journal article (Sund et al., 2020) on the distribution of selective teaching traditions and excerpts from unpublished data in a new study from 2020 (Sund & Gericke, 2020). This part of the chapter also offers some information about teacher groups other than science in order to increase the overall interest in research on teaching traditions related to ESD.

The national report from 2002 (Education, 2002) shows that lower secondary teachers in the Swedish compulsory school (a total of 67 language, science and social science teachers who all claim to be teaching EE/ESD) are distributed in the three selective teaching traditions as follows: fact-based tradition 11%, normative tradition 67% and pluralistic tradition 22% (Table 3.1). The report does not show the distribution of science teachers in all the traditions, but states that 12% (Education, 2002. p 122) of the science

Table 3.1 The distribution of teachers in lower secondary school (aged 13–16 years), 67 teachers in total, from different subjects and science (separate row) in selective teaching traditions (Education, 2002)

Lower secondary teachers	Fact tradition	Normative tradition	Pluralistic tradition
Science, social science and language together	11%	67%	22%
Science teachers only	12%	No data available	No data available

Table 3.2 The distribution of teachers from different subject areas in selective teaching traditions (Sund et al., 2020)

Subject area	Fact tradition	Normative tradition	Pluralistic tradition
Science	54% (8)	33% (5)	13% (2)
Social science	9% (1)	9% (1)	82% (9)
Language	10% (1)	60% (6)	30% (3)

Percentage (number of teachers)

teachers in lower secondary school follow the fact-based tradition. This report used enquiries and follow-up interviews.

The study from 2020 (Sund et al., 2020) compares lower secondary school teachers in different subject areas and the distribution of teachers in the different selective teaching traditions for each subject area (Table 3.2). This study uses written enquiries and follow-up individual interviews with 36 teachers.

A comparison between the national report in 2002 (Education, 2002) and the study from 2020 (Sund et al., 2020) shows that there is a shift towards the fact-based tradition concerning lower secondary science teachers. Although the number of teachers involved is few ($N = 36$), the tendency is clear. This comparison is strengthened by a second study (Sund & Gericke, Submitted) that also includes group interviews about ESD in science and starts with a specific question: *In your experience, which social changes have put pressure on you to change your teaching in the last 10 years?* Here, the last 10 years is the interval between 2007 and 2017 and includes the launch of the latest curriculum

in 2011 (Education, 2011). This period was deliberately chosen by the research group in order to discern changes due to the new curriculum. This was also communicated to the lower secondary school teachers in the ten teacher groups (four groups from science, three from social science and three from language). The strongest societal change pressure is the digitalisation of education, and the second strongest pressure the new national curriculum. In the transcribed data, there are clear oral statements from all four science groups (15 science teachers in total). Two out of the four groups point to the pressure of an overcrowded science curriculum. When comparing before and after the introduction of the new curriculum, a member of the third group claimed:

> There are now more concepts in the curriculum. We could go deeper in concepts earlier, but now the overall knowledge in science has been reduced [an explicit comparison before and after the latest curriculum, remark].

To offer more data and further strengthen the fact that science teachers in Sweden moved more towards the fact-based tradition after the latest curriculum change, some individual voices from the group discussions (Sund & Gericke, Submitted) are presented below. These voices refer to overcrowded curricula, core content, facts and fewer possibilities for group work. Examples from all the four science teacher groups are presented. The teachers in each group work at the same school and know each other well, which means that the interview situation is socially secure and enables them to speak freely about their current situation.

In the following, T = individual teachers and I = interviewer.

Group 1

T 1: One pressure [a response to the start question, above] is the massive amount of core content in course plans, there is actually quite a lot. In physics there are an enormous number of concepts, and also in biology.

………

I: Form of work – have pupils worked more individually or more in groups over the ten last years?

T 2:   The forms are more varied, I would say. There is more small talk (small buzz), turning round and checking with your neighbour. Group work as it used to be will probably never happen again until I try to organise it. There is not enough time for it…

This brief discussion with two teachers shows that the increased amount of core content reduces the possibilities for group work by limiting the time for activities other than those that are teacher led and controlled in the fact-based tradition.

The next excerpt is a reminder from a teacher to other peers to keep to the general purposes of chemistry education and not risk drowning in the amount of core content. The focus on teaching guided by subject purposes that was stressed in the former curriculum in 1994 made this experienced teacher remind herself and her peers about what is important in teaching.

Group 2

T1:   I think that the summary of the purposes with the course [the teaching aims in the curriculum], you cannot forget to keep them in focus. You should not be caught up in the core content only. These three or four purposes are actually the reason why they [pupils] have chemistry at all in my classroom. You need to read them again now and then, otherwise you get suffocated by the core content…

In this excerpt, two teachers compare the old and new curriculum. According to one teacher, the former curriculum was too 'woolly' and contained no core content at all, and each school subject was goal-oriented through well-described subject purposes in the course plan. The other teacher reflected on and considered this former 'woolliness' as a lost opportunity to adapt to and harness the pupils' own interests.

Group 3

T1:   But you need only to make a comparison between core content in the 2011 curriculum and that in 1994.

T2   In the 1994 curriculum it was very woolly.

T1:   Yes, but at the same time it was much freer, you could stop at a topic that your class was interested in, such as nuclear power. We could dig deep down into knowledge. But now, we have the core content, all knowledge requirements [grading criteria, remark]. We do not have the time to stop and think. We just have to continue dutt, dutt, dutt [makings sounds, remark]

The next excerpt shows that cross-curricular collaborations could be part of the solution to working with overcrowded curricula. This kind of extra-curricular collaboration between teachers of different school subjects could benefit ESD teaching in school and is also emphasised in the new 2011 curriculum (Sund & Gericke, 2020). This can be regarded as a way of strengthening ESD teaching in Swedish schools. Unfortunately, the crowded curriculum seems to consume what is regarded as valuable time for the development of collaborative work.

Group 4

TI: The new design in the steering documents, it is like a pendulum that swings back and tells you not just to work on enhancing abilities, but that you need to ensure that they [pupils] know some facts as well. The amount of subject matter has increased, which forces you to make priorities in your teaching.

I: More subject matter, how has it increased?

T2:   By the core content in the 2011 curriculum [the new curriculum].

T1:   With the added core content [compared to the earlier curriculum, remark] it feels as though there is more content in general. There is a lot of discussion about needing to include all the subject matter in the core content in each subject. Then we would need to work in a cross-disciplinary way and try to fuse our subjects together to save time, otherwise we will never succeed in getting through our expected teaching. We need to have more cross-disciplinary collaborations in our teaching, but I do not think we have done that yet. I think it is about needing time to plan at the beginning, when people are starting a collaboration, and it is difficult to organise that in practice. Not enough time is set aside for these types of development.

3.3.1 The Result of the Comparison Due to Curriculum Change

The comparison of the change in the distribution of teaching traditions between the two studies (Education, 2002; Sund et al., 2020) and the unpublished excerpts from Sund and Gericke (2020) shows the consequences of the introduction of the new curriculum on selective teaching traditions and pupils' possibilities to participate in their education. The shift towards the fact-based tradition and teachers' voiced experiences of overcrowded curricula due to the adaption to the new curriculum that was developed in response to earlier results in PISA studies (Wahlström & Sundberg, 2015) all show that the possibilities to undertake a pluralistic and democratic ESD teaching decreased in Swedish schools after 2011.

3.4 Discussion

The results of the comparison of reports and studies (Education, 2002; Sund & Gericke, Submitted; Sund et al., 2020) before and after the implementation of the new curriculum in 2011 show that when teaching issues relate to EE and ESD, science teachers tend to adopt the fact-based tradition.

The problem with the fact-based tradition is that this type of teaching easily turns sustainability and development issues into a question of a lack of objective and neutral knowledge. The risk is that teaching will ignore values and conflicting interests that are intimately connected to how the problems are perceived and how they can best be remedied. Further, in general, pupils are not given opportunities to practise participating in discussions in which different perspectives and stances are voiced, critically evaluated and challenged. The problem with the normative tradition is that that teaching approach delivers ready-made answers to value-laden questions. There is therefore a risk that teaching will become a political tool in the creation of a specific and pre-defined society. The teaching in this tradition can impede an education that embraces democracy and a pluralism of ideas and opinions. It can, in other words, become an education *for* a specific goal in the future (Jickling & Wals, 2008) or one that threatens pupils' possibilities to encounter resistance and be challenged in a plurality of alternative answers (Biesta, 2009). In comparison, the pluralistic tradition teaching approach seems to include many advantages. For example, it includes the political dimensions of sustainability and development issues and strives to avoid the risk of indoctrination by supporting pupils' critical thinking and strengthening their democratic action competences. Through questions and comments, teachers can guide pupils in their decision-making, help them to evaluate and compare alternatives and enable them to explore their arguments in discussion and debates (Rudsberg & Öhman, 2010).

What are the objections to a pluralistic teaching approach? One possible objection is that democracy is time-consuming. A teaching that builds on the examination of many and sometimes conflicting development alternatives needs time and space for group discussion. An overcrowded curriculum therefore becomes an immediate challenge for teachers in the classroom. It is difficult also in a pluralistic approach to produce quick and easy solutions to societal challenges. The consequences of a pluralistic teaching are that it does not offer clear and effective solutions for the conversion of the society towards sustainability. In addition, public voices in the media claim that we do not have sufficient time to wait for the best possible actions concerning, e.g. climate change, but that 'we have to act now according to the best known knowledge'. There is not enough time to wait for democratic processes to generate reliable answers. Democracy takes (too much) time, and it is difficult to know which decisions, quick or democratic, will create the most sustainable world in the long run. Democracy takes time. Further, overcrowded curricula do not offer time for democratic and pluralistic group discussions in schools, and there is less room for teaching that focuses on subjectification (Hasslöf & Malmberg, 2015). National curricula that aim to become competitive in PISA studies are constructed with the qualification function of education in mind (Biesta, 2009, p. 37). Earlier research has shown that young people want to be an active part of education, participate in the development of the future society (Jensen, 2000) and be part of a sustainability education that communicates hope (Ojala, 2013). Subjectification, as a function of education that aims at students' independent thinking and empowerment, aligns with their own ideas of an education that enables them to be part of a positive societal development.

3.5   Conclusion and Suggestions

Curriculum change as an attempt to solve knowledge problems in science and increase the results in PISA may create problems in education by preventing teachers from teaching about sustainability issues in a democratic way. The strong belief in factual knowledge conveyed by lecturing and instruction can lead to a reduction in pluralistic teaching and the educational function of subjectification. Knowledge about science is important (qualification), as is understanding the values of and in a democratic society (socialisation). These functions are prerequisites for a subjectification function of an education in which democracy is a life form (Dewey, 1916/1966). The balance between knowledge and values for participation and the room for pupils' subjectification (possibilities to be subjected-to others) for empowerment and emancipation are important

for research to continuously study and raise questions like 'what is education for?'

For teachers to be able to open space for student's subjectification, there are some things teachers can reflect on in their daily practice. Less educational content is often more learning opportunities for students. They have time to reflect and deepen their knowledge. Teachers use authentic real-world problems instead of having students practise on simplified 'school issues' in textbooks. Good teaching is about including the learners. Teachers should listen to students' interests and invite them to open discussions about important future issues. In these discussions, students learn and develop their argumentation abilities. They are able to develop their factual and value groundings for their decision-making abilities in important issues for the future.

Acknowledgements This research was supported by ROSE (Research On Subject-specific Education) at Karlstad University and by Stockholm University.

References

Biesta, G. (2007). Why "what works" won't work: Evidence based practice and the democratic deficit in educational research. *Educational Theory, 57*(1), 1–22.

Biesta, G. (2009). Good education in an age of measurement: On the need to reconnect with the question of purpose in education. *Educational Assessment, Evaluation and Accountability, 21*(1), 33–46.

Biesta, G. (2012). The future of teacher education: Evidence, competence or wisdom? *Research on Steiner Education, 3*(1), 8–21.

Callahan, B. E., & Dopico, E. (2016). Science teaching in science education. *Cultural Studies of Science Education, 11*(2), 411–418.

Dewey, J. (1916/1966). *Democracy and education: An introduction to the philosophy of education.* Free Press.

Dewey, J. (1922). *Human nature and conduct: An introduction to social psychology.* Holt.

Dewey, J. (1929/1958). *Experience and nature* (2nd ed.). Dover.

Dewey, J. (1938/1997). *Experience and education. Touchstone.* Simon and Schuster.

Education, T. S. N. A. f. (1994). The curriculum for the (Swedish) compulsory school, preschool and school-age [Läroplan för det obligatoriska skolväsendet, förskoleklassen och fritidshemmet Lpo 94]. http://ncm. gu.se/media/kursplaner/grund/Lpo94.pdf. Accessed 10 May 2021.

Education, T. S. N. A. f. (2002). Sustainable development in school. *Skolverket*. http://www.skolverket.se/ publikationer?id=925. Accessed 1 May 2021.

Education, T. S. N. A. f. (2011). Curriculum for the compulsory school, preschool class and the leisure-time centre 2011. *Skolverket*. http://www.skolverket. se/2.3894/in_english/publications. Accessed 15 May 2021.

Education, T. S. N. A. f. (2019). PISA 2018. 15-åringars kunskaper i läsförståelse, matematik och naturvetenskap [PISA 2018. 15-teen years old knowledge in reading, mathematics and science]. *Skolverket*. https:// www.skolverket.se/publikationsserier/rapporter/2019/ pisa-2018.-15-aringars-kunskaper-i-lasforstaelse-matematik-och-naturvetenskap. Accessed 20 May 2021.

Fensham, P. (1988). Familiar but different: Some dilemmas and new directions in science education. In P. Fensham (Ed.), *Development and dilemmas in science education* (pp. 1–26). The Falmer Press.

Goodson, I. (1987). *School subjects and curriculum change. Studies in curriculum history*. Falmer Press.

Hasslöf, H., & Malmberg, C. (2015). Critical thinking as room for subjectification in Education for sustainable development. *Environmental Education Research, 21*(2), 239–255. https://doi.org/10.1080/13504622.20 14.940854

Jensen, B. B. (2000). Participation, commitment and knowledge as components of pupils' action competence. In B. B. Jensen, K. Schnack, & V. Simovska (Eds.), *Critical environmental education and health education* (pp. 219–238). Research Centre for Environmental and Health Education. The Danish University of Education.

Jensen, B. B., & Schnack, K. (1997). The action competence approach in environmental education. *Environmental Education Research, 3*(2), 163–178.

Jickling, B. (1992). Why I don't want my children educated for sustainable development. *Journal of Environmental Education, 23*(4), 5–8.

Jickling, B., & Wals, A. E. J. (2008). Globalization and environmental education: Looking beyond sustainable development. *Journal of Curriculum Studies, 40*(1), 1–21.

Leicht, A., Heiss, J., & Won Jung, B. (2018*).* Issues and trends in education for sustainable development. *UNESCO*. https://unesdoc.unesco.org/ark:/48223/ pf0000261445.

Lundegård, I., & Wickman, P.-O. (2007). Conflicts of interest: An indispensable element of education for sustainable development. *Environmental Education Research, 13*(1), 1–15.

Öhman, J. (2004). Moral perspectives in selective traditions of environmental education. In P. Wickenberg, H. Axelsson, L. Fritzén, G. Helldén, & J. Öhman (Eds.), *Learning to change our world* (pp. 33–57). Studentlitteratur.

Öhman, J. (2008). Environmental ethics and democratic responsibility- A pluralistic approach to ESD. In J. Öhman (Ed.), *Values and democracy in education for sustainable development - contributions from Swedish research* (pp. 17–32). Liber.

Ojala, M. (2013). Emotional awareness: On the importance of including emotional aspects in education for sustainable development. *Journal of Education for Sustainable Development, 7*(2), 167–182.

Östman, L. (1995). *Socialisation och mening: No-utbildning som politiskt och miljömoraliskt problem* [Meaning and socialisation. Science education as a political and environmental-ethical problem]. Almqvist & Wiksell International.

Roberts, D. A. (1982). Developing the concept of "curriculum emphases" in science education. *Science Education, 66*, 243–260.

Roberts, D. A. (2007). Scientific literacy/Science literacy. In S. K. Abell & N. G. Lederman (Eds.), *Handbook of research on science education* (pp. 729–780). Lawrence Erlbaum Associates.

Roth, W.-M., & Désautels, J. (2002). *Science education as/for sociopolitical action*. Peter Lang.

Rudsberg, K., & Öhman, J. (2010). Pluralism in practice - experiences from Swedish evaluation, school development and research. *Environmental Education Research, 16*(1), 95–111.

Sandell, K., Öhman, J., & Östman, L. (2005). *Education for sustainable development*. Studentlitteratur.

Scott, W., & Gough, S. (2003). *Sustainable development and learning*. RoutledgeFalmer.

Sund, P. (2008). Discerning the extras in ESD teaching: A democratic issue. In J. Öhman (Ed.), *Values and democracy in education for sustainable development - contributions from Swedish research* (pp. 57–74). Liber.

Sund, P. (2016). Discerning selective traditions in science education: A qualitative study of teachers' responses to what is important in science teaching. *Cultural Studies of Science Education, 11*(2), 387–409.

Sund, P., & Gericke, N. (2020). Teaching contributions from secondary school subject areas to education for sustainable development – A comparative study of science, social science and language teachers. *Environmental Education Research*. https://doi.org/10 .1080/13504622.2020.1754341

Sund, P., & Gericke, N. (Submitted). External pressures – major change forces on teaching through teachers' everyday adaptations. *Scandinavian Journal of Educational Research*.

Sund, P., Gericke, N., & Bladh, G. (2020). Educational content in cross-curricular ESE teaching and a model to discern teacher's teaching traditions. *International Journal of Education for Sustainable Development*. https://doi.org/10.1177/0973408220930706

Sund, P., & Wickman, P.-O. (2011). Socialization content in schools and education for sustainable development - II. A study of students' apprehension of teachers' companion meanings in ESD. *Environmental Education Research, 17*(5), 625–650.

Wahlström, N., & Sundberg, D. (2015). *En teoribaserad utvärdering av läroplanen Lgr 11, RAPPORT 2015:7*. [A theory based assessment of the national curriculum for compulsory school 2011, Report 2015:7]. Institutet för arbetsmarknads- och utbildningspolitisk utvärdering. https://www.ifau.se/globalassets/pdf/se/2015/r-2015-07-teoribaserad-utvardering-av-laroplanen-lgr11.pdf. Accessed 20 May 2021.

van Poeck, K., & Vandenabeele, J. (2012). Learning from sustainable development: Education in the light of public issues. *Environmental Education Research, 18*(4), 541–552.

van Poeck, K., Östman, L., & Öhman, J. (2019). *Sustainable development teaching*. Routledge.

Wickman, P.-O. (2004). The practical epistemologies of the classroom: A study of laboratory work. *Science Education, 88*(3), 325–344.

Wickman, P.-O. (2012). How can conceptual schemes change teaching? *Cultural Studies of Science Education, 7*, 127–136.

Wickman, P.-O., Hamza, K., & Lundegård, I. (2018). Didaktik och didaktiska modeller för undervisning i naturvetenskapliga ämnen [Didactic models and how they can be produced through didactic modelling]. *NorDiNa, 14*(3), 239–249.

Williams, R. (1973). Base and superstructure in Marxist cultural theory. *New Left Review, 82*, 3–16.

Zeidler, D. (Ed.). (2003). *The role of moral reasoning on socioscientific issues and discourse in science education*. Kluwer Academic Publishers.

Per Sund is a docent in science education at the department of teaching and learning at Stockholm University, Sweden. He has been a Research fellow at Karlstad University 2017-20. Per's research interest is science education and environmental and sustainability education from a teacher's perspective. He is involved in several research projects, including Research on subject education, ROSE. He is the former the link-convenor (2013-2017) of the Environmental and Sustainability Education Research Network, ESER (no: 30) collaborating within the European Educational Research Association, EERA. Per trains and supervise teachers and student teachers at national and international level.

Outdoor Education for Sustainability with Systems Thinking Perspective

4

Güliz Karaarslan-Semiz

Abstract

In order to achieve the goals of ESD and build a sustainable society, we need a new line of thinking. This new way of thinking is called systems thinking, which is a key component of ESD. It is clear that we need more systems thinkers to find sustainable solutions to today's complex problems: systems thinking should be integrated into school curriculum, and learning activities should be designed to introduce systems thinking to both students and teachers. In this chapter, I offer outdoor education for sustainability as a pedagogy to develop young people's systems thinking skills. Outdoor education provides a diverse learning environment and helps the youth understand the interactions between the Earth systems and build a strong relationship with our planet. In the chapter, I first examine the theoretical foundations and practical applications of systems thinking within the ESD context and describe some systems thinking components in the science and sustainability context. Then, I explain the relationship between outdoor education, ESD, and systems thinking. I conclude this chapter with an outdoor learning activity example which can be applied in an urban environment. Teachers can use or adapt this activity to improve students' systems thinking skills.

Keywords

Systems thinking · ESD · Sustainability · Outdoor education · Young people

4.1 Introduction

We live in an age of crisis today. Social, ecological, economic, and health crises impact the life support systems of Earth and lead us to think about our relationship with Earth again. Lönngren (2014) notes that sustainability issues such as climate change, poverty, biodiversity loss, and global health are not only technical problems. Instead, they are highly complex and contested problems including multiple causes and multiple solutions, and they cannot be solved with generic principles or linear thinking (Blackman et al., 2006). For example, the Covid-19 pandemic is a global crisis associated with many problems such as the destruction of forests, weakened ecosystems, and climate crisis. Filho et al. (2020) examined the impacts of the Covid-19 crisis on the social, environmental, and economic aspects of sustainability. The authors argued that the main impacts of the Covid-19 crisis on sustainability are related to income reduction, increasing pov-

G. Karaarslan-Semiz (✉)
Department of Mathematics and Science Education, Ağrı İbrahim Çeçen University, Ağrı, Türkiye
e-mail: gkaraarslan@agri.edu.tr

erty, food crisis, health problems, inequalities, unemployment, and limited access to learning. The Covid-19 crisis has shown us that a change in one part of the Earth's system can cause diverse problems in other systems and can threaten human lives. Moreover, the pandemic verified that we need a holistic view to understand how the Earth systems work and how the consequences of our actions impact our planet (Vasconcelos & Orion, 2021). The major problems we encounter today cannot be resolved with our current way of living and traditional thinking (UNESCO, 2014a; Tilbury, 2007). We need to learn how to live sustainably and change our old ways of thinking (Sterling et al., 2005; UNESCO, 2020). Stibbe (2019) notes that we need to be aware of the stories that are not working and that contribute to unsustainability and we should create new stories to build a sustainable society.

In order to deal with today's complex problems, individuals should be equipped with necessary skills and competencies, which may be possible through a transformation toward ESD (Wiek et al., 2014). ESD is a well-established framework to prepare young people to cope with the current problems of the world and to build a sustainable future (UNESCO, 2020). To reach the goals of ESD and understand the complex systems and interactions, we need a new line of thinking, which is called systems thinking, a key component of ESD (UNECE, 2011) and an important framework to realize the complex systems and relationships and to see the big picture (Sterling, 2003; Tilbury & Cooke, 2005). One of the ways to achieve a sustainable society is to understand how social, environmental, and economic issues are linked to each other (Sterling et al., 2005). Therefore, in the current times, there is a need to develop students' ability to understand sustainability problems in a holistic and integrative manner (Lönngren & Svanström, 2015). For this reason, systems thinking skills should be developed in schools as students can understand the complex and dynamic relationships and they can feel a part of the solution (Schuler et al., 2018). Moreover, when students have systems thinking skills, they can make important decisions to create healthy ecosystems,

sustainable economies, and equal social systems for humanity (Booth-Sweeney, 2017). Students should understand how human systems and ecological systems should be in balance, and this understanding can be developed with the help of ESD-competent teachers. It is important to strengthen the capacity of teachers, educators, trainers, and change agents to create educational change toward sustainability (UNESCO, 2014a). Particularly, teachers are perceived as the key agents to develop the abilities of new generations for a sustainable world (UNESCO, 2014b). Teachers could design and facilitate learning environments to develop students' systems thinking skills (Strachan, 2012). They could assist their students in developing systems thinking skills in primary and secondary schools.

In this chapter, I explore the theoretical foundations and practical applications of systems thinking within the ESD context. Then, I explain the outdoor education for sustainability approach as a learning strategy to foster the systems thinking skills of students. I end the chapter with an outdoor learning activity which can be useful for primary and secondary school teachers to develop their students' systems thinking skills.

4.2 Systems Thinking as an Important Skill for Dealing with Today's Complex Problems

Before describing systems thinking within the ESD context, there is a need to explain the historical and theoretical foundations. Systems thinking was unearthed as a reaction to reductionism. According to the reductionist approach, systems work in a static, mechanical, closed, and linear manner (Zhang & Ahmed, 2020). As systems are complex, interconnected, and dynamic, reductionism does not describe how complex systems work. Today, this old view of the world is unsuccessful in solving complex problems. Meadows (2008) noted that hunger, poverty, environmental degradation, economic instability, unemployment, wars, and diseases are all systemic problems because the system structure produces these problems. Therefore, we need to see the world in a different way and change our perception to understand how systems work. Systems thinking helps us change our perceptions. Before defining systems thinking, we should first understand what "system" means. Meadows (2008) states that a system is not a collection of things. According to the author, a "system is an interconnected set of elements that is coherently organized in a way that achieves something" (p.11). Based on this definition, system includes three important dimensions which are *elements, interconnections*, and *a function* (Meadows, 2008). For example, a school is a system; Earth is a system; the national economy is a system; a tree or an animal is a system (Meadows, 2008). Some corporations see forest from a reductionist perspective and cut down and burn forests easily. Forest is not merely a collection of trees; it is a large system including trees, animals, plants, microorganisms, and the interconnections between these organisms. System functions are significant because a small change in the functions of a system can lead to significant changes in the whole system (Meadows & Wright, 2009). The story of the wolves in Yellowstone National Park is a good example to understand how ecosystems work in relationships. The disappearance of the wolves in Yellowstone after the 1930s changed the flora and fauna of the park and affected the entire system. The reintroduction of the wolves as apex predators in the 1990s helped to restore the overall ecosystem in the park. A systems thinker can understand how the complex systems work and how the change in one part of the system impacts the whole system in a similar way to natural ecosystems.

In the 1930s, systems thinking was conceptualized by biologists, psychologists, and ecologists. As a biologist, Ludwig von Bertalanffy proposed general systems theory (Capra & Luisi, 2014). von Bertalanffy (1969) explained the unity of science; defined living organisms as open, dynamic, and interconnected systems; and for the first time listed the characteristics of systems thinking (as cited in Zhang & Ahmed, 2020). When the literature is examined, it is difficult to find a unique definition of systems think-

ing. Richmond (1993) described systems thinking as a holistic paradigm to understand the complexities in the world, and he defined a systems thinking model including seven tracks which are *dynamic thinking, closed loop thinking, generic thinking, structural thinking, operational thinking, continuum thinking*, and *scientific thinking*. These seven tracks indicate the multidimensional nature of systems thinking, and each should be developed to create a meaningful learning (Richmond, 1993). On the other hand, Peter Senge, one of the leaders of systems thinking, defined the term as "a discipline for seeing wholes, as a framework for seeing interrelationships rather than things, for seeing patterns of change rather than static snapshots" (Senge, 2006, p. 68). Senge contributed to understanding the language of systems thinking (Zhang & Ahmed, 2020). In the literature, systems thinking is also defined as a framework to recognize the whole systems, to understand the interrelationships, and to see longer-term solutions and the bigger picture (e.g., Capra, 1999; Sleurs, 2008; Sterling, 2003; Tilbury & Cooke, 2005).

Systems thinking has been accepted as a significant skill that needs to be developed in schools. Logan (2018) maintained that school curricula should have a framework to develop students' systems thinking skills. Karaarslan-Semiz (2021) argued that systems thinking research can focus more on identifying and developing the systems thinking skills of primary and secondary school students. Moreover, the author highlighted that both indoor and outdoor learning strategies in science and sustainability education can be used to develop students' systems thinking skills. In the literature, we can find systems thinking inquiries in different contexts such as the Earth system, biology education, education for sustainability, and climate change education. In these studies, various systems thinking components were explained (see Table 4.1). The systems thinking components are as follows: *recognizing components in a system and complex relationships, recognizing feedback loops, dynamic thinking, cyclic thinking, recognizing hidden dimensions in a system, time dimension,*

Table 4.1 Some systems thinking components and explanations

Components of systems thinking	Explanations
Identifying components in a system and complex relationships (Ben-Zvi-Assaraf & Orion, 2005; Karaarslan & Teksöz, 2020; Stave & Hopper, 2007; Sterling et al., 2005)	Understanding the whole system and how the parts of the system are related to each other. An alteration in one part of the system can cause chain reactions within interrelating systems
Dynamic thinking (Batzri et al., 2015; Ben-Zvi-Assaraf & Orion, 2005; Booth-Sweeney & Sterman, 2007; Richmond, 1993)	Thinking that systems are not static and they can change over time. For example, the Earth systems are dynamic and continue to react to changes
Cyclic thinking (Ben-Zvi-Assaraf & Orion, 2005; Batzri et al., 2015; Richmond, 1993)	Thinking that the Earth systems work in cycles and there is not waste in this system. For example, in a sustainable system, there should be a circular economy
Recognizing feedback loops (Batzri et al., 2015; Booth-Sweeney, 2017)	Recognizing the cause-effect relationships in a system and the negative or positive feedback loops that influence the system. Feedback loops are the key elements of dynamic and cyclic processes
Recognizing own responsibility in the system (Sleurs, 2008; UNECE, 2011)	Realizing one's personal role in the system and taking respon sibility for the choices made in daily life. Being part of the solutions to global problems
Considering the relationship between the past, present, and future actions (time dimension) (Ben-Zvi-Assaraf & Orion, 2005; Sterling et al., 2005)	Thinking about the relationship between the past, present, and future actions. Instead of short-term thinking, long-term thinking is important to understand the long-term impacts of the incidents
Developing empathy (Karaarslan & Teksöz, 2020; Sleurs, 2008; Sterling et al., 2005)	Looking into events from other people's perspectives and understanding their needs behind their actions and also building empathy with non-human beings
Recognizing hidden dimensions in a system (Ben-Zvi-Assaraf & Orion, 2005, 2010)	Recognizing the non-obvious parts of the system which are not seen at first glance
Identifying the key aspects of sustainability and analyzing interactions (Karaarslan & Teksöz, 2020)	Identifying the social, economic, and environmental aspects of sustainability and the relationship among them
Seeing nature as a system (Karaarslan & Teksöz, 2020)	Understanding the complexity of nature and describing human-nature relationship from a holistic perspective

recognizing own responsibility in the system, developing empathy, identifying key aspects of sustainability and analyzing interactions, and *seeing nature as a system.*

4.2.1 The Importance of Systems Thinking in ESD

Many scholars maintain that the root of complex problems is related to not seeing things systemically; thus, seeing issues from a holistic view or systems thinking perspective is a solution to the current problems of the world (Capra, 2005; Nolet, 2009; Orr, 1992; Sterling, 2003). Orr (2004) stated that in order to change our way of thinking, first we need to change our education

system that shapes it. ESD is a holistic and transformational approach to reshape our knowledge, skills, and values (UNESCO, 2020). Education plays a key role in transforming societies toward sustainability, and learners need to think critically and systemically to develop values and attitudes toward sustainability (UNESCO, 2018). In this context, systems thinking is a core tool to resolve the complex problems of today and create a sustainable planet. Sterling (2003) emphasized the need for a paradigm change across education, which is more transformative and relational. The author described the importance of systems thinking for the change in education culture and introduced the concept of "whole systems thinking." Furthermore, Sterling (2009) noted that: "If we want the change of a sustainable future, we need to think relationally" (pg.1). Systems thinking is considered to be a key element of ESD for changing the unsustainable system and our old perceptions which impede sustainability. Therefore, we need to focus on developing students' systems thinking skills both in and for ESD.

In the ESD literature, the components of systems thinking were defined in various ways, and diverse pedagogical tools were developed. For example, Sterling et al. (2005) created linking thinking teaching materials to introduce whole systems thinking to students and teachers. The authors described several principles of linking thinking to build a sustainable society. Some of these principles are listed below:

- The whole is greater than the sum of the parts.
- You can never do only one thing (the simpliest action can have consequences).
- Everything is connected but not equally strongly (we need to look beyond our boundaries to understand the wider influences in our systems).
- A change in the parts affects the whole (and vice versa) (systems have a degree of resilience in the face of change).
- Complex systems show delayed response (if you try to change a person, an organization, or an ecosystem, you might get unexpected responses) (pg.24).

Nolet (2009) also identified *systems thinking and interdependence* as a component of sustainability literacy. According to the author, systems thinking does refer not only to the interconnections between natural systems but also to the interconnections between environmental, social, and economic systems. Today, it is better understood that a variation in the interrelated feedback loops of Earth's systems can influence environmental, social, and economic systems (Nolet, 2009). The impact of climate change on ecosystems and people's life is an example of this interconnection between social and environmental systems. Sustainability allows social, environmental, and economic systems to work in balance.

Wiek et al. (2011) identified systems thinking as one of the key sustainability competencies. The authors defined systems thinking as the ability to analyze complex systems across various areas (environment, society, economy, etc.) and described systemic structures such as feedback loops, cascading effects, and complex cause-effect chains related to sustainability issues. Similar to Wiek et al. (2011), the United Nations Economic Commission for Europe (UNECE) (2011) identified the core competencies for ESD. In the report the Commission prepared, a holistic approach with integrative thinking; the connections between the past, present, and future; and achieving transformation were determined as the essential characteristics of ESD (UNECE, 2011). Systems thinking was placed at the heart of these competencies. UNECE (2011) reported that systems thinking is a valuable tool to achieve an integrative approach. Through systems thinking, individuals can understand the interconnections between social, natural, and economic systems, appreciate different cultures and worldviews, and feel connected to other people both locally and globally (UNECE, 2011). Sleurs (2008) also noted that systems thinking helps individuals change their perspectives on how to act for a sustainable planet. It is a useful tool to reshape our values and ethical considerations to understand sustainability issues (Hofman-Bergholm, 2018; Sleurs, 2008). On the other hand, Riess and Mischo (2010) conceptualized systems thinking as "the ability to recognize, describe, model and explain complex aspects of reality as systems." Schuler et al. (2018) agreed on this definition and designed a competence model to measure and develop the systems thinking skills of students and student teachers. According to this model, abilities for systems thinking were determined as "declarative system knowledge" (knowledge of different system properties), "system modeling" (understanding complex systems through system models), "solving problems using system models," and "evaluation of system models."

Recently, Karaarslan and Teksöz (2020) elaborated on 12 systems thinking skills derived from

Fig. 4.1 Specific
systems thinking
components in the ESD
context

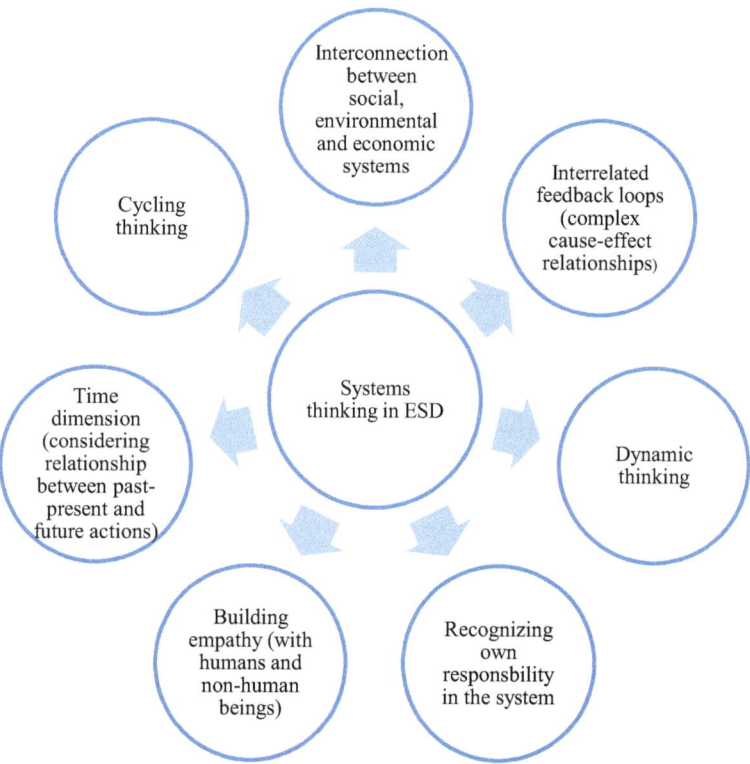

the relevant literature. The authors described some of these skills as "identifying the components of a system and relationships," "recognizing hidden dimensions," "analyzing the interconnections among the social, economic and environmental aspects of sustainability," "recognizing personal role in the system," "the cyclic nature of the system," and "developing empathy with people and non-human beings." The authors emphasize that outdoor education for sustainability is a valuable tool to develop the systems thinking skills of pre-service science teachers. In line with the literature in this chapter, specific systems thinking components that can be developed through learning activities related to ESD are suggested (Fig. 4.1). In Table 4.1, there are more systems thinking components derived from the literature, but, in this chapter, we focus on several of them and present a learning activity to incorporate these components.

Sterling (2003) emphasized the role of education in a whole system shift as being a more participatory, dynamic, and active learning process

to generate knowledge and focus on real-world problem-solving. In this chapter, I offer outdoor education for sustainability as an active, transformative, and experiential learning approach to foster systems thinking skills. In the following sections, I describe the rise of outdoor education in the ESD context and its role in developing systems thinking skills and propose an outdoor learning activity to develop students' systems thinking skills. This activity can be seen as a practical guide for teachers in order to implement systems thinking within the ESD context.

4.3 Outdoor Education for Sustainability Approach to Foster Systems Thinking Skills

ESD embraces an interactive, action-oriented, exploratory, interdisciplinary, and transformative pedagogy to inspire individuals to act for sustainability (UNESCO, 2014a). In order to achieve sustainability goals, formal, non-formal (such as nature centers), and informal education (such as media sector) should work collaboratively (McKeown, 2002). At this point, outdoor education plays an important role in supporting these interdisciplinary and transformative approaches related to ESD (Lugg, 2007). The incorporation of sustainability into outdoor education has been addressed in the literature (e.g., Aksland & Rundgren, 2020; Beames et al., 2012; Higgins, 2009; Hill, 2012; Prince, 2017). Aksland and Rundgren (2020) pointed out that schools and centers promoting outdoor experiences could be learning spaces where students can see what a sustainable lifestyle means and outdoor educators could be a role model to encourage sustainable living. Outdoor education provides opportunities to develop our relationship with the planet to build a sustainable future (Beames et al., 2012). Orr (2004) emphasized the importance of learning outside the classroom for sustainability. The author argued that people need to develop their connection with nature in order to better understand our relationship with nature and its social, ecological, and aesthetic value. Lugg (2007) also stated that outdoor education can make a significant contribution in terms of raising sustainability-literate citizens who can see the interconnections between the social, economic, and environmental aspects of issues. Hill and Brown (2014) discussed how the intersections between sustainability, transformation, and place might be included in outdoor education. The authors explored that outdoor education has a critical potential for transforming society and individuals by developing our relationship with place and learning outside the classroom enables us to understand the relationship between nature and society.

Outdoor education can foster a holistic understanding of sustainability, our sense of connection with nature and with the places we live in, and also our recognition of our place on the planet. The importance of local places in outdoor education has been emphasized in the relevant literature (e.g., Beames et al., 2012; Gruenewald & Smith, 2008; Wattchow & Brown, 2011). For instance, Beames et al. (2012) highlighted that

outdoor education should focus on the story of local places and develop students' connection with place. Higgins (2009) argued that developing a sense of connection with place enables individuals to understand the consequences of their actions and build an ethic of citizenship and care.

Outdoor education is related not only to learning in nature but also to learning in urban places and cultural learning (Drexler, 2019). Bellino and Adams (2017) emphasized that while making observations in urban environments, young people can understand the social and ecological complexities of urban spaces and ponder an improved life for all humans and non-humans. At the end of this chapter, readers will find an outdoor learning activity related to water systems in an urban environment. This activity can help students realize complex sustainability issues and develop systems thinking skills.

What is the relationship between outdoor education and systems thinking skills? How can outdoor education develop students' systems thinking skills? In the literature, a number of studies indicated that outdoor education contributes to developing students' systems thinking skills such as understanding complex systems and interactions, temporal thinking, and dynamic and cyclic thinking (e.g., Ben-Zvi-Assaraf & Orion, 2005, 2010; Keynan et al., 2014). Ben-Zvi-Assaraf and Orion (2010) demonstrated that through outdoor learning activities, fourth grade students explored the water cycle components in a natural environment, identified the relationships among these components, and understood that water cycle is a dynamic system. Thus, outdoor learning environments helped students explore the components and relationships in a system in a concrete way and fostered their systems thinking skills (Ben-Zvi-Assaraf & Orion, 2010).

Karaarslan and Teksöz (2020) worked with pre-service science teachers and revealed that pre-service science teachers' systems thinking skills could be developed through an outdoor-based ESD course. The authors prepared some learning activities for the pre-service teachers related to examining a lake system from different perspectives (ecosystem, human use, and water quality monitoring). It was seen that through outdoor learning activities, the pre-service teachers developed their systems thinking skills such as identifying and analyzing different aspects of sustainability, seeing nature as a system, and recognizing their own responsibility in a system.

In a report related to the educational outcomes of learning for sustainability in Scotland, Christie and Higgins (2020) described systems thinking as an outcome of learning for sustainability. The authors highlighted that interdisciplinary learning and outdoor education provide many opportunities to adopt the systems thinking approach and support learners to take transformative actions. Logan (2018) also underlined that outdoor education is a key strategy to strengthen young people's systems thinking skills and their interconnection with living and non-living things in the Earth systems. The author stated that the Australian science curricula separate disciplines and the Earth systems are conceptualized as a resource for the benefit of human beings. Although the Australian curricula incorporate sustainability as a cross-curriculum priority, sustainability is not elaborated in different curriculum areas, and the environmental aspect of sustainability is generally embedded into science curriculum (Logan, 2018). Turkish primary and secondary school science curriculum which was revised in 2018 incorporated sustainable development as a separate unit, and it was not comprehensively addressed in the curriculum subjects. Sustainable development was solely linked to recycling and the efficient use of resources, which does not necessarily promote students' systems thinking skills. Sustainability-related learning objectives from the Turkish science curriculum are listed below:

- Designs a project related to the sustainable use of resources
- Discusses the importance of recycling of solid wastes and its contribution to national economy and provides solutions related to this issue
(Ministry of National Education (MoNE), 2018)

The Turkish science curriculum does not have a specific goal to foster the systems thinking skills of students, and sustainability is not a priority in the curriculum. Generally, environmental and economic aspects of sustainability were embedded in the science curriculum. Moreover, Earth cycles such as the water cycle and carbon cycle are explained separately, which does not guide students in terms of seeing the connections between the cycles.

School curricula should incorporate the systems thinking approach; thus, young people can understand Earth as a dynamic system and see the connections between the components of the Earth system such as biological, geological, atmospheric, and hydrological systems (Logan, 2018). Moreover, through systems-based science education, students can understand how ecological systems are related to each other and how human activities impact these systems (Logan, 2018). From a sustainability perspective, Aksland and Rundgren (2020) found that teachers mostly do not integrate sustainability into outdoor education courses; therefore, their professional knowledge to connect sustainability and outdoor education should be improved, and their school practices should be increased. Logan (2018) also pointed out that most of the teachers do not understand education for sustainability in a holistic way and they have difficulty integrating sustainable practices into their lessons effectively. In terms of sustainability and the systems thinking approach, curriculum descriptions should be more explicit and comprehensive; hence, teachers can understand and integrate these approaches into their lessons. In the following section, a learning activity incorporating outdoor education, sustainability, and the systems thinking approach is presented in primary and secondary level.

4.4   Incorporating Systems Thinking in Outdoor Education for Sustainability: A Learning Activity Example

In this part of the chapter, a learning activity that aims to develop primary and secondary school students' systems thinking skills through an outdoor education for sustainability approach is presented. The subject of the activity is dynamic water systems. More specifically, in this activity, the hidden rivers issue is introduced to the students, and the possibility of daylighting rivers (returning a culverted river to open water) is discussed. UNESCO (2014a) described the ESD learning areas including critical themes which are *climate change, disaster risk reduction, sustainable consumption and production,* and *biodiversity.* There is a need to integrate these critical subjects into school curricula and develop students' core competencies such as critical thinking and systems thinking (UNESCO, 2018). The hidden rivers issue was chosen as it is related to the abovementioned critical sustainability themes, in particular, climate change. In line with this activity, students can explore many subjects such as urban development and its impact on water quality and the future of water resources, economic growth, population growth, altering river ecosystems, human relationship with water, and also the impact of climate change on rivers. Students can discuss the connections between these subjects from a sustainability perspective. Therefore, the hidden rivers issue is an interdisciplinary subject that can be integrated into many disciplines like science, sustainability, geography, history, and mathematics. In Turkish primary and secondary science curriculum, it is difficult to find subjects that are related to water systems from a systemic approach. The curriculum includes only two learning objectives which are related to the cycles of matter, including the water cycle, carbon cycle, and nitrogen cycle. The relevant learning objectives from the Turkish science curriculum are as follows:

- Describes the cycles of matter and demonstrates them on a figure
- Questions the importance of the cycles of matter for life
(MoNE, 2018)

In the following part, a local example about water systems is presented. Teachers can adapt this activity according to the environmental characteristics of their region, and they can redesign it according to different age groups. For primary school students, a short version of this activity can be prepared.

Hidden Rivers Case
Subjects: Science, sustainability, geography, mathematics, and language
Materials: Water quality measurement devices (PH meter, DO meter, turbidity meter, photometer, worksheets, PC, Internet, camera/smartphones, recycled materials)
Duration: 5 hours
Learning Objectives
1. Exploring the change in river systems from past to today
   - Exploring the impact of urban development on river management
   - Discussing the impact of short-term solutions on the environment and society
2. Discussing the impact of climate change on water systems and the water cycle
3. Analyzing the hidden rivers issue from the social, environmental, and economic aspects of sustainability
4. Investigating how daylighting rivers is possible for the sustainability of rivers and society
5. Recognizing one's personal role in the system
6. Developing a strong connection with hidden rivers

Systems Thinking Connection
The systems thinking components as shown in Fig. 4.1 can be integrated into this activity.

Cyclic thinking, dynamic thinking, and interrelated feedback loops Students explore how water systems work in cycles and how river management impacts the whole system, and they examine the interrelationship between climate change, river management, and urban development.

Time dimension Students explore how rivers were covered in the past and the consequences of culverted rivers for the environment today. They also learn how it is possible to revitalize rivers in the future. Students establish a connection between past, present, and future and develop their ability to think forward and backward.

Interconnections between social, environmental, and economic systems Students recognize the economic, social, and environmental values of rivers and explore sustainable solutions for the future.

Recognizing own responsibility in the system Students explain their vision for their city, create a city plan for daylighting rivers, and prepare an informative poster about hidden rivers to engage with their community. They find inspiring ideas to become change makers for creating a sustainable future.

Building empathy Students build empathy with the non-human world and develop a strong connection with hidden rivers.

Background of the Topic
The impact of urbanization on water systems can easily be seen in cities. Most of the river beds in urban cities have been channeled, drained, diverted, or culverted. Many cities around the world have rivers or streams flowing under the roads. Ankara, the capital of Turkey, is one of these cities. It is generally known as a dry and gray city, but this is not true. It was established near the Ankara River, and there were many streams around the city in the past. Through the years, the city population has grown up, and now six million people live in the city. A growing population and urbanization changed the river systems. After the 1950s, the rivers of Ankara were mostly culverted for several reasons such as river pollution, flood risk, and urbanization (Semiz, 2019). Today, most of these rivers are flowing under roads, and they are highly polluted because of sewage and industrial pollutants.

In summer months, the city has limited water because of the semi-arid climate of the region (Kaymaz, 2019). Therefore, the streams and riv-

ers of Ankara are crucially important for the future. Water systems are essential for all living organisms, and they have many ecological, economic, and cultural values (Kaymaz, 2019). In this activity, students will learn about how rivers changed over the years, the current problems about the water systems, and how the hidden rivers can be revitalized. They will make a connection between the past, present, and future. Students will analyze water systems through a systemic approach, and they will consider the relationships between social, environmental, and economic systems.

They will explore sustainable solutions for the future. This will be an outdoor learning activity in an urban environment. Three teaching strategies, namely, problem-based learning, project-based learning, and place-based education, can be used in this activity.

Introduction to activity At the beginning of the lesson, the teacher asks students the following questions about rivers. Students work in groups, write their answers, and discuss in the class.

- Why were most of the cities around the world established near water resources?
- What is the importance of water for the environment and society?
- Do you know any river or stream flowing in your city? How many rivers are there in your city?
- Do you know any floods that occurred in the past in your city?
- Have you heard about hidden rivers in your city?

Later, students are shown the documentary named *Under the Road, the River*, which is about the hidden rivers of Ankara (Semiz, 2019), and they discuss the issues in the documentary and learn more about the hidden rivers in the city. Students can watch different documentaries about the lost rivers as well.

Field trip The teacher organizes a walking tour to visit the rivers touched upon in the documentary. During the visit, students follow a city map indicating the hidden rivers. They work in groups and collect data about the environmental and cultural story of the rivers. The teacher can ask some questions as presented below:

- How were rivers used in the past?
- What is the social, environmental, and economic value of rivers?
- How has the biodiversity of the river ecosystem changed from past to present?
- What can you say about the water pollution problem in the rivers?

Students also take photographs of the areas where rivers are culverted and make interviews with local people who remember the rivers in the past. Furthermore, they make a research plan about the water quality of the rivers they follow. They write a hypothesis to investigate the water pollution problem in the rivers. They conduct scientific measurements about water quality (PH, DO, turbidity, phosphate, and nitrate) near the open sides of the rivers.

Discussion and solutions After the field trip, each group organizes their data and prepares a presentation about the walking tour. The teacher makes a link between the hidden rivers and climate change so that students can understand how climate change impacts water cycle and how culverting rivers accelerates the problem. Some of the questions that can be asked to students are as follows:

- How does climate change affect Earth's water cycle?
- What is the relationship between river management (culverting rivers), urbanization, climate change, and water cycle (thinking of the cause-effect relationship)?

The teacher explains feedback loops and how positive feedback loops worsen the climate change problem and how the hidden rivers issue amplifies this process. The water vapor feedback loop can be explained to students as an example that makes the climate warmer (Climate Reality Project, 2020) (Fig. 4.2).

Students make a search on the Internet and investigate how the precipitation regime has changed in the city from past to present. They can show this change on a graph. The teacher explains how climate change causes dramatic changes in water availability. Urban rivers and streams create a "cool island" and regulate the climate in cities. Rivers enable a microclimate effect in the city centers (Hathway & Sharples, 2012). Covering rivers and urbanization accelerate the impact of climate change. Thus, water cycle is affected, and the risk of heavy rains and extreme droughts increases. Therefore, daylighting rivers is an important solution to mitigate the impact of climate change and create more sustainable cities. While elaborating on the solutions, some examples about daylighting rivers around the world can be presented. In this way, students can see how cities are changed in a sustainable way after daylighting rivers such as reduction of traffic jam, increasing green areas, and changing the climate in the city.

A sustainable city plan At the end of this activity, students can work on a project. They can organize some field trips around the city. Based on the city map, they can decide which river can be revitalized to create a sustainable city plan. A city planner can be invited to the class to guide students. Students can be asked to consider how to restore the area after opening a hidden river. They can think about more green areas, cycling paths, and walking roads around the river. They can design their city model by using a variety of recyclable materials such as some art materials, plastic bottles, waste cardboards, and natural materials.

Students can also prepare a poster to increase the awareness of the community (family, friends, neighbors) about hidden rivers. In this way, they can engage with the community to express themselves.

At primary level, students can write a letter to a friend who lives a long way away, and they can tell about the hidden river, where it is and what they think about it while they are sitting outside near the hidden river. The teacher can choose one

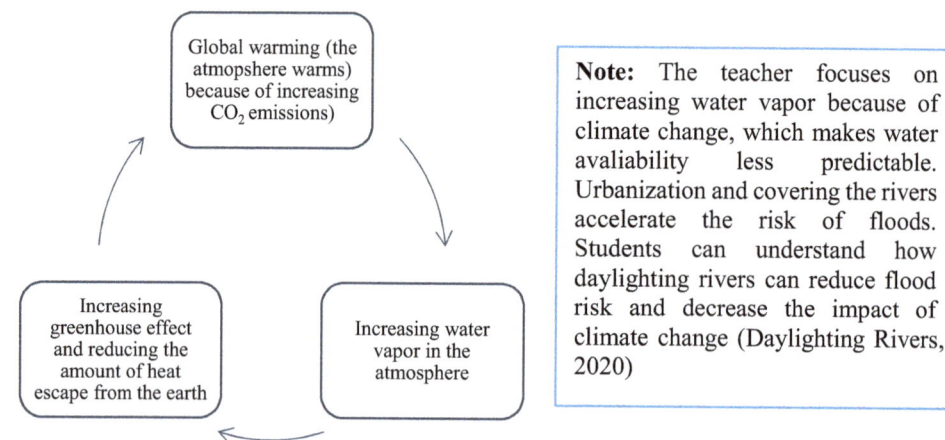

Fig. 4.2 Positive feedback loop related to climate change

of these activities according to the grade level of students.

Evaluation

The teacher can ask the following questions after the activity:

1. What mostly influenced the students in this activity?
2. How did their thoughts about rivers change?
3. How do you define your role in this hidden rivers issue?
4. Please draw a concept map showing the relationship between the hidden rivers issue, climate change, and sustainability.

4.5   Conclusion and Suggestions

Sustainability is related to seeing the wholeness between social, environmental, and economic systems. It involves understanding the fact that the change in our way of life with urbanization is also linked to the disappearance of rivers in a city, water pollution, flood risk, drought, forest fires, change in our social life, and climate crisis. In order to understand these complex relationships and uncertainties regarding the current challenges of the world, we need sustainable citizens who have competencies for building a sustainable future (UNESCO, 2017). Systems thinking is one of these key sustainability competencies. If we want to achieve a sustainable society, we need

more systems thinkers. We need to guide our students so that they can realize and understand the systemic connections between people, places, issues, rivers, and environment. In this way, they can make their own decisions to create a balance between healthy ecosystems, equitable sustainable economy, and equal social systems for all people (Booth-Sweeney, 2017). A systems thinker can look at the problems from a wider perspective considering multiple impacts and trying to develop systemic solutions. He/she can also understand the importance of long-term instead of short-term thinking while dealing with the sustainability issues. As Sterling (2005) stated, "some solutions just produce more problems. Instead, we need to develop solutions that generate further solutions" (p.15).

In this chapter, I have proposed outdoor education for sustainability as a pedagogical approach to improve students' systems thinking skills. Outdoor education provides multifarious learning environments to understand how natural systems are dynamic and work in cycles, to recognize our place in this system, and to develop a strong connection with the human and non-human world. Outdoor education can help us understand the complex structure of our planet and develop a deep connection with the places we live in (Beames et al., 2012). Developing our connection with Earth and understanding the complex relationships are significant while shaping our decisions for sustainability. At the end of this chapter, I have shared a learning activity related to the hidden rivers issue in an urban environment. This outdoor activity can be linked to many subjects in the curriculum and help young learners to understand complex sustainability issues. While using outdoor education in ESD, we can transform the worldviews of young people and develop their systems thinking skills. In the future, teachers can use or adapt this outdoor learning activity and evaluate its impact on their students' systems thinking skills.

References

Aksland, C., & Rundgren, S. C. (2020). 5th–10th-grade in-service teachers' pedagogical content knowledge (PCK) for sustainable development in outdoor environment. *Journal of Adventure Education and Outdoor Learning, 20*(3), 274–283. https://doi.org/10.1080/14729679.2019.1697713

Batzri, O., Assaraf, O., Cohen, C., & Orion, N. (2015). Understanding the earth systems: Expressions of dynamic and cycling thinking among university students. *Journal of Science Education and Technology, 24*(6), 761–775. https://doi.org/10.1007/s10956-015-9562-8

Beames, S., Higgins, P., & Nicol, R. (2012). *Learning outside the classroom. Theory and guidelines for practice*. Routledge.

Bellino, M. E., & Adams, J. E. (2017). A critical urban environmental pedagogy: Relevant urban environmental education for and by youth. *The Journal of Environmental Education, 48*(4), 270–284. https://doi.org/10.1080/00958964.2017.1336976

Ben-Zvi-Assaraf, O., & Orion, N. (2005). The development of system thinking skills in the context of earth system education. *Journal of Research in Science Teaching, 42*, 1–43. https://doi.org/10.1002/tea.20061

Ben-Zvi-Assaraf, O., & Orion, N. (2010). System thinking skills at the elementary school. *Journal of Research in Science Teaching, 47*(5), 540–563. https://doi.org/10.1002/tea.20351

Blackman, T., Greene, A., Hunter, D. J., McKee, L., Elliott, E., Harrington, B., & Williams, G. (2006). Performance assessment and wicked problems: The case of health inequalities. *Public Policy and Administration, 21*(2), 66–80. https://doi.org/10.1177/0952076706021002006

Booth-Sweeney, L. (2017). All systems go. Developing a generation of systems smart kind. In E. Assadourian & L. Mastny (Eds.), *Rethinking education on a changing planet* (pp. 141–331). Island Press.

Booth-Sweeney, L. B., & Sterman, J. D. (2007). Thinking about systems: Student and teacher conceptions of natural and social systems. *System Dynamics Review, 23*, 285–311. https://doi.org/10.1002/sdr.366

Capra, F. (1999). *Ecoliteracy: The challenge for education in the next century. Liverpool Schumacher Lectures, 20*.

Capra, F. (2005). Speaking nature's language: Principles for sustainability. In M. K. Stone & Z. Barlow (Eds.), *Ecological literacy. Educating our children for a sustainable world* (pp. 18–29). Sierre Club Books.

Capra, F., & Luisi, P. L. (2014). *The systems view of life: A unifying vision*. Cambridge University Press.

Christie, B. & Higgins, P. (2020). *The educational outcomes of learning for sustainability: A brief review of literature*. https://www.gov.scot/publications/educational-outcomes-learning-sustainability-brief-review-literature/pages/3/ . Accessed 20 Dec 2020.

Climate Reality Project (2020). *How feedback loops are making the climate crisis worse*. https://www.climaterealityproject.org/blog/how-feedback-loops-are-making-climate-crisis-worse. Accessed 21 Jan 2021.

Daylighting Rivers (2020). *Themes and learning units*. https://www.daylightingrivers.com/themeslearningunits/. Accessed 25 Dec 2020.

Drexler, S. (2019). *Exploring the idea of an outdoor primary school* (Master's thesis, Linköping University). Diva Digital Archive. https://www.diva-portal.org/smash/record.jsf?pid=diva2%3A1324134&dswid=9348

Filho, W. L., Brandli, L. L., Salvia, A. L., Rayman-Bacchus, L., & Platje, J. (2020). COVID-19 and the UN sustainable development goals: Threat to solidarity or an opportunity? *Sustainability, 12*(13). https://doi.org/10.3390/su12135343

Gruenewald, D., & Smith, G. A. (2008). *Place-based education in the global age: Local diversity*. Lawrence Erlbaum Associates.

Hathway, E. A., & Sharples, S. (2012). The interaction of rivers and urban from in mitigating the urban island effect: A UK case study. *Building and Environment, 58*, 14–22. https://doi.org/10.1016/j.buildenv.2012.06.013

Higgins, P. (2009). Into the big wide world: Sustainable experiential education for the 21st century. *The Journal of Experimental Education, 32*(1), 44–60. https://doi.org/10.1177/105382590903200105

Hill, A. (2012). Developing approaches to outdoor education that promote sustainability education. *Australian Journal of Outdoor Education, 16*(1), 15–27.

Hill, A., & Brown, M. (2014). Intersections between place, sustainability and transformative outdoor experiences. *Journal of Adventure Education and Outdoor Learning, 14*(3), 217–232. https://doi.org/10.1080/14729679.2014.918843

Hofman-Bergholm, M. (2018). Could education for sustainable development benefit from a systems thinking approach? *System, 6*(43), 1–12. https://doi.org/10.3390/systems6040043

Karaarslan, G., & Teksöz, G. (2020). Developing the systems thinking skills of pre-service science teachers through an outdoor ESD course. *Journal of Adventure Education and Outdoor Learning, 20*(4), 337–356. https://doi.org/10.1080/14729679.2019.1686038

Karaarslan-Semiz, G. (2021). Systems thinking research in science and sustainability education: A theoretical note. In E. Jeronen (Ed.), *Transitioning to quality education. Transitioning to sustainability series 4* (pp. 39–61). MDPI.

Kaymaz, I. (2019). Lost streams of Ankara. A case study of Bent Stream. *IOP Conf. Series. Materials Science and Engineering, 603*. https://doi.org/10.1088/1757-899X/603/5/052040

Keynan, A., Assaraf, O. B. Z., & Goldman, D. (2014). The repertory grid as a tool for evaluating the development of students' ecological system thinking abilities. *Studies in Educational Evaluation, 41*, 90–105. https://doi.org/10.1016/j.stueduc.2013.09.012

Logan, M. (2018). Challenging the anthropocentric approach of science curricula: Ecological systems approaches to enabling the convergence of sustainability, science, and STEM education. In A. C. Mackenzie, K. Malone, & E. B. Hacking (Eds.), *Research handbook of childhood nature* (pp. 1–28). Springer.

Lönngren, J. (2014). *Engineering students' ways of relating to wicked sustainability problems* (Licentiate Thesis). Chalmers University of Technology, Gothenburg: Chalmers.

Lönngren, J., & Svanström, M. (2015). *Systems thinking for dealing with wicked sustainability problems: Beyond functionalist approaches*. http://umu.diva-portal.org/smash/get/diva2:1305363/FULLTEXT01.pdf. Accessed 10 May 2021.

Lugg, A. (2007). Developing sustainability literate citizens through outdoor learning possibilities for outdoor education in higher education. *Journal of Adventure Education and Outdoor Learning, 7*(2), 97–112. https://doi.org/10.1080/14729670701609456

McKeown, R. (2002). *ESD toolkit*. http://www.esdtoolkit.org/esd_toolkit_v2.pdf. Accessed 15 June 2021.

Meadows, D. (2008). *Thinking in systems: A primer*. Earth Scan. https://wtf.tw/ref/meadows.pdf. Accessed 16 June 2021.

Meadows, D. H., & Wright, D. (2009). *Thinking in systems: A primer*. Earthscan.

Ministry of National Education (MoNE). (2018). *Turkish science curriculum (Grades 3-8)*. http://mufredat.meb.gov.tr/ProgramDetay.aspx?PID=325. Accessed 10 Oct 2020.

Nolet, V. (2009). Preparing sustainability literate teachers. *Teachers College Record, 111*(2), 409–442.

Orr, D. W. (1992). *Ecological literacy: Education and the transition to a postmodern world*. State University of New York Press.

Orr, D. W. (2004). *Earth in mind: On education, environment, and the human prospect*. Earth Island Press.

Prince, H. E. (2017). Outdoor experiences and sustainability. *Journal of Adventure Education and Outdoor Learning, 17*(2), 161–171.

Richmond, B. (1993). Systems thinking: Critical thinking skills for the 1990s and beyond. *System Dynamics Review, 9*, 113–133.

Riess, W., & Mischo, C. (2010). Promoting systems thinking through biology lessons. *International Journal of Science Education, 32*(6), 705–725. https://doi.org/10.1080/09500690902769946

Schuler, S., Fanta, D., Rosenkraenzer, F., & Riess, W. (2018). Systems thinking within the scope of education for sustainable development (ESD) – a heuristic competence model as a basis for (science) teacher education. *Journal of Geography in Higher Education, 42*(2), 192–204.

Semiz Y. (2019) *Asfaltın Altında Dereler Var!* [Under the road, the river]. Luwi Film.

Senge, P. (2006). *The fifth discipline: The art and practice of the learning organization* (p. 1990c). Doubleday/Currency.

Sleurs, W. (2008). *Competences for education for sustainable development (ESD) teachers. A framework to integrate ESD in the curriculum of teacher training institutes*. Commenius 2.1 Project.

Stave, K., & Hopper, M. (2007). What constitutes systems thinking? A proposed taxonomy. In *Proceedings of the 25th International Conference of the System Dynamics Society*, Boston, MA, 29 July–2 August 2007. https://pdfs.semanticscholar.org/506d/f8001a8b9190a6f9b22abf-7c1495e96de72d.pdf. Accessed 20 Jan 2021.

Sterling, S. (2003). *Whole systems thinking as a basis for paradigm change in education: Explorations in the context for sustainability* (Unpublished doctoral dissertation). University of Bath, UK.

Sterling, S. (2005). Linking thinking, education and learning: An introduction. In S. Sterling, P. Maiteny, D. Irvine, & J. Salter (Eds.), *Linking thinking – new perspectives on thinking and learning for sustainability*. WWF.

Sterling, S. (2009). Ecological intelligence. In A. Stibbe (Ed.), *The handbook of sustainability literacy: Skills for a changing world* (pp. 77–83). Green Books Ltd.

Sterling, S., Maiteny, P., Irving, D., & Salter, J. (2005). *Linking thinking: New perspectives on thinking and learning for sustainability*. WWF.

Stibbe, A. (2019). Education for sustainability and the search for new stories to live by. In J. Armon, S. Scoffham, & C. Armon (Eds.), *Prioritizing sustainability education: A comprehensive approach* (pp. 233–243). Routledge.

Strachan, G. (2012). *WWF-professional development framework of teacher competences for learning for sustainability*. WWF.

Tilbury, D. (2007). Learning based change for sustainability: Perspectives and pathways. In A. E. J. Wals (Ed.), *Social learning towards a sustainable world* (pp. 117–132). Wageningen Academic Publishers.

Tilbury, D., & Cooke, K. (2005). *A National review of environmental education and its contribution to sustainability in Australia: Frameworks for sustainability*. Australian Government Department of the Environment and Heritage and Australian Research Institute in Education for Sustainability.

UNECE (United Nations Economy Commission for Europe). (2011). *Learning for the future. Competences in education for sustainable development*. https://unece.org/DAM/env/esd/ESD_

Publications/Competences_Publication.pdf. Accessed 1 Feb 2021.

UNESCO. (2014a). *Road map implementing global action program on education for sustainable development*. UNESCO.

UNESCO. (2014b). *Shaping the future we want. UN decade education for sustainable development (2005–2014) Final Report*. UNESCO.

UNESCO. (2017). *Education for sustainable development goals. Learning objectives*. UNESCO.

UNESCO. (2018). *Issues and trends in education for sustainable development*. UNESCO.

UNESCO. (2020). *Education for sustainable development. A roadmap. ESD for 2030*. UNESCO.

Vasconcelos, C., & Orion, N. (2021). Earth science education as a key component of education for sustainability. *Sustainability, 13*, 1316. https://doi.org/10.3390/su13031316

Von Bertalanffy, L. (1969). *General system theory: Foundations, development, application* (revised ed.). George Braziller.

Wattchow, B., & Brown, M. (2011). *A pedagogy of place: Outdoor education for a changing world*. Monash University Publishing.

Wiek, A., Withycombe, L., & Redman, C. L. (2011). Key competencies in sustainability: A reference framework for academic program development. *Sustainability Science, 6*, 203–218.

Wiek, A., Xiong, A., Brundiers, K., & Van Der Leeuw, S. (2014). Integrating problem-and project-based learning into sustainability programs: A case study on the school of sustainability at Arizona State University. *International Journal of Sustainability in Higher Education, 15*(4), 431–449.

Zhang, B. H., & Ahmed, S. A. M. (2020). Systems thinking—Ludwig Von Bertalanffy, Peter Senge, and Donella Meadows. In B. Akpan & T. Kennedy (Eds.), *Science education in theory and practice*. Springer. https://doi.org/10.1007/978-3-030-43620-9_28

Güliz Karaarslan-Semiz is an assistant professor in the Department of Mathematics and Science Education at Ağrı İbrahim Çeçen University in Türkiye. She studied science education (BS) and then received her master degree (focus on environmental education). She completed her PhD at Middle East Technical University in Ankara (focus on ESD). Her study was awarded as the best PhD thesis of the year at Middle East Technical University in 2016. Her research interests are education for sustainable development, systems thinking, and recently whole school approach to sustainability. She has worked on national and international projects related to ESD and science education. She is a member of the Environmental and Sustainability Education Research (ESER) network in Europe and recently she has worked as a visiting researcher at Stockholm University in Sweden.

Real-World Learning as a Frame for Sustainability in Education

5

Daniela Conti and Richard Dawson

Abstract

This chapter explores the role outdoor education and learning can play not only in learning about the natural world but also in developing behaviours, which care and sustain the natural world. It draws on extensive research across Europe and presents a practical model for practitioners to use within their teaching work. Topics explored include the role of values and frames in influencing behaviours and how science can be framed in different ways to increase opportunity for sustainable social change. The chapter offers a specific enquiry-based approach to using science to address urban sustainability issues.

Keywords

Enquiry-based learning · Authentic learning · Outdoor learning; STEM · Sustainability competences

5.1 Introduction

Nobody can discover the world for somebody else. Only when we discover it for ourselves does it become common ground and a common bond and we cease to be alone. Wendell Berry, A Place on Earth

This chapter presents a reflection on the potential for authentic learning settings to encourage positive social change in learning for sustainability and to empower learners to take action. It presents suggestions and practical tools to envision how to embed real-world learning in teaching outdoors and how it can be utilised for enhancing learning. We draw upon good practice with teachers and educators developed through the experimental activity of two projects in the frame of the European Comenius and Erasmus Plus programmes – Real World Learning Network (2012–2015) and Urban Science (2017–2020). Each project examined the interrelations between the use of real-world learning approaches and the development of sustainability competences in formal and non-formal education. The Real World Learning Network developed a model to guide the process of planning and reflecting on learning, while Urban Science improved the use of the city environment as the perfect setting where sustainability learning can take place. The question driving our work was to identify the elements which can support a substantial paradigm shift and transition in learning and teaching, one

D. Conti (✉)
Centre for Environmental Research, Documentation and Education CREDA, Monza, Italy
e-mail: daniela.conti@creda.it

R. Dawson
Wild Awake, Shrewsbury, UK

which moves learning from linear didactics where it is not easy to find the possibilities to work on sustainability into a more attentive and comprehensive vision for the construction of knowledge, skills and attitudes as a cultural and social process supporting learning towards a more sustainable world.

5.2   Learning Together in Real World

Real-world learning entails the use of education in authentic situations to trigger a change in knowledge, abilities, mindsets, behaviours and values. Learning emerges when students are called to act in a reflective way in a situation which displays complex and difficult problems related to real-life situations, to which straightforward answers are insufficient to find an explanation or solution. In such authentic learning situations, students learn by doing, acquiring knowledge and understanding while developing life skills (Sala et al., 2020) and competences such as critical thinking, problem-solving, thinking out of the box, observation, researching, collaboration, writing, reading and presenting. These are competences that are not only transversal in all subject areas but that will be determinant throughout a student's life outside of school.

The two projects revealed that a real-world learning process needs planning and an intention that goes far beyond creating a setting where content can be better visualised as a representation of reality. It is a concrete and first-hand experience that should add value to the student's tangible engagement, the authenticity of the task, the sense of a shared purpose and, most of all, the construction of knowledge which is meaningful for the learner. An illustration of how to implement a real-world situation and of the inner possible learning potentialities which can be embraced is shown in Fig. 5.1.

In this example, the instruction on how to calculate the area of a quadrilateral is traditionally taught through calculation using a geometrical formula. This is merely an application of a piece of information which can reveal essentially if students can remember or not how to calculate an area of a particular polygon. A teacher might put this task into a context, for example, suggesting a problem where students need to calculate the area of a wall which needs to be painted in a hypothetical situation. This might result in more effective engagement by students as a clearer purpose for this knowledge is presented. Better though, when the instruction becomes essential and urgent for a real task that, if planned carefully, may end up being a powerful learning setting. This could be an assignment for refurbishing a wall in the science laboratory of the school which students always wished to improve. In this case, the teacher may not even suggest to pupils that the area of the wall needs to be calculated and, instead, let them experience what is essential to know and do and how to best proceed and, finally, to find the tools to measure physically and for real the two dimensions of the wall to calculate its area, experimenting errors as steps they need to overcome for the sake of fulfilling the task.

Other valuable, but not less important, matters might emerge during the learning process, for example, estimating the required resources, reporting to the school board, selecting the type of painting to buy – an eco-friendly one? seeking collaboration with a larger community than the school for the help, the process to make a collective decision on the new colour of the wall with school friends and with the principal and teachers of the school. As a result, in authentic learning, students are required to use what they are learning, motivating students to learn beyond the usual and often ineffective to acquire enough knowledge on a subject for the purpose of getting a good grade.

The critical point of learning triggered by a real-world issue is that students and teachers are not pretending; together, they live a concrete experience where contents, subjects and disciplines make sense and are relevant to them in that precise moment (Humberstone & Stan, 2012; Lundegård, 2018). The teacher and educator role is to look for the thresholds where numerous subjects of the everyday school curriculum start to live and weave together for understanding the

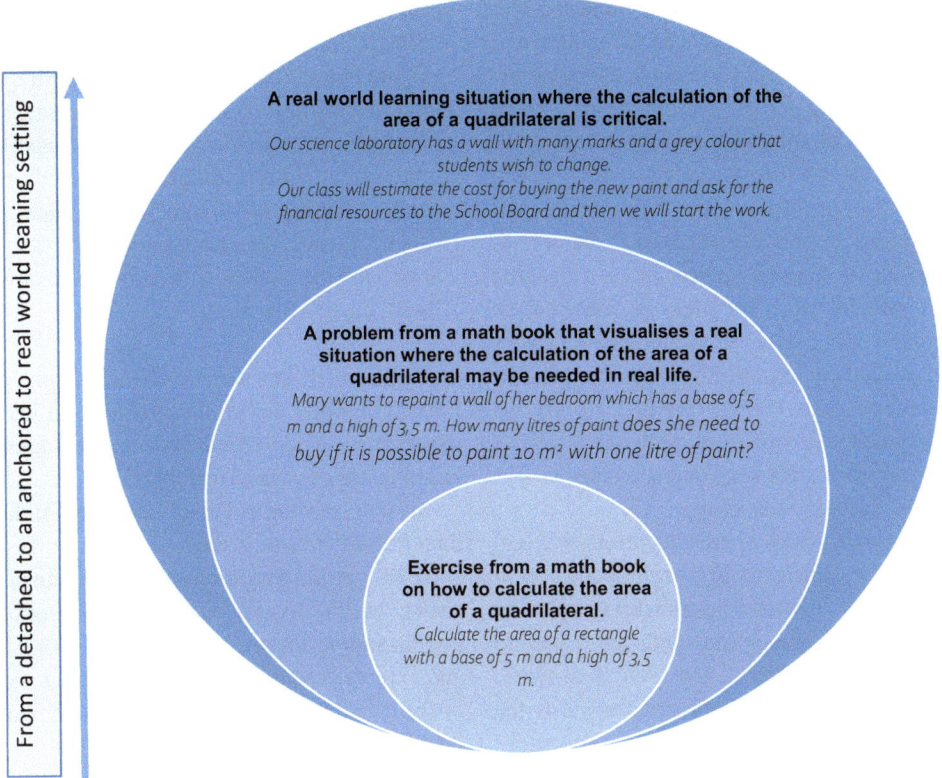

From a detached to an anchored to real world leaning setting

A real world learning situation where the calculation of the area of a quadrilateral is critical.
Our science laboratory has a wall with many marks and a grey colour that students wish to change.
Our class will estimate the cost for buying the new paint and ask for the financial resources to the School Board and then we will start the work.

A problem from a math book that visualises a real situation where the calculation of the area of a quadrilateral may be needed in real life.
Mary wants to repaint a wall of her bedroom which has a base of 5 m and a high of 3,5 m. How many litres of paint does she need to buy if it is possible to paint 10 m² with one litre of paint?

Exercise from a math book on how to calculate the area of a quadrilateral.
Calculate the area of a rectangle with a base of 5 m and a high of 3,5 m.

Fig. 5.1   The same task of calculating the area of a quadrilateral through the lens of real-world learning approach

world around us. In this sense, real-world learning helps to reconstruct curriculum content to offer teachers a concrete opportunity to work together, to plan the complexity of a learning setting which is inherently transversal to context, disciplines and competences.

5.3   Sustainability and Real-World Learning

Modern human activity is so disruptive it is affecting how the biosphere supports life on the Earth. We are changing the geological course of the planet itself. After 11,700 years of relative environmental stability following the last glacial era, the planet seems to have entered in a new geological era, the so-called Anthropocene (Crutzen, 2002), where humans have become a geological agent able to catalyse this change: land, structural and climate changes are modify-

ing the timing of geological phenomena such as erosion; sedimentation; the cycles of carbon, nitrogen and phosphorus; and the biosystems of the planet (Steffen et al., 2015; Waters et al., 2016; Zalasiewicz et al., 2017). The challenge we urgently face is to respond wisely to the environmental and social problems triggered by these recent and rapid changes made by humanity. Climate crisis, biodiversity loss, energy and food production, social injustice, poverty, pollution and health issues, while in some cases improving, are not changing fast enough to prevent huge problems affecting all humans and the natural systems we depend on. There is a clear need to share new approaches to addressing sustainability since our current attempts are insufficient.

We need an overall strategy which not only considers the political, economic and scientific dimensions of the challenges we face but also delivers effective cultural and educational answers to promote action and to embrace the transformational potential of sustainability to build a better world for all. If education has been recognised as a crucial element to deliver sustainable development (UNESCO, 2012, 2014), education institutions still struggle to fill the gap between the easier fostering students' environmental awareness, which seems however insufficient, and the more compelling nurturing to empower students to be able to make decisions towards sustainability. Engaging young people to become change-makers (Rieckmann et al., 2017) in the big challenges of our time, so compelling described in the Sustainable Development Goals (SDGs) set in 2015 by the United Nations General Assembly (UN, 2015), seems to be still an unknown and strenuous task to fulfil in the everyday schooling.

In these whirlpools, where education strives to be effective in transforming curricula and teaching approaches towards sustainability, a real-world learning approach may indicate a way to navigate the complexity of the task. Real-world learning offers a grounded practice towards the educational challenges of sustainability problems, considering real and meaningful issues as starting points for teaching sustainability and integrating curricula subjects. Learning from a real-life setting turns upside down our point of view, from teaching *about* sustainability to promoting learning *for* sustainability and *as* sustainability (Vare & Scott, 2007): from a subject whereby students might understand what could be done to improve life on the planet to a subject to live by, where students learn through experience what sustainability means for them. To address this goal, teachers and educators, within the Urban Science project, who planned and piloted learning modules to explore sustainability and to develop action competences while teaching in urban settings have found crucial:

- The necessity to focus attention on the pedagogy, to design learning processes for the development of knowledge, mindsets, competences and values which emerge when we are called to live a real situation where sustainability is critical.
- The need to select learning contexts which offer the potential to disclose and address *all*

together the issues which are entailed in the 17 SDGs. The choice of the learning situation should be carefully designed to easily show the spectrum of possibilities to consider while exploring a particular task; the dimensions of the self, and the community; the needs, rights and wellbeing of all species; an attention to social justice, gender equality and cooperation; and the quality of the environment where students live and study.

Some other key characteristics found to be important when designing a real-world learning experience while focusing on sustainability learning are described in Table 5.1.

In such learning contexts, students will have a direct experience (e.g. they will monitor the presence of plastic and microplastic in their local river), outline issues that are meaningful for understanding its implied elements, then search and explore resources, ask questions to members of the local community, share ideas and plan and decide which action to address the situation with. In the meantime, the teacher will facilitate to reflect and to form concepts considering different dimensions of sustainability. As a result, students are engaged to change the ways they see and think about the world while deepening their understanding of it (Mezirow, 2000; Slavich & Zimbardo, 2012). They co-create knowledge with others, learn to deal with societal and environmental problems that are complex, uncertain and without a clear solution, and to make wiser choices possible in that moment, and personal and collective decision-making for a better world for all living beings.

5.3.1 A Model for Planning Learning Experience About Sustainability in the Real World

Between 2012 and 2015, the Real World Learning Network brought together outdoor educators from across Europe to explore how science and sustainability can link towards behaviour change for a sustainable future. After

Table 5.1 The characteristic of a learning environment where students can 'breathe' sustainability

Characteristic	Explanation
Facilitating the learning process	A teacher's role is to guide the learning experience in a manner which is meaningful and where the pleasure of the discovery and of understanding involves not only students but also teachers
Sensibility to all the SDGs' dimensions	The learning situation should be sufficiently complex to let students embrace, potentially, all SDGs' dimensions. In such a favourable situation, students are supported to consider different points of view and to look at the matter they are exploring from a range of perspectives to enlarge their opinions and to connect different dimensions and implications of sustainability to what they are studying
Exploring interdisciplinary and transdisciplinary learning	The issue and approach provide opportunities for students to discover autonomously how different disciplines and competences are needed for a comprehensive understanding and how they can contribute to formulate an idea, a solution or a model for the required task. This attention challenges students to synthesise learning without simplifying complex issues and to find links which might be new to both the students and the teachers. This type of learning setting helps to overcome the linearity of curriculum planning, often characterised by the progression of one content into another. It offers a structural vision of the learning at the intersection of disciplines and allows learners the possibility to create new knowledge
Embedding curricula subjects	Students and teachers develop or review curriculum contents which make sense to clarify, understand, elaborate and evaluate a sustainability issue students are addressing. A real-world situation works as a sliding door where curricula contents are identified and take part in the learning for their meaning in relation to the sustainability task that students are living
Problem-based learning	Learning for sustainability benefits from a learning setting which is grounded in a relevant issue of sustainability and is meaningful to students. Problem-based learning is a powerful way not only to engage students into the learning process but also to build agency and responsibility to face the problem. It is a natural constructivist approach to understanding what we do not yet know about the world (Bell, 2010)
Place- and community-based learning	Outlining a real-world setting with a strong link to the place and community where students live develops a sense of belonging and responsibility, they concretely experience some of the elements of the issue and can contextualise their actions and solutions. The involvement of the community in the learning path represents an essential resource to foster school-community collaboration (Smith & Sobel, 2010)
Inquiry-based learning	The inquiry approach gives the opportunity to enrich the learning process, developing curiosity, creativity and the will to investigate, understand and do. The learning process begins from student's authentic questions about the problem or issue where sustainability is the critical element. Used especially in science education, inquiry-based learning is an applicable approach which can suit and be effective with the learning of all disciplines (Rocard, 2007)
Open-ended questions	The learning process needs to give students the possibility to mirror the complexity and ambiguities that characterise sustainability problems where there are no clear right and wrong answers or single solution, but rather complex situations with many linked elements that need to be monitored to tackle the problem (Rittel & Webber, 1973)
Tuning into values	Students are encouraged to express and discuss perspectives, doubts, necessities, interests and ethics related to sustainability issues they are exploring, recognising why values matter to achieve sustainability and which values can guide behaviours for sustainability (Holmes et al., 2011; Schwartz, 1992)
Promoting action experiences	A crucial step is to help students take decisions and action in their life. The learning process includes opportunities for students to feel commitment, to take action and to develop action competences, among them courage and a sense of responsibility (Jensen & Schnack, 1997)
With the future in mind	The possibility to envision thinkable and alternative futures in relation to the sustainability issue will encourage students to practise creativity, to imagine solutions taking into account impact and evidence that comes from the understandings of the past and to be engaged already in the present (UNECE, 2011)

3 years of collaborative work within the network, in consultation with experts across Europe and through European conferences and workshops, the Network launched the RWL Model (Real World Learning Network), its vision for a truly embodied holistic approach to learning for sustainability (see-www.rwlnetwork.org/rwl-model.aspx). The model seeks to be a compass for teachers and practitioners in their thinking about and planning, delivery and evaluation of learning experiences. It is a tool with which to play and explore within their education work and to challenge their thinking and practice and a way of deepening learning experiences to the level of sustainability.

The RWL Model is based upon an experiential educational perspective and proposes that learning is the result of a construction of knowledge and understanding through direct experience. The model emerged from a complex interaction between the partners, through sharing their own experiences and asking themselves 'what works in our experience'. In this sense, the model is based on the experience of the individuals taking part not on a conscious review of existing models and frameworks. Although there was undoubtedly influence from other approaches, this liberating approach allowed the model to emerge from practised experience rather than academic research. The learning in the model is seen as a journey rather than a fixed destination, and within this journey, the learner is encouraged to build connections between, and make meaning from, their experiences, rather than simply remembering facts. This way of seeing education is particularly useful when encouraging discussion around sustainable futures, as to deal with some of the most challenging issues of our time, it is important that our education system equips learners with the ability to make meaning within their lives. Not only do learners need to be able to make meaning from the situations they find themselves in, but in order to move beyond unsustainable behaviours, learning must become 'transformative' – becoming aware of and challenging our deeply held assumptions about the world.

The model brings together six interconnecting areas, each contributing to the whole. We can visualise these as fingers on a hand; each can operate independently but function for more effectively as a whole. The fingers represent understanding, transferability, experience, empowerment and values, linked through the palm representing frames. A brief description is found in Textbox 5.1. The planning and delivery of teaching and learning experiences are personal to the individual; however, the RWL Model offers some helpful guidelines for planning. The model offers a degree of 'structure' to be used as a fluid, flexible guide rather than a fixed pro forma.

Textbox 5.1 Real-World Learning

The Real World Learning Network was established to explore and share successful approaches to real-world learning through the outdoor classroom that leads to action for sustainable development. It was a consortium of seven partners from six countries across Europe. During the project, an additional 56 outdoor learning and education organisations joined the network in this exploration of effective outdoor learning. The core achievement of the network was to synthesise learning and practical experience in outdoor learning into a unifying model – the RWL Model. When developing the model, the challenge of delivering learning that leads to sustainable behaviour change was held in mind. The result, therefore, is not so much a model for outdoor learning but a model for transformative learning. Quotes from project participants:

I've recognized that this way of learning has helped me think much more outwardly.

I've realised that the hand model is really very inspirational for everyday work of tutors and verifying what we are already doing and planning for our centre's future.

I changed – Thoughts, Relationships, Self – Reflection; probably there was change in everybody.

> **Textbox 5.2 Description of the RWL Model Elements (Real World Learning, 2014)**
>
> *Understanding – Are scientific concepts of life involved?*
>
> Scientific concepts, like cycles or change, infuse all areas of life. Understanding these concepts means to understand the complex interplay of processes and patterns that sustain life. However, true understanding comes from combining a scientific approach with emotions, values and humanity. Exploring scientific concepts of life in this holistic way develops thinking and action for sustainability.
>
> *Transferability – Are different areas of life included?*
>
> Sustainability goes through all areas of life. It is important to transfer learning of, for instance, understanding of scientific concepts, with experiences that learners have had, actions taken or values held. This allows learners to make connections between themselves, their communities, global society and the non-natural and natural environment.
>
> *Experience – Do learners get in touch with outdoor settings?*
>
> By getting in touch with an outdoor setting, learners can experience real life with their head, heart and hands, follow their curiosity, become sensitive to the complexities and interconnections around them and recognise that they are a part of a bigger system. This intensity of experience is held and lifted by the other aspects of the model.
>
> *Empowerment – Are learners empowered to shape a sustainable future?*
>
> Empowerment brings learners to the centre of the learning experience: it's about recognising and realising their own humanity and their own ability to take action for positive change. Empowering learners enables them to cooperate and to take ownership of their learning. Everybody can make a change. To experience this can help learners to shape the future in a sustainable way.
>
> *Values – Are self-transcendence values promoted?*
>
> Values represent our guiding principles, our broadest motivations, influencing the attitudes we hold and how we act. Self-transcendence values support bigger-than-self thinking and action. Being concerned about the wellbeing of others and the planet is essential for sustainability. For more on self- transcendent values, see Holmes et al. (2011).
>
> *Frames – Is there a frame providing a connecting story?*
>
> Frames play a powerful part in how we understand and interpret the world around us. For example, when we hear the word 'nature', subconsciously a bundle of different memories, emotions and values are activated. Such associations, often leading to strong narratives under the surface of our awareness, are called 'frames'. In our model, the frames are in the palm of the hand as they ensure that values, empowerment, experience, transferability and understanding are connected, leading to a deeper sustainability learning experience (Lakoff & Johnson, 2003; Lakoff, 2008).

5.4   Working with the RWL Hand Model

The planning and delivery of teaching and learning experiences are personal to the individual; however, the RWL Model offers helpful guidelines for planning an experience, whether it is a single session/ lesson, a day or an entire course. The model offers a degree of 'structure', and it is best used as a fluid, flexible guide rather than a fixed pro forma. This is especially the case when presented with content which has already been planned, outcomes which are predetermined or a favoured approach which has been already decided. Any model which is to be used in education must be able to accommodate these demands since teachers and educators will not always have a 'blank canvas' upon which to start their planning. The following entry points may be useful when considering the applicability of the model to their teaching and planning process:

- *The 'Blank Canvas': Starting with Elements of the Model*

If there is the luxury of designing a teaching experience from the 'ground up', whether as part of a taught course or standalone lesson, this model will provide useful guidelines for planning about where the experience sits with the learners. For example, what is the learners' socio-cultural background, what views or experiences might already shape their thinking, and what are their specific learning needs? Whatever the case, teachers and educators might like to look at what that experience offers in relation to the elements of the model.

Teachers can ask themselves: Will the learners benefit from some self-directed learning to prompt empowerment, are activities able to link their home and community lives with the contents of study to transfer understanding, could they link all of this into the universal principles of life, and what type of frame can be used for this experience which helps to bring in bigger-than-self values and connects the different elements of the model?

- *A Structured Approach: Teaching Topics and Methods*

In the majority of cases, curricula contents will already have a familiar structure and have attached to them a certain set of expectations. In this case, the approach might be to 'retro fit' elements of the model into the lesson/curricula as appropriate. The way in which to go about this will vary and very much depend on circumstances such as the support of other staff, the need to resource the lesson/teaching activity, time and personal style and approach of teaching. In this case, the model can be used as a think-piece for thoughts and reflections. It is likely that a frame can be used either explicitly or implicitly to effectively bring in elements of the model. The overall format of the lesson will not change dra-

matically, but the approach will determine the path that learning takes throughout. Even the way in which questions are asked of learners or the emphasis placed upon activities and the freedom given to undertake them will affect the learning outcomes and the values which the experience embodies. Using the model for this approach also allows reflection on where the opportunities and 'gaps' are.

- *Outcomes: Boundaries and Directions*

The increasing push within education towards higher 'standards' and preparing young people for the world of work can lead to overly deterministic outcomes. The RWL Model provides a handrail to move away from this and into a more experiential, empowering and creative form of learning. However, it is recognised that while attempting to modify teaching practice to take into account the elements of the model, we will inevitably come up against the deterministic nature of education, be it pre-defined learning objectives, exam skills or time restrictions. When faced with this apparent stand-off between open and closed learning, it is easy to say 'this isn't appropriate here' of the RWL Model. However, to do so would be to turn our back on some essential thinking and understanding about learning for sustainability. So, we encourage educators to make use of the model in this scenario as a guiding light from which to be able to select elements which fit and may help to link the experience into a more holistic and values-based context.

5.5 Bringing Together the 'In' and the 'Out' of the School for Learning Towards a Sustainable World

Despite the fact that we are used to thinking of the classroom as the main location where teachers work and students learn, teaching and learning happen effectively everywhere. The school itself and the spaces outside have enormous educational potentiality to trigger learning processes because of their inherent characteristic of not being neutral spaces. Whatever non-classroom locations we are considering– school entrance or street, the institute canteen or plant for city waste management, courtyard, community garden or parking lot, the local square or the Council meeting room – they are all authentic situations to be considered in education and which potentially have powerful insight to design learning settings to address sustainability (Quay, 2015).

If it is the real world that relevant problems of sustainability challenges occur, is it not obvious to go outside to have a direct learning experience of them which is authentic? Real-world settings provide the opportunity to understand the complexity of sustainability and to experience relationships, interconnections, correlations and dependencies, to make sense of diversity and dissonance and to have space where social, historical and physical phenomena can be vividly displayed   (see   www.rwlnetwork.org/media/75352/approaches_methods.pdf   for review of approaches). Reading, listening, asking, researching, experimenting and debating find their best conditions outside where students can immerse themselves in situations which reveal how all the sustainability elements are relevant, critical and connected while considering how personal values can influence views and behaviours, analysing alternative visions or solutions and, finally, deciding and making choices in relation to the experience they lived and the future they want to live in. 'I believe that education is a process of living and not a preparation for future living', Dewey (1897) asserted. The outside classroom, as advocated by the American philosopher and psychologist of the active and experiential learning approach, is the critical place where students experience the mutual, active and transformative exchanges between people and the environment.

Creating conditions for students exploring sustainability with an authentic and outdoor approach fosters a transformation of the school itself that goes beyond the simple organisation of part of the teaching lesson considering just the educative potentialities of the territory where it is settled. The ideal situation is to adopt a model of school able to frame sustainability coherently in its inherent and external processes and operating as a link with its environment and the community. That implies an attention to values and principles of sustainability in all school functions, considering the school as a whole system and not only reorganising curriculum and the pedagogy but also taking into account internal procedures, organisation of spaces, activities and facilities, decisions on purchases and assets and planning of collective and collaborative actions in and by the school community (Henderson & Tilbury, 2004). As a result of this process, sustainability should be breathed by teachers, students, administrative workers, families and community. The school should reinforce partnerships to bring community programmes and resources inside and, in the other way around, school resources and facilities into the community to support and make vital contributions to their communities, becoming a vibrant and dynamic node of the network for better life conditions. In this perspective, the thoughtful planning of the thresholds between the 'in and out' of the classroom and of the school as authentic open learning spaces represent the special areas where a school can make the difference to enhance the opportunities to learn sustainability as a process of social learning (Wals, 2007) in which all stakeholders, from students to the elders of the community, are involved and can actively contribute. In such a perspective, teachers can also consider exploring above and beyond the school boundaries and investigating the city around to find how students, people, families and citizens can support healthy cities.

5.6   The City Around the School Is a Living Hub for Sustainability and Science Learning

Located just outside the school, there is an accessible open-air laboratory waiting for students and teachers to examine and experiment with authentic challenges to creating sustainable cities.

We need not just think of parks areas of cultural, historical or natural heritage as significant places for field study.

The urban environment around the school is also an extraordinary educational resource, a living lab. Firstly, it is near to school and available even for short sessions of work which can overcome logistical issues. Secondly, it represents a valid means for students to discover the area where they spend a considerable time of their life, strengthening their connections with the community and developing their sense of place. Finally, the identification and the direct analysis of the challenges of their town increase students' exposure to problems which undermine their urban environment, enabling them to recognise that they have a role to play and decisions to make for changing what interferes with health, justice and wellbeing in their cities and for shaping better living settings for all.

Textbox 5.3 Urban Science-Integrated Learning for Smart Cities

Urban Science focused on how science can develop solutions to urban issues, so motivating pupils to view the positive benefits of science to the urban environment. Our aim was for pupils to explore solutions to urban issues not just the issue itself. This placed a greater emphasis on creative thinking and problem-solving skills and ensuring what the science pupils learn is seen to be practically applied. The project developed 35 enquiry-based learning modules piloted with 383 teachers, 46 schools and 1602 pupils.

What have students loved of this experience: the possibility to do science lesson in a different way, they appreciate the fact to relate a content of their book to a real-life situation of their school territory. (Italian teacher)

Materials developped in the project allow to do research, make conclusions based on the real life situations which usually does not happens in the ordinary lessons. (Latvian teacher)

Provokes questions, creativity, decision-making and engagement. (Hungarian teacher)

What is surprising that students who had been 'outsiders' in my science classroom found their motivation and started to develop self-efficacy in inquiry-based science, and it was a really rapid and profound change. (Hungarian teacher)

Between 2017 and 2020, a consortium of organisations together with primary and secondary teachers joined in Urban Science (see www.urbanscience.eu, 2017–2020a). Together they explored the correlation between real-world learning in urban settings and the improvement of scientific and sustainability competences. This involved an experimental phase in schools to pilot learning resources and evaluate student learning development through the design of specific rubrics. The objective was to relate learning for sustainability to the everyday lives of learners to envision and develop healthy cities. The project developed a framework grounded in the inquiry-based approach to guide teachers' work (Bybee et al., 2006; European Commission, 2007; Minner et al., 2010), and it created learning resources integrating STEM and sustainable development.

Five were the pillars we considered:

- *Strong focus on the potential for authentic learning in urban areas addressing the significance of sustainability.* Challenges chosen by students and teachers on the 17 SDGs should display strong links with the urban area where students live. This choice recognises the fact the cities are critical to the lives of young people and cities account for the majority of environmental impacts directly or indirectly. Changes within cities can have greater traction; there is an urgent need for cities to reinvent themselves to be future fit. For example, students can choose a problem related to the low air quality of the place where they live or consider the abandoned places near their school, the difficulties to move around in cities, the shortage of areas to meet or to practise sports and music, extreme weather phenomena such as floods or heat islands, the lack of good food in the local canteen or the development of a commercial area nearby.
- *Sustainability challenges as the 'red thread' to develop both scientific and sustainability knowledge and competences.* The central idea was to engage students to analyse a sustainability problem of their city or territory around the school, to develop key competences related to sustainability (Rieckmann, 2018)

and content knowledge which are necessary to understand the observed phenomena and to envision actions to design a healthier and more liveable place, embracing a constructive approach to learning. The learning pillars considered in the project are summarised in Table 5.2, while an example of a matrix to

Table 5.2 Content knowledge and competences considered in the Urban Science project

Urban Science learning pillar	Descriptors
Develop knowledge and understanding of key Urban Science issues	State observable features
	State or use a classification system
	State relationships between variables
	Show understanding of scientific theory
Learn how to	Identify equipment
	Use equipment
	Describe a standard procedure
	Carry out a standard procedure
Develop an understanding of scientific inquiry	Propose a question
	Plan a strategy
	Evaluate risk
	Collect relevant data
	Present data effectively
	Process data
	Interpret data
	State a conclusion
	Evaluate a conclusion
The ability to understand and apply systems thinking: inputs, outputs, connections, loops and feedback	Able to connect different elements within an urban environment
	Seeing how they relate to each other
	Recognising that all actions are part of a system and often have multiple consequences positive or negative
An understanding of how natural systems function ecological limits and resource constraints	Understanding how natural systems work within limits and use strategies to adapt, optimise and flourish
	Understanding how human activity that exceeds ecological limits or capacity has negative effects
	Understanding how sustainable systems balance resource and use within a fixed carrying capacity
The ability to think in time – to forecast, to think ahead and to plan	Develop, understand and evaluate ideas for alternative futures
	Able to predict the consequences of actions today on future choices and their ability to act
The ability to think critically about value issues	Identify behaviours and values that reinforce a sustainable future
	Able to apply a values perspective to decision-making; integrating scientific knowledge with personal and societal values in making choices
The ability to separate number, quantity, quality and value	Being able to distinguish between actions which improve quality of life versus quantitative changes in material standards
The capacity to move from awareness to knowledge to action	Able to take responsibility to develop and implement plans and evaluate their success
The capacity to develop an aesthetic and compassionate response to the environment	Having a sense of connection beyond self
	See the needs of others
	Demonstrate compassion and sympathy for others and the natural world
The capacity to use processes (knowing, inquiring, acting, judging, imagining, connecting, valuing and choosing)	Being able to integrate a range of technical and emotional capacities
	Know which capacities to apply to a given situation

Table 5.3 An example of rubric for the assessment of students' learning process of Urban Science learning modules (Urban Science, 2017–2020b)

Area of learning	Novice	Beginner	Practising	Advanced
A. Develop knowledge and understanding of key Urban Science issues	Need to develop understanding of scientific thinking in the context of urban environment	Able to identify elements of scientific thinking and to identify urban challenges	Apply some elements to understand challenges	Able to apply scientific thinking to understand challenges
B. Able to use scientific methods for inquiry in Urban Science	Need to learn methods of science inquiry	Need to practise methods of science inquiry in urban context	With support, use scientific methods in urban context	Able to use scientific methods in urban context
C. Carry out inquiry science activities in urban context	Need to acquire inquiry activities in urban context	With scaffolding, able to complete an inquiry activity in urban context	With support, able to use science inquiry in urban context	Able to apply science inquiry autonomously in urban context
D. Understand the basic features of sustainable urban systems, using science knowledge	Need to know more about the basic features of sustainable urban systems	Able to identify some basic features	Able to distinguish between sustainable and unsustainable urban systems	Understand the basic features of sustainable urban systems
E. Understand alternative futures in urban environments, using science knowledge	Need to develop time-related thinking in urban context	Able to understand the basics of forecast alternative scenarios using scientific thinking	Able to develop ideas using scientific thinking	Develop ideas and understand alternative futures based on scientific thinking
F. Identify behaviours and values that reinforce a sustainable future	Need to learn more about values in a sustainability context	Able to identify some behaviours and values	Able to identify some behaviours and values that act towards a sustainable future	Able to distinguish between behaviours and values that act towards a sustainable future
G. Agency and responsibility supported by scientific thinking	Need to develop responsibility and ownership for actions	Understand evidence supporting responsibility for actions in urban environment	Need to develop agency but express responsibility for actions in urban environment	Ready to practise evidence-based responsibility for actions in urban environment
H. Evaluate success of proposed interventions in cities based on scientific thinking	Need to practise how to evaluate consequences of actions in systems such as cities	Possess skills to evaluate actions in systems	Able to apply scientific thinking to evaluate success of interventions in cities	Able to evaluate success of (proposed) interventions in cities based on scientific thinking

assess competence and content knowledge development is illustrated in Table 5.3.

- *Inquiry-based approach.* The flow of learning should encourage students to ask questions linked to the sustainability urban challenge. The main idea was to develop learning journeys in search of answers and solutions to those questions which can be investigated directly in an urban environment. We decided to use a simplified IBSE (Inquiry-Based Science Education) approach because of its inherent characteristics of students-centred

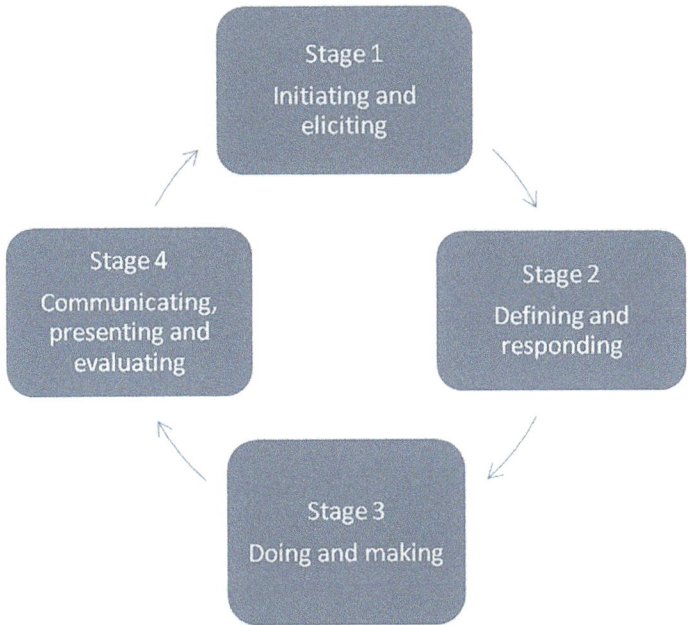

Fig. 5.2 The Enquiring Minds flow of learning (Morgan et al., 2015) used in Urban Science project. *Stage 1*: eliciting the knowledge, interests, ideas and motivation of students about sustainable challenges in their city. The teacher's role is to help them draw on their own lives and experiences to discover things that interest them, make them curious and want to ask questions. *Stage 2*: shaping, defining and focusing an idea or question and making plans to research it further. The teacher's role is ensuring students can advance their inquiries meaningfully, providing frameworks and learning so that they can organise their research. *Stage 3*: research, design and construct activities, in order to make a contribution in the chosen enquiry, during which students engage in a variety of tasks depending on the nature of their enquiry. Teachers encourage students to manage their time, identify clear goals and monitor their progress. *Stage 4*: students communicate, share and present their new knowledge and understanding with others

learning and actively engaging learners and developing their curiosity to understand the world around them and to make informed decisions. The model which has inspired our planning of learning modules and resources was Enquiring Minds (Morgan et al., 2015). A description of the approach and how to use it for designing activities is simplified and described in Figs. 5.2 and 5.3.

- *Transforming STEM teaching*. Our aim was to explore how contextual understanding is

critical in learning science for understanding complex problems which interrogates society and the way we live on the planet to plan a sustainable future. We aimed to identify if using locally generated content focused on the urban environment could make learning more meaningful and applicable for students, thus influencing their motivation to learn. We aimed for a move from knowledge being held by authority figures to one where knowledge is dynamic, collaboratively

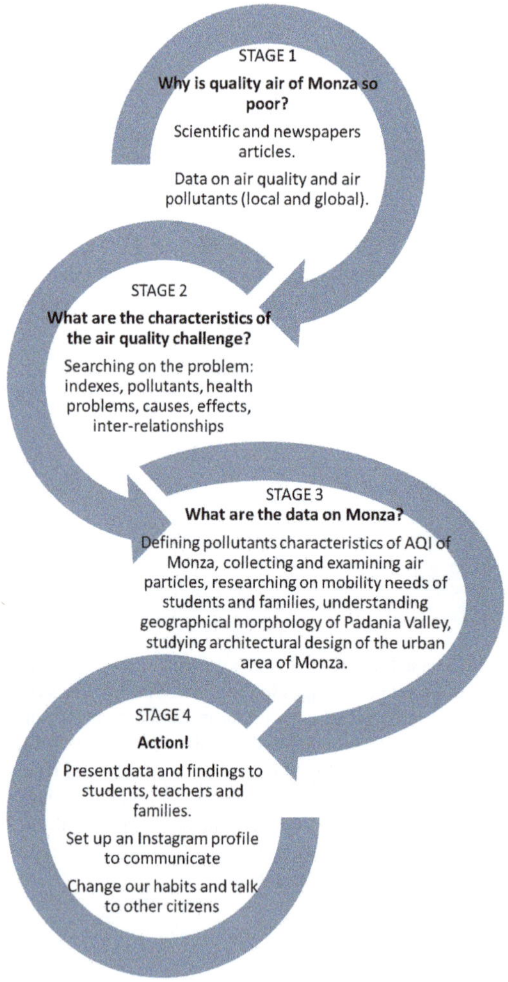

STAGE 1

Why is quality air of Monza so poor?

Scientific and newspapers articles.

Data on air quality and air pollutants (local and global).

STAGE 2

What are the characteristics of the air quality challenge?

Searching on the problem: indexes, pollutants, health problems, causes, effects, inter-relationships

STAGE 3

What are the data on Monza?

Defining pollutants characteristics of AQI of Monza, collecting and examining air particles, researching on mobility needs of students and families, understanding geographical morphology of Padania Valley, studying architectural design of the urban area of Monza.

STAGE 4

Action!

Present data and findings to students, teachers and families.

Set up an Instagram profile to communicate

Change our habits and talk to other citizens

Fig. 5.3 Example of learning flow planned and piloted on city air quality in the Urban Science project

developed and social context matters, where teachers and students becoming co-learners (Fig. 5.4).

- *Align science with values and future-thinking to create sustainability.* The relationship between science and society is vital and sensitive. In this learning process, students should experience how scientific research and tech-

nological innovation are instruments at the service of the needs of the community and how science may help in integrating people perspectives on sustainability issues into a common understanding by linking concepts with shared social values (Lang et al., 2017; Hall et al., 2017) and identifying points of leverage to designing well-integrated sustainability solutions.

This learning module was developed with a team of 3 teachers and trialled with 52 students in a secondary school in a city in the north area of Milan, Italy, whose Air Quality Index (Akimoto, 2003) exceeded in 2018 standard values for a noticeable number of days/year (European Parliament, 2016). This learning module works through the inquiry-based flow of learning. The urban area near school becomes a living laboratory which helped students to explore the complexity of the air quality issue and its scientific understanding in a real context. It facilitated the envisioning of a sustainable and healthy city. At the end of learning experience, students decided to open an Instagram page; they experimented with different ways to travel to school and decided then to promote these habits to school friends.

During the trialling, it was found that all students were able to research and elaborate scientific information, if guided, for debating the issue of air pollution. Moreover, they were able to relate findings to their everyday life with a call to actions for themselves, school friends and families. From student assessment, it was found that the possibility of inquiring into questions they want to explore and to understand step by step their misunderstandings and misconceptions on air quality, air pollution and climate change were the most important keys for turning on curiosity and a will to know more.

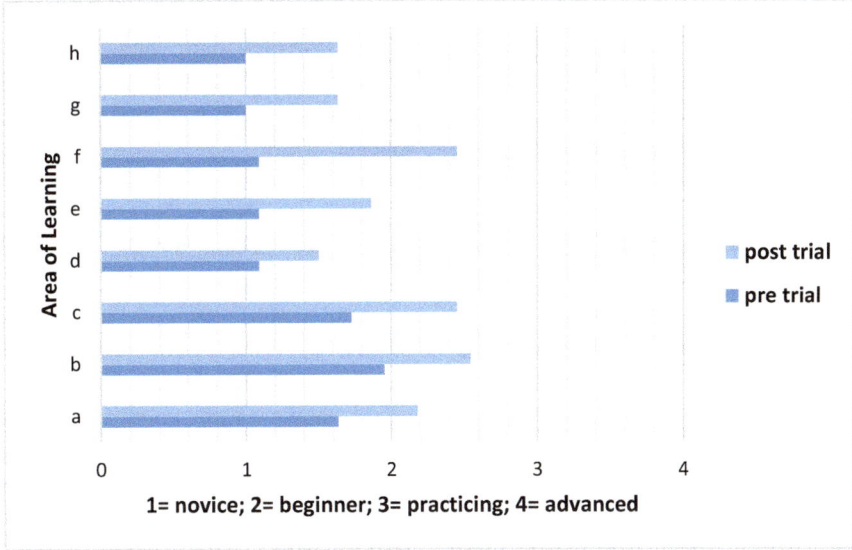

Fig. 5.4 Assessment of the learning process of students using the rubrics described in Table 5.3 during the Urban Science piloting of the Air Quality learning module

5.7   In Conclusion: Why Is There Not More Real-World Learning?

We strongly advocate for the benefits of real-world contexts and settings in learning, extending already familiar pedagogies including inquiry and place-based learning. It is relevant to ask, if such learning has tremendous merits, why this discussion is necessary. There are several educational issues limiting responses to sustainable development. Firstly, the education sector is not providing an adequate response to fully understanding the causes of unsustainability or providing attractive solutions (e.g. see increase in climate anxiety; Marks et al., 2021). There is a gap in what we know about human behaviours creating environmental challenges and our beliefs about what we want to do. There still exists a dominant frame that 'everything will turn out fine', 'there's nothing I can do about it' and 'technology will always find a solution'. More broadly, this is rooted in the dominant human condition of greed, underpinned and promoted by the consumption-based economic system, and unwillingness to change for the risk this apparently brings (Scharmer & Kaufer, 2013; Lent,

2017, 2021). Secondly, learning is not sufficiently connected to the real world and is often abstract. Climate change taught through PowerPoint can never truly engage the emotions strongly enough to motivate change. Content-driven learning rarely encourages the curiosity that comes from closely observing a real-world situation and questioning what is happening (Greer & Glackin, 2021). Neither does it allow learners to reflect on what has importance for them in their lives or even ask learners what matters. As a result, sustainability topics such as climate change and biodiversity loss are seen as just another subject rather than learning integral to key life choices. In England, for example, climate change is not even mentioned within the primary curriculum and so not seen as important until a learner reaches 13 years of age. Fundamentally, we forget we are a part of nature, completely dependent upon it for our survival, and that we can at best influence but never control it.

Education needs to provide an effective response for Europe to meet its sustainable development targets. It needs to support young people in developing the skills to actively build the green economy and society. Results from the RWL Network and Urban Science partners suggest that outdoor learning through first-hand experience is increasing; however, there is a huge potential for it to contribute more meaningful messages for sustainability. Learning within the Urban Science project was closely aligned with competences for working scientifically. A total of 35 learning modules were developed and piloted with 1602 pupils engaged during the project benefiting from improved knowledge and understanding of their urban environment, improved methodology to investigate their urban environment and new perspectives of how science can support them in addressing the challenges and opportunities of urban sustainability. Partners collected feedback from pupils through questionnaires and evaluation of teachers. Data of the competence levels measured during and after learning modules show that pupils performed a progression; indeed, some of the pupils started their journey from the beginner level but ended with more confidence in all the IBSE and sustainability competences. The core achievement of the RWL Network was to synthesise learning and practical experience in outdoor learning into a unifying model – the RWL Model. The model emerged from a need to create a coherent way to communicate the outputs and outcomes of the working groups. The RWL Model brings together the elements of five working groups into a meaningful whole, thereby providing educators with an overview of the components of outdoor learning and entry points to deeper understanding. When developing the model, the challenge of delivering learning that leads to sustainable behaviour change was held in mind. The results, therefore, is not so much a model for outdoor learning but a model for transformative learning.

There are of course other approaches to science and sustainability such as e-learning and classroom-based approaches; however, neither engages the learner directly in the subject matter. Sustainability issues are often large scale and can seem too distant for learners to grasp; theoretical approaches simple magnify this. Real-world learning makes problems real and direct, reducing them to a scale whereby the learner can 'see' the issue and react in a meaningful way. We need to rebalance learning from classroom-based to real-world learning. It is clear that learning which simply provides more and more information about sustainability is not changing minds and actions fast enough and may even be having the opposite effect.

Can we succeed? We proposed at the start of this chapter that real-world learning can support a substantial paradigm shift and transition in learning, one which moves learning from linear didactics into a more attentive and comprehensive vision supporting learning towards a more sustainable world. The practice and evidence above provide ample starting points for teachers to shift the perspective of teaching and learning towards sustainability: to see that sustainability is not additional to 'core' learning but is in fact the basis for all learning that really matters.

Can we afford to fail?

References

Akimoto, H. (2003). Global air quality and pollution. *Science, 302*, 1716. https://doi.org/10.1126/science.1092666

Bell, S. (2010). Project-based learning for the 21st century: Skills for the future. *The Clearing House, 83*(2), 39–43.

Bybee, R., Taylor, J. J. A., Gardner, A., Van Scotter, P., Carlson, J., Westbrook, A., & Landes, N. (2006). *The BSCS 5E instructional model: Origins and effectiveness*. BSCS.

Crutzen, P. J. (2002). Geology of mankind. *Nature, 3*(6867), 415. https://doi.org/10.1038/415023a

Dewey, J. (1897). My pedagogic creed. *School Journal, 54*, 77–80.

European Commission. (2007). *Science Education NOW: A renewed Pedagogy for the Future of Europe* (EUR series 22845). Office for Official Publications of the European Communities.

European Parliament. (2016). Directive (EU) 2016/2284. https://eur-lex.europa.eu/legal-content/EN/TXT/?uri=uriserv%3AOJ.L_.2016.344.01.0001.01.ENG. Accessed 15 June 2021.

Greer, K., & Glackin, M. (2021). What counts' as climate change education? Perspectives from policy influencers. *School Science Review, 103*(383), 15–22.

Hall, D. M., Feldpausch-Parker, A., Peterson, T. R., Stephens, J. C., & Wilson, E. J. (2017). Social-ecological system resonance: A theoretical framework for brokering sustainable solutions. *Sustainability Science, 12*(3), 381–392.

Henderson, K, & Tilbury, D. (2004). *Whole-school approaches to sustainability: An international review of sustainable school programs*. Report prepared by the Australian research Institute in Education for Sustainability (ARIES) for the Department of the Environment and Heritage, Australian Government.

Holmes, T., Blackmore, E., Hawkins, R., & Wakeford, T. (2011). *The common cause handbook*. Public Interest Research Centre.

Humberstone, B., & Stan, I. (2012). Nature and well-being in outdoor learning: Authenticity or performativity. *Journal of Adventure Education & Outdoor Learning, 12*(3), 183–197.

Jensen, B., & Schnack, K. (1997). The action competence approach in environmental education. *Environmental Education Research, 3*(2), 163–178.

Lakoff, G. (2008). The political mind: A cognitive scientist's guide to your brain and its politics. Penguin Group.

Lakoff, G., & Johnson, M. (2003). *Metaphors we live by*. University of Chicago.

Lang, D. J., Wiek, A., & von Wehrden, H. (2017). Bridging divides in sustainability science. *Sustainability Science, 12*(6), 875–879. https://doi.org/10.1007/s11625-017-0497-2

Lent, J. (2017). *The patterning instinct*. Prometheus Books.

Lent, J. (2021). *The web of meaning*. Profile Books.

Lundegård, I. (2018). Personal authenticity and political subjectivity in student deliberation in environmental and sustainability education. *Environmental Education Research, 24*(4), 581–592. https://doi.org/10.1080/13504622.2017.1321736

Marks, E., Hickman, C., & Pihkala, P., Clayton, S., Lewandowski, E. R., Mayall, E. E., Wray, B., Mellor, C., & van Susteren, L. (2021). Young people's voices on climate anxiety, government betrayal and moral injury: A global phenomenon. https://ssrn.com/abstract=3918955. Assessed 05 Feb 2021.

Mezirow, J. (2000). Learning to Think like an Adult. Core Concepts of Transformation Theory. In J. Mezirow, & Associates (Eds.), *Learning as Transformation. Critical Perspectives on a Theory in Progress* (pp. 3–33). San Francisco, CA: Jossey-Bass.

Minner, D. D., Levy, A. J., & Century, J. (2010). Inquiry-based science instruction-What is it and does it matter? Results from a research synthesis years 1984 to 2002. *Journal of Research in Science Teaching, 47*(4), 474–496.

Morgan, J., Williamson, B., Lee, T., & Facer, K. (2015). *Enquiring minds* (p. 38). Future Lab/Microsoft. www.academia.edu/3270700/Enquiring_Minds_Guide. Accessed 4 July 2021.

Quay, J. (2015). *Understanding life in school: From academic classroom to outdoor education*. Palgrave Macmillan.

Real World Learning. (2014). The hand model guidelines. https://www.rwlnetwork.org/media/86467/the_hand_model_guidelines__v.4_pdf. Accessed 05 May 2021.

Real World Learning Network (RWL). (2014). Real World Learning. https://www.rwlnetwork.org/. Accessed 05 May 2021.

Rieckmann, M. (2018). Chapter 2 - Learning to transform the world: key competencies in ESD. In A. Leicht, J. Heiss, & W. J. Byun (Eds.), *Education on the move. Issues and trends in education for sustainable development* (pp. 39–59). United Nations Educational, Scientific and Cultural Organization.

Rieckmann, M., Mindt, L., & Gardiner, S. (2017). *Education for sustainable development goals, learning objectives*. UNESCO.

Rittel, H., & Webber, M. M. (1973). Dilemmas in a general theory of planning. *Policy Sciences, 4*, 155–169.

Rocard, M. (2007). *Science education NOW: A renewed pedagogy for the future of Europe*. European Commission.

Sala, A., Punie, Y., Garkov, V., & Cabrera Giraldez, M. (2020). *LifeComp: The European framework for personal, social and learning to learn key competence* (EUR 30246 EN). Publications Office of the European Union, Luxembourg. Accessed 10 May 2021.

Scharmer, O., & Kaufer, K. (2013). *Leading from the emerging future: From ego-system to eco-system economics*. Berrett-Koelher Publishers Inc..

Schwartz, S. H. (1992). Universals in the content and structure of values: Theoretical advances and empirical tests in 20 countries. In M. P. Zanna (Ed.), *Advances in experimental social psychology* (pp. 1–65). Academic Press.

Slavich, G. M., & Zimbardo, P. G. (2012). Transformational teaching: Theoretical underpinnings, basic principles, and core methods. *Educational Psychology Review, 24*(4), 569–608.

Smith, G., & Sobel, D. (2010). *Place- and community-based education in schools*. Routledge.

Steffen, W., Richardson, K., Rockström, J., Cornell, S. E., Fetzer, I., Bennett, E. M., Biggs, R., Carpenter, S. R., Vries, W., Wit, C. A., et al. (2015). Planetary boundaries: Guiding human development on a changing planet. *Science, 13*, 347. https://doi.org/10.1126/science.1259855

UNECE. (2011). *Learning for the future. Competences in education for sustainable development*. UNECE.

UNESCO. (2012). *Shaping the education of tomorrow. Report on the UN decade of education for sustainable development*. UNESCO.

UNESCO. (2014). *Roadmap for implementing the global action programme on education for sustainable development*. UNESCO.

United Nations (UN). (2015). *Transforming our world: The 2030 agenda for sustainable development* (A/RES/70/1). United Nations.

Urban Science. (2017–2020a). *Engaging science, creating sustainable cities*. https://urbanscience.eu/

Urban Science. (2017–2020b). *Framework for Urban Science*. https://urbanscience.eu/uk/about/reports-and-outputs/. Accessed 03 July 2021.

Vare, P., & Scott, W. (2007). Learning for a change: Exploring the relationship between education and sustainable development. *Journal of Education for Sustainable Development, 1*(2), 191–198.

Wals, A. E. J. (2007). *Social learning towards a sustainable world: Principles, perspectives, and praxis*. WAP. https://doi.org/10.3920/978-90-8686-594-9

Waters, C. N., et al. (2016). The Anthropocene is functionally and stratigraphically distinct from the Holocene. *Science, 351*(6269), aad2622.

Zalasiewicz, J., et al. (2017). The working group on the Anthropocene: Summary of evidence and interim recommendations. *Anthropocene, 19*, 55–60.

Daniela Conti, MS in biological sciences, is currently in charge of the research area in the Centre for Environmental Research, Documentation and Education at the Regional Park of Monza in Italy. She is the project manager for several projects, handling the design and management of educational programs for schools, teacher professional development and citizens on environmental and science issues related to sustainability. Her research focuses on the transition from science education to sustainability education, and how to use authentic context and active and participatory learning to develop science and sustainability competences.

Richard Dawson has developed a reputation for bringing fresh insights to education and learning over 25 years. His focus is on helping teachers, schools, and organizations to improve the quality of their learning, create learning for a sustainable future, and enhance their capacity to deliver learning effectively with lasting benefits. He has worked in over 30 countries with NGOs, government, business, and civil society organizations. He is the director of Wild Awake, a not-for-profit social enterprise whose purpose is to develop and provide learning which inspires change towards a more sustainable planet, and support people to live healthy and happy lives which respect natural limits.

The Environment: A Question of Justice?

6

Ben Ballin

Abstract

This chapter is framed by two tragic incidents where the environment has impacted on human beings, raising complex questions about social, economic and environmental injustice. It then takes a critical look at equity-related Sustainable Development Goals, before outlining the history of the environmental justice movement. It goes on to highlight two intertwined educational strands: people's right-to-know about the issues that affect them and the difficulties often involved in realising this. Drawing on discussions with teachers and young people, it considers how ESD has tended to approach such potentially contentious content and the resulting need for criticality and multiple perspectives. In doing so, it argues that sustainable development itself can and should be considered as a learning agenda. Taking air pollution and the climate crisis as examples that relate to the two incidents with which the chapter began, it offers substantial case studies of educational strategies for addressing the human and environmental questions involved, at scales from the macro (community and whole school) to the micro (specific teaching activities).

Keywords

Environmental justice · Air pollution · Climate crisis and criticality

6.1 Introduction: The Power Line

On 15 February 2013, a 9-year-old girl called Ella Kissi-Debrah died of respiratory failure following an asthma attack. Her family lived less than 30 metres from London's busy and congested South Circular Road. Ella has already been hospitalised 28 times following similar episodes. A report from the University of Southampton (UK) 'found that the times Ella was rushed to the hospital corresponded with times when air pollution spiked around her home' (Rosane, 2020).

In December 2020, the inner South London coroner ruled that Ella's death had been caused by illegal levels of air pollution. They added that the failure to provide her mother with information about the risks had been a contributory factor (Laville, 2020).

On 7 February 2021, a Himalayan glacier burst in the Indian state of Uttarakhand, flooding a huge area and destroying the dam at the Rishiganga Hydroelectric Project. At least 200 people were left missing or dead. BBC News reported on the causes the following day: 'experts say one possibility is that massive ice blocks broke off the gla-

B. Ballin (✉)
Geography and Sustainability Education Consultant, Birmingham, UK

© The Author(s), under exclusive license to Springer Nature Switzerland AG 2022
G. Karaarslan-Semiz (ed.), *Education for Sustainable Development in Primary and Secondary Schools*, Sustainable Development Goals Series, https://doi.org/10.1007/978-3-031-09112-4_6

cier due to a temperature rise, releasing a huge amount of water' (Khadka, 2021).

Three years earlier, researchers from the University of Geneva, studying 150 years of avalanche occurrence in another part of the Himalayas, had warned that 'climate warming observed in recent decades has been accompanied by increase in the occurrence of avalanches … posing a risk to the people of the region' (Choudhary, 2018).

A whole 14 years before that, environmental journalist Mark Lynas photographed glacier retreat as one of the 'fingerprints of global warming' (Lynas, 2004). So, while the BBC reported, the day after the devastation in Uttarakhand, that 'it is not yet clear what caused the glacial burst' (Choudhary, 2018), the general pattern of global heating is very clear.

There is something depressingly familiar about these stories of untimely death and environmental disaster. We have become familiar with them through news reports and via star-studded movies, from 'Erin Brockovich' to 'Dark Waters' (children, too, are familiar through films like 'Fern Gully' and picture books like 'Rainforest') (Cowcher, 1988). There is a trope that runs from the North Dakota pipeline to Union Carbide in Bhopal, from mercury poisoning in Minamata to Shell's misadventures in the Niger Delta: a few brave individuals or communities take on big corporations and governments to seek redress for grim environmental injustices. In many ways, the idea of environmental justice is about precisely such stories. But 'justice' is not only about addressing injustices: it is also about asserting that there is a more just and proper way in which things might be done. And not every brave individual emerges victorious. In 1980, the political theorist Ambalavaner Sivanandan asserted that 'the colour line is the power line is the poverty line' (Samarasinghe, 2018). In the 2020s, we might confidently add to this 'the environmental line'. The concept of environmental justice recognises that those who find themselves on the less powerful side of other social and economic faultlines often experience the worst environmental impacts. Ella Kissi-Debrah was a Black child in South London; those who died or

went missing in Uttarakhand were mostly power station workers. These faultlines – or power lines – can run within and between communities, generations and nations. Too often, as with Ella's mother, the power line is also about access to essential knowledge and information.

As teachers and educators, these are central concerns to us. If we lack reliable access to the things that we and young people need to know, then that creates for us 'a problem of knowing' (Ballin, 2016a), something which is both a problem of power and an educational problem. We will return to this later, but I will first set out more detail on some key contexts for environmental justice and where the idea comes from.

6.1.1 Social Justice and the Sustainable Development Goals

The 17 Sustainable Development Goals (UN, 2015) were agreed by almost all the world's nations at the UN Sustainable Development Summit in 2015. They bring both socio-economic and environmental considerations together. It is not accidental that the first ten of them are addressed explicitly to areas of persistent global inequality, kicking off with headliners SDG1 'No Poverty' and SDG2 'Zero Hunger' and then moving on to:

- SDG3 'Good Health and Well-being (… *for all at all ages)*'
- SDG4 '(inclusive and equitable) Quality Education (… *for all)*'
- SDG5 'Gender Equality' (*for 'all women and girls'*)
- SDG6 'Clean Water and Sanitation (… *for all)*'
- SDG7 'Affordable and Clean Energy (… *for all)*'
- SDG8 'Decent Work and Economic Growth (*Promote sustained, inclusive and sustainable economic growth, full and productive employment and decent work for all)*'
- SDG9 'Industry, Innovation and Infrastructure' (including '*inclusive industrialisation*')

And, just in case you hadn't got the idea yet …

- SDG10 'Reduced Inequalities (Reduce inequality within and among countries)'

One could be forgiven for thinking that those repeated 'for all's and such a weighting of the goals towards equality and inclusion are more-or-less a manifesto for social, economic, political and environmental justice. When I drill down into the 169 targets associated with the goals, however, it is not quite so simple.

Let us take SDG1 as a telling example. This pegs the level of 'absolute poverty' at US $1.25 per day (2015 rates) and measures its apparent successes accordingly. Many critics have pointed out that this level is well below what many people would need simply to survive and that it wholly ignores the problem of over-consumption by the rich (e.g. Hickel, 2015). Indeed, a copiously evidenced United Nations report, produced 6 years before the Sustainable Development Goals, suggested that there was a serious problem with these calculations: the 'poverty line' miscalculated inflation, ignored a huge range of contextual factors and overlooked significant areas of necessary expenditure: 'There is evidence to suggest that the [UN and World Bank] poverty lines underestimate the actual extent of poverty' (DESA, 2010).

Why does this matter? In a nutshell, a higher figure (say, US $2) 'would have meant many millions more people being counted as living in "extreme poverty"' (Tide~ global learning, 2016). Instead, the needs of those 'many millions' risked invisibility. If they were not counted, perhaps they did not count? There are plenty of other such examples for this and other Sustainable Development Goals. To critique them, however, is not to render them worthless or invalid: only to invite a scrutiny of what lies beneath surface appearances. This is an ethical point which points towards critical pedagogies, as we shall explore later.

6.1.2   What Is 'Environmental Justice' and Where Does It Come from?

I first came across the concept of environmental justice through the work of Agyeman (2000), who stated at the time that 'environmental and sustainability policy discourses and claims are beginning to be re-framed. Instead of being firmly allied to a "green" agenda, these discourses are being refocused around notions of justice and equity'. Agyeman (2000) pointed out that 'Environmental justice concerns have been around since the Conquest of Columbus in 1492'. He located the origins of the modern-day movement in the 1980s and the struggles of working-class communities in the USA, especially people of colour, against dramatically disproportionate levels of toxic waste dumping and other health-threatening practices in their communities. By the early 1990s, this had become:

> a fully-fledged environmental justice movement. It occurs from Alaska to Alabama and from California to Connecticut, driven by the grassroots activism of African-American, Latino, Asian, Pacific Islander, Native American and poor white communities who are organising themselves around LULUs (Locally Unwanted Land Uses) such as waste facility siting, and other issues such as lead contamination, pesticides, water and air pollution, workplace safety, and transportation. (Agyeman, 2000, p. 9)

In October 1991, the People of Color Environmental Leadership Summit established a set of criteria for environmental and social justice. Some successes followed, including the establishment of a federal Office of Environmental Justice (within the Environmental Protection Agency), a National Environmental Justice Advisory Council and the US Emergency Planning and Community Right-to-Know Act.

People in other countries also began to explore these ideas, and Boardman et al. (1999) as cited in Agyeman (2000) offered ten 'Proposed Principles of Environmental Justice' in the

UK. These ten 'proposed principles' addressed both local and global inequities, and, while all remain pertinent, it is particularly worth considering the first two:

1. Environmental problems are a component of social exclusion and an issue of social justice. Most environmental pollution is unevenly distributed: even in rich countries like Britain, it is normally the poor and disadvantaged who suffer most as a result. Even where the effects are more even, impacts are uneven as the rich can more easily respond.
2. Communities and individuals should have a right to know and the ability to respond to distributed environmental hazards. This means that government should … incorporate environmental objectives in area-based regeneration initiatives, work to strengthen participation and fully involve communities in locally-based strategies. (p. 14)

In 2012, the United Nations established an advisory council with environmental justice as a part of its remit (UNEP, 2012). Its remit is founded on clear evidence that 'increasing environmental pressures from climate change, biodiversity loss, water scarcity, air and water pollution, soil degradation, among others, contribute to poverty and to growing social inequalities'. However, without the 'right-to-know' asserted as a principle by Boardman et al. (1999), those experiencing such social and economic consequences have limited opportunities to address the unjust situations that significantly affect them.

6.2 Environmental Justice: Some Educational Implications

This right-to-know is enshrined in US law and has clear educational implications. In effect, the principles of environmental justice assert an entitlement for all (young and old) to know about the factors that influence their lives (and sometimes their literal survival). We have however also noted some of the difficulties involved in knowing about such things, be they access to balanced and meaningful information or the need to engage critically with grand-sounding statements such as those related to the Sustainable Development Goals.

Vare and Scott's (2008) distinction between 'ESD 1 – learning *for* sustainable development' and 'ESD 2 – learning *as* sustainable development' is perhaps helpful to us here. ESD 1 is about information provision and promoting 'positive behaviours, which can be helpful where the need is clearly identified and agreed'. ESD 2 recognises that not all information is straightforward or uncontested and therefore highlights criticality and creativity, aiming 'to build capacity to think critically about (and beyond) what experts tell us and test sustainable development ideas, exploring the contradictions inherent in trying to do the "right" thing'. Vare and Scott (2008) emphasise that these two forms of Education for Sustainable Development (with their different epistemological starting points) are mutually supportive, rather than exclusive or competing. I shall later offer practical examples of both.

As with Vare and Scott (2008), Boardman et al.'s (1999) ten 'proposed principles' also caution sensitivity when employing 'behaviour change' strategies, some of which are a familiar part of many (ESD 1) environmental education programmes. For example, asking people who are struggling on low incomes to consume less might be wholly inappropriate, as might telling them how well-off people are in the 'rich world' when compared to other places (a wise friend once memorably described this as trying to 'obliterate one injustice with another injustice') (C. Cooper, personal communication, 2000). As Principle 6 reminds us, 'mechanisms designed to change individual or corporate behaviour … can be socially regressive if applied in isolation' (Boardman et al. (1999) as cited in Agyeman (2000, p. 14).

Vare and Scott's emphasis on 'testing ideas' is particularly important. I would argue that any effective process of sustainable development is *in itself* a social learning process, where strategies and ideas are developed in response to the challenges of sustainability, tested and reconsidered as necessary. New and improved strategies and ideas then arise. Without this cyclical and reflexive element, action on sustainable development at any scale becomes a form of unthinking activity. So what are the implications of such considerations for ESD? What might they mean for us as teachers and educators?

6.2.1 Environmental Justice and ESD

The Sustainable Development Goals brought together both environmental and socio-economic considerations and (in their headline statements, at least) significantly emphasised equity and inclusion. This has something in common with environmental justice. However, the starting point for environmental justice is not universality and holding the environment in equilibrium with other factors, but a framing of the issues around the specific injustices of a situation.

A similar state of affairs could be described in relation to Education for Sustainable Development, which in one sense is an attempted synthesis of Environmental and Development Education: the former traditionally emphasising field studies or natural history and the latter, human development and especially social justice, equality and power. Education for Sustainable Development holds the two paradigms in dialectical tension. Such syntheses have been proposed many times (e.g. Greig et al., 1987, 1989; Belk et al., 1992; Huckle & Sterling, 1996; Fien & White, 2002–2009; Webster & Johnson, 2008). The abundance of related 'adjectival' descriptions (e.g. world studies, education for sustainability, global learning, global citizenship, global education, etc.) has further led to frequent confusion over terminology and paradigms (Ballin, 2016b). Whatever the wishes of the educational theorists, however, the more liberal values of balance and universality have in practice tended to have the upper hand over the more socially radical framing of environmental justice. Readily comprehensible programmes such as Eco Schools and its many international counterparts (Eco Schools Global, 2021) have often proved more adoptable and adaptable than more challenging and contentious alternatives. In short, there has in general been more of an emphasis on 'ESD 1' than 'ESD 2'.

It was against this background that three UK organisations with an interest in both equality and the environment decided to hold a consultative meeting with teachers on Environmental Justice and Education: Black Environment Network, the Environmental Law Foundation and the teachers' network Tide~. The event took place in 2001 in Birmingham and was attended by 12 teachers and educators (including myself) plus a prominent environmental lawyer. As part of this, teachers presented some questions and issues raised by primary-age children (ages 5–11). This event highlights some crucial issues about environmental justice and ESD. Table 6.1 (below) highlights some of these and matches issues raised by teachers with those from children. It could form a useful starting point for a staff planning session in school. What would you prioritise or change as a teacher? What issues are students raising in your school? How do the teachers' and the students' agendas match up or differ? Why?

The list from teachers clearly highlights some of the pedagogical dilemmas and potential pitfalls involved in addressing potentially contentious subject matter. The teachers raise several quite proper concerns about possible bias, the need for a supportive whole school approach and the relationship between environmental justice and both the formal curriculum and the 'hidden curriculum' of a school's values, practices and embodied relationships. One thing that strikes (and heartens) me about the children's issues in the table is how practical, concrete and to-the-point they generally are. Rather than worry about policies and institutions, they emphasise that they would like to be listened to as there is a job of work that needs to be done and there are things that they will need to know in order to deal with it. This emphasis on positive practical solutions as an antidote to environmental anxiety corresponds with the findings of educationalist Alexander (2010) and neurobiologist McAndrews (2018). All this is not to pretend that these issues are simple for teachers to deal with. Apart from their questions about why the needs of some people appear to be prioritised over those of others, the children's list is however more aligned with 'ESD 1' (where knowledge is clear and uncontested) than 'ESD 2' (where it is less so) (Vare & Scott, 2008).

Table 6.1 Teachers' and primary-age children's issues about environmental justice

Teachers	Children
Children's rights. There are serious environmental justice issues which directly affect many children, for example, children affected by asthma due to airborne pollutants. (Do we help children see that crime is not only about people stealing their belongings but also about people stealing their health?) Safety. How do we create safe environments? What roles are needed for children to feel safe?	Who will listen to us? As children, do we have rights? Can we really get things done? Our playground sometimes smells really chemically and makes us feel sick. Should this be permitted? Is the smell a real health hazard for us?
Local issues as an entry point for young people's understanding of the world – this includes issues about the school itself.	Do traffic-calming measures create more pollution? Why is the nearby factory covered in black soot? Who is responsible for cleaning it? Is the electricity pylon so close to our school a problem for us? Who should we contact to find out about this siting? Graffiti – how can we stop it? Where can we get security cameras from? What do they cost? Litter – where do we get more bins from? Who will pay? Empty buildings – Why build more office blocks and houses when so many are already empty? Who gives permission for these buildings to be built and open land used up? This leaves less space for people to use. What about the homeless? Could they not be put into these buildings?
Power, empowerment and citizenship. The relative powerlessness of children within adult environments Real citizenship leads to children challenging authority – what is the role of a school in relation to this? Where does it intervene, hold back, facilitate, prevent, etc.? Are children really listened to when environmental justice/local issues come up? Are their views acted upon? What can they do about it if they are not? Student voice: having more say in the curriculum, ownership of learning and the justice agendas they pursue. ('We'll strike if we don't!') Appropriate pedagogies: 'justice' is something that a child has to discover for themselves: it is not something that can be taught through transmission. The risk of tokenism. Certain issues keep on coming back without being resolved. Should we be creating 'false successes'? If so, do we sometimes disempower children when we are trying to empower them?	Why were only adults present at this consultation? How can we be sure we are being listened to? Why do some areas get problems like graffiti sorted out quickly and others not? Do the police take crime seriously in poorer areas?
The need for a degree of political literacy A desire to avoid propagandizing. How do we deal with contentious questions? (See Davies et al., 2005)	We know what could be done, but who can help? Access to services and information – where do you go to deal with graffiti? litter? pollution?
An emphasis on a supportive whole school ethos for such work. Vision. Environmental justice embracing and connecting a subject-led curriculum, as a key purpose of the learning, rather than merely supporting the subjects.	

Source: Consultative meeting on Environmental Justice and Education (2001)

6.2.2   Environmental Justice and the Problem of Knowing

Because a right-to-know is such a central tenet for environmental justice, we will now explore a little bit more about the problem of knowing in this context. My example analyses the knowledge implications following a terrible fire that killed large numbers of people in Sinai, a Nairobi township, in 2011 (Ballin, 2016a). It identifies four problematic areas related to knowing:

- *Access to knowledge.* In this case, under-reporting of the fire: very little print or broadcast news in the UK mentioned the event, and little time or print space was given over to it. It was therefore hard for many people in the UK to even know that it had happened, let alone find out any detail about it. (In comparison, the deaths of two British tourists in Kenya on the same day were widely reported.)
- *Lack of data.* Sinai was a marginal community and many residents were undocumented. When it comes to the real death toll (let alone information on longer-term impacts), 'the real figure may well be *unknowable*'.
- *Conflicted knowledge.* There were contradictory accounts of the event at the time and especially about its causes. Who and what to believe? How to weigh up these different perspectives? This is particularly important in an era of 'fake news': What linguist Noam Chomsky has called 'the conflict of epistemologies' (Polychroniou, 2020) and political commentator Peter Oborne, 'a nightmare epistemological universe' (2021).
- *Framing.* Such events are often viewed from countries like the UK through the prism of an objectivist 'single story' about African countries (Adichie, 2009; Andreotti, 2013; Wainaina, 2005). Because they fit into a familiar narrative that is widely believed to be true, they can reinforce a partial or biased view and distort understanding.

The problem of knowing requires citizenly and pedagogical care. The example from Nairobi highlights a need for learners (and teachers):

- To be conscious of dominant frames when tackling environmental justice stories.
- To realise that significant investigation may be needed in order to reveal hidden or partially obscured knowledge – it may well not be 'handed to us on a plate'.
- To engage critically with what is presented (e.g. inviting even very young learners to look hard at images or texts and imagine what lies 'beyond the frame').
- To engage with multiple perspectives (and indeed to recognise that often there may be no single 'correct' authoritative view of a particular issue; sometimes we will need to hold such perspectives in tension with each other or – to borrow a metaphor from music – hear them 'contrapuntally') (Said, 1993).

The example of the Sinai fire emphasises that 'we need both "felt understandings" and cool analysis' when engaging with such stories and issues and the importance of narrative and imaginative modes of knowing in order to humanise our understanding of them (Bruner, 1996; Ballin, 2016a). In the following pages, we offer some suggestions for what all this might mean in practice. We opened this chapter with the stories of Ella Kissi-Debrah and the Uttarakhand flood. These two stories were chosen because they frame air pollution and the climate crisis as environmental justice issues. It is to these two issues that we will now turn.

6.3   Teaching About Environmental Justice

6.3.1   Issue 1: Air Pollution

The next few pages describe attempts at systematic and systemic change involving whole school and classroom approaches to air pollution and environmental justice in the UK. 'The numbers are chilling. Globally, air pollution cuts short 7 million lives every year: about 40,000 in Britain, some 100,000 in the United States, and upward of a million each for China and India' (Gardiner, 2019). As the Sustainable Development

Goal headlines earlier reminded us, it is investigating what lies behind these figures that most sharpens our focus on the power lines of class, race and environment. Here are some figures from the early 2000s that highlight environmental justice aspects of UK air pollution.

- 'The poorest families are twice as likely to be in a neighbourhood with a polluting factory as the most wealthy families' (Friends of the Earth, 2000).
- 'Wards in the most deprived decile provide the location for five times as many sites and authorisations and seven times as many emission sources as wards in the least deprived decile. Out of the 3.6 million estimated people living within 1km of an IPC [Integrated Pollution Control] site, there are 6 times more people from the most deprived decile compared to the least deprived' (Walker et al., 2003, p. 2).
- 'IPC sites are also disproportionately clustered together in deprived wards. As site and emission clusters become more concentrated, the bias towards the more deprived deciles becomes more acute' (Walker et al., 2003, p. 2).
- 'Analysis of emission levels from IPC sites for particulates and carcinogenic emissions to air, show a disproportionate concentration of emissions in more deprived areas' (Walker et al., 2003, p. 3).

Air pollution also highlights international faultlines. One has only to compare Gardiner's mortality figures for India and China with those for the UK and the USA. 'It is said that being in Beijing on one of their bad air days is equivalent to smoking 40 cigarettes' (Scott & Vare, 2018). A look at real-time air pollution mapping at the world air pollution website (https://waqi.info) quickly highlights where the 'hotspots' tend to be found: above all, in industrialising, industrialised and densely populated parts of the Global South. As an illuminating exercise during a Geography lesson, students could use these online maps and data sets (or the 'city dial' at https://breathe-life2030.org) to compare the air quality of places

around the world with that in their own area, reporting back on what is different and why this might be.

The following draws significantly on interviews with two organisations, who between them have been creating system-wide approaches to addressing air pollution injustice in schools. These range from the macro (e.g. national policy), through local authorities and communities, to the whole school and the individual classroom. We will use this macro-to-micro framework for describing their work, although there are significant differences between the organisations and their approaches.

Danielle Kennell is Air Quality Education Officer for the East Midlands city of Leicester. Her Twitter name of 'Clean Air Girl' says something about the hands-on and child-friendly approach that she has adopted. It helps frame the issue as something positive and solution-based: about what a just or good life might look like, rather than an unjust one.

Leicester is one of two UK cities where Black and Minority Ethnic residents are in a majority. The city prides itself on its 'green' credentials and the local authority continues to maintain a dedicated Environmental Education Team (2021) with whom the Air Quality Education Officer works closely. The Air Quality Education Programme offers 'a *free* tailored programme of activity to suit your school's needs and links into the national curriculum'. This includes an impressive array of free downloadable resources created for students aged 4–16 (Leicester Air Quality Education, 2021a, b).

Skips Educational is a Birmingham-based Social Enterprise founded by 'accidental author' Ash Sharma. Skips provides resources with supporting CPD programmes to several local authorities around England, often tapping into contractors' corporate social responsibility budgets to help fund behaviour-change-orientated activities. One of its most distinctive offerings is workbooks for children to read with their parents, thereby influencing a great many families. Its website claims that 2,750,000 family members were reached this way by February 2021 (Skips Educational, 2021). This includes those deemed 'hard to reach' by local authorities: itself a somewhat problematic term (Manchester Research Hive, 2018). The Skips programme, 'Clean Air Cops' (Sharma, 2018), directly addresses air pollution. In 2020, Skips helped establish The Westminster Commission for Road Air Quality, including a working group on Education (Westminster Commission on Road Air Quality, 2021). For this chapter, I interviewed Ash Sharma and his colleague Lorraine Cookson.

6.3.1.1 Addressing the Big Picture: The Westminster Commission for Road Air Quality

The Westminster Commission is 'trying to take a joined-up approach to the issues', linking together education with research, policy, smart monitoring, health outcomes, etc. These different sectors are 'bound by a deep-rooted belief that it is every citizen's inalienable right to breathe clean air', according to Commission Chair Barry Sheerman MP (2021). In his interview, Ash Sharma said that working as part of the Commission has allowed them to open the agenda up and 'push government to do more'. This includes resisting attempts to over-emphasise individual behaviour change while evading governmental responsibility. As Lorraine Cookson explained:

> they can't keep pointing the finger at people and saying it's the individual that has to make the difference: there are things that people can do, but there also have to be some huge things going on, like replacing fleet vehicles with electric cars or hybrids.

Ash Sharma added in his interview that business also needs holding to account, whether this means regulating the location of social housing or addressing malpractice by car manufacturers and vehicle testing centres (Client Earth, 2021). In this, the Commission is also explicitly about people's right-to-know: 'They are saying that is vitally important for the nation to know what they are up against with air pollution'. However, Skips recognises that there can be blocks, whether poor communications strategies from local authorities and decision-makers, academics wishing to protect their work or the consequences of commer-cial interests. With the other parties in the Commission, they are piloting creative ways of overcoming these hurdles.

6.3.1.2 Leicester City: A Local Authority with Clean Air Equalities on the Agenda?

Leicester's Air Quality Education Programme is embedded within a city-wide Action Plan that prioritises action on transport emissions while also contributing to the city's Climate Emergency Strategy. The Action Plan states that air pollution levels from nitrous oxide in the city are decreasing but still a cause for concern: 'monitoring shows that levels of pollutants have decreased from 80 $\mu g/m^3$ to 60 $\mu g/m^3$ in the past few years but still not enough to meet EU thresholds' (i.e. 40 $\mu g/m^3$) (Leicester City Council, 2015). The plan also highlights diesel particulates as an issue. An earlier Action Plan especially targeted an inner-city 'Air Quality Management Area' which 'correspond[s] to areas of elevated social deprivation' (Davies & Scott, 2011). The 2015 plan focuses explicitly on health inequalities:

> People who live in more deprived areas are more affected than people living in less deprived areas even if they are exposed to the same levels of pollution. Those who are already in poor health are more affected by pollution than those who are healthy. Air pollution is thus an equality issue and tackling it will help to address Leicester's health inequalities. (Leicester City Council, 2015)

The 2011 plan had described educational interventions as '"soft" initiatives' whose impacts are 'not directly quantifiable'. Regardless of such qualms, the Air Quality Education Officer post was brought into being in 2018, and both Action Plans have highlighted the contribution of schools. The following quote from the 2011 Action Plan is worth highlighting: it clearly chimes with ones from the climate justice example at the end of this chapter and speaks resonantly to a sense of environmental injustice:

> Disadvantaged people tend to contribute least to atmospheric emissions and also tend to be the group least able to take action to address them. (Davies & Scott, 2011)

The Air Quality Education Programme is situated within the Transport section of the council, which tends to orientate its work towards traffic pollution, although the programme does also look at emissions from wood-and-coal burning (a key source of $PM_{2.5}$ particulate emissions) and industrial pollutants. In interview, the Education Officer helped explain this focus: 'school congestion is one of the big contributors to poor air quality outside schools'. Schools are therefore prioritised in relation to local congestion levels: in practice, this tends to overlap with areas of high housing density and deprivation. It is interesting to set against this the perception from Skips that the development of 'Clean Air Cops' was sometimes steered by nervous local authority officers away from issues such as indoor pollution, wood burners (Bland, 2021), industrial and waste-related pollution. As the interviewees explained:

> The Transport people said, 'we don't want it, we just want cars, cut it out' … we had a bit of a battle to say it should be in (Lorraine).
> The Health people said, 'don't scare the public, don't tell them it's a poison playground' (Ash).

(Compare Table 6.1, where children asked explicitly if their playground was 'poison' – the 'scare' was already present in their minds.) Ash related these concerns to the short-term anxieties of councils and electoral cycles:

> So we couldn't mention the airport because it was a polluter, we had to make it that it was the passengers in their cars on the way to the airport that was the problem.

All this not only has an impact on people's right-to-know but also adds to their problem of knowing.

6.3.1.3 Air Pollution: Community Education and Whole School Approaches

In the Skips interview, the right-to-know element of air pollution was central:

> The messages local authorities were giving out didn't make sense to the average person on the street. They just thought it was another campaign, not something that was happening to them already (Ash).

Skips cited the common experience of slow carbon monoxide poisoning as experienced by motorway drivers:

> what is alarming is the lack of understanding about the health issues that air pollution causes (Lorraine).

With that in mind, while the Skips Clean Air Cops programme works with children, its ultimate intention is about communicating to adults through providing materials that parents and children will read together:

> It's too late if we are waiting for primary children to grow up and take action, but nobody wants their child to have bad health and it's the parents who are going to be driving the car or deciding to take the bus (Lorraine).

This model is intended to go beyond the widely adopted behaviour change strategy of 'pester power'. Instead, parents and children become co-learners, coming to understand the issues and possible responses together:

> We've got to get away from the idea that because we've taught the child, the pester power will do … and stop thinking that education is only something that happens in school (Lorraine).

While also adopting community-scale approaches, Leicester's Air Quality Education Programme especially emphasises change at the whole school level. It always starts with a whole school or phase assembly, before moving on to action-focused projects within the curriculum or through extra-curricular EcoTeams. These teams, involving both students and staff members, are widespread in the city, linked to its active promotion of the Eco Schools Programme. The programme makes strong links to Eco Schools themes of Transport and Healthy Living, as well as explicit links to SDGs 3 (Good Health and Wellbeing), 4 (Quality Education) and 13 (Climate Action). There is support from the programme for schools to develop their own Clean Air Plan. The Education Officer also presents to schools governors, who in England have an oversight responsibility for the whole school.

6.3.1.4 Air Pollution: Curriculum Responses

In the interview, Skips recognised that teachers often want support on issues like air pollution but can lack time. Offering something that teachers can 'run with' was therefore crucial and allowed their programme to get beyond the narrow band of existing enthusiasts, the 'eco teachers', and also engage their colleagues. Recognising a similar need, the Leicester programme includes activities, workshops, assemblies and resources that can be delivered by a visiting presenter or a teacher (e.g. as an 'off-the-shelf' lesson plan or assembly presentation). Many of these activities, slides and plans can be freely downloaded from their website (Leicester Air Quality Education, 2021b).

In interview, the Air Quality Education Officer related how, when leading sessions in person, she was keen to lead by example: dressing in a tracksuit to highlight active living and travelling to school by bicycle, walking or public transport. Students often asked her about these things and expected her behaviour to match up to what was being said. The thematic structure for the Leicester programme (Leicester Air Quality Education, 2021a, b) runs across all age groups, from 4–5 to 16+, as follows:

1. *Basic knowledge* – what air pollution is (including seasonal as well as human impacts on air quality; understanding air pollution sources).
2. *Understanding impact* – how it affects our health and environment.
3. *Action* – based on that knowledge and understanding, what you can do about it.

As part of *(1) 'Basic knowledge'*, there is a focus on sources of air pollution, including activities which sort out natural from human causes and then consider what can be done about them. As the Education Officer says, 'there are so many pollutants! I say to the kids, What about volcanoes? Paint? Dust? Sea Salt? Sandstorms?'. Older children look at how air pollution can be measured, so as to better understand, for example, the different significance of a temporary local 'spike' in pollutants (such as occurs in England after 'fireworks night' on 5 November) and longer-term trends.

The Education Officer found *(2) 'Understanding impact'* the hardest element to resource: 'you really struggle to find anything that's easy to digest, although there are some quite hefty scientific papers on it'. From a 'right-to-know' perspective, it is highly significant that the programme is making such information accessible and taking it beyond the realm of technical specialists and policymakers: addressing precisely the sort of information deficit that was highlighted in the death of Ella Kissi-Debrah.

(3) 'Action' includes popular 'Park and Stride' and 'Switch off your engine (anti-idling)' campaigns and a schools award scheme. As the Education Officer explains, 'it's teaching people about the important things that we can do something about'.

An annual Clean Air Day features street closures around a selected school (Fig. 6.1). It is perhaps noteworthy that the street closure outside an inner-city school (on a through-route) had more impact that that outside a suburban school (situated on a cul-de-sac, where parking was largely displaced to neighbouring streets).

One programme innovation is the creation of Walking Maps that use isochrones. These show how far someone can walk in a particular time period (Fig. 6.2). In the interview, the Education Officer outlined how it works:

> you have the school in the centre and you can map your house if you live nearby, map a route to school, challenge yourself and your friends: a 10-minute walk, then up to a 15-minute walk and so on.

The Education Officer explained that actions in the programme are organised like a gradient scale: starting with one thing that you could do right now, like switching something off and reducing pollution from a power station, and then something you could do next week, like Park and Stride or finding a different way of getting to school, and then in a term, like a Walk to School Day. In doing this, students are also moving from small-scale individual actions to collective whole school and community responses. There is also

Fig. 6.1 Clean Air Day street closure at a Leicester Primary School. (Photo taken from Leicester City Council with their permission)

an implicit progression in learning complexity within the programme: what and how students learn as they get older. This could to some extent be seen as a movement from more ESD 1-type activity to integrating some elements of ESD 2 which invite a greater degree of critical reflection and creativity and which touch on questions about the deeper underlying causes at play (and thus more explicitly on questions about environmental justice). Programme interventions with Infant children (ages 4–7) have moved away from formal presentations to using props. Language development and basic understandings are key. Having introduced a puppet called 'Clean Air Clive' (Fig. 6.3), the Education Officer reflects that:

> what was fascinating was how articulate [young] children could be about poor air quality and air pollution after they had engaged with a puppet, the amount they had retained and could talk about, retell it in their own words.

Building up to this, children complete unfinished sentences, such as 'Clean Air Clive says

…', and then explore Clive's 'Top Tips for Clean Air'. They go on to make their own posters and to stage a social media photo shoot. As students move into Junior and Secondary classes (ages 7 plus), the programme begins to focus more on the formal curriculum: Science, Geography and English (e.g. letter-writing to parents or the mayor). A module on the History of air pollution 'gets a really good take-up … let's talk about the Romans and how they were polluting the air with lead' (Education Officer). Activities aimed at older students include more sophisticated investigations (Where does pollution come from? How do we measure air quality? How do we improve air quality? Studying and measuring lichens for evidence of air quality). An 'air pollution catcher' Geography/Science activity is very popular with students aged 11–14. For this, students draw something that represents clean air (e.g. a plant, a cloud) on a sheet of waterproof paper. The picture is then coloured in with petroleum jelly. Students predict where most pollution will be coming from,

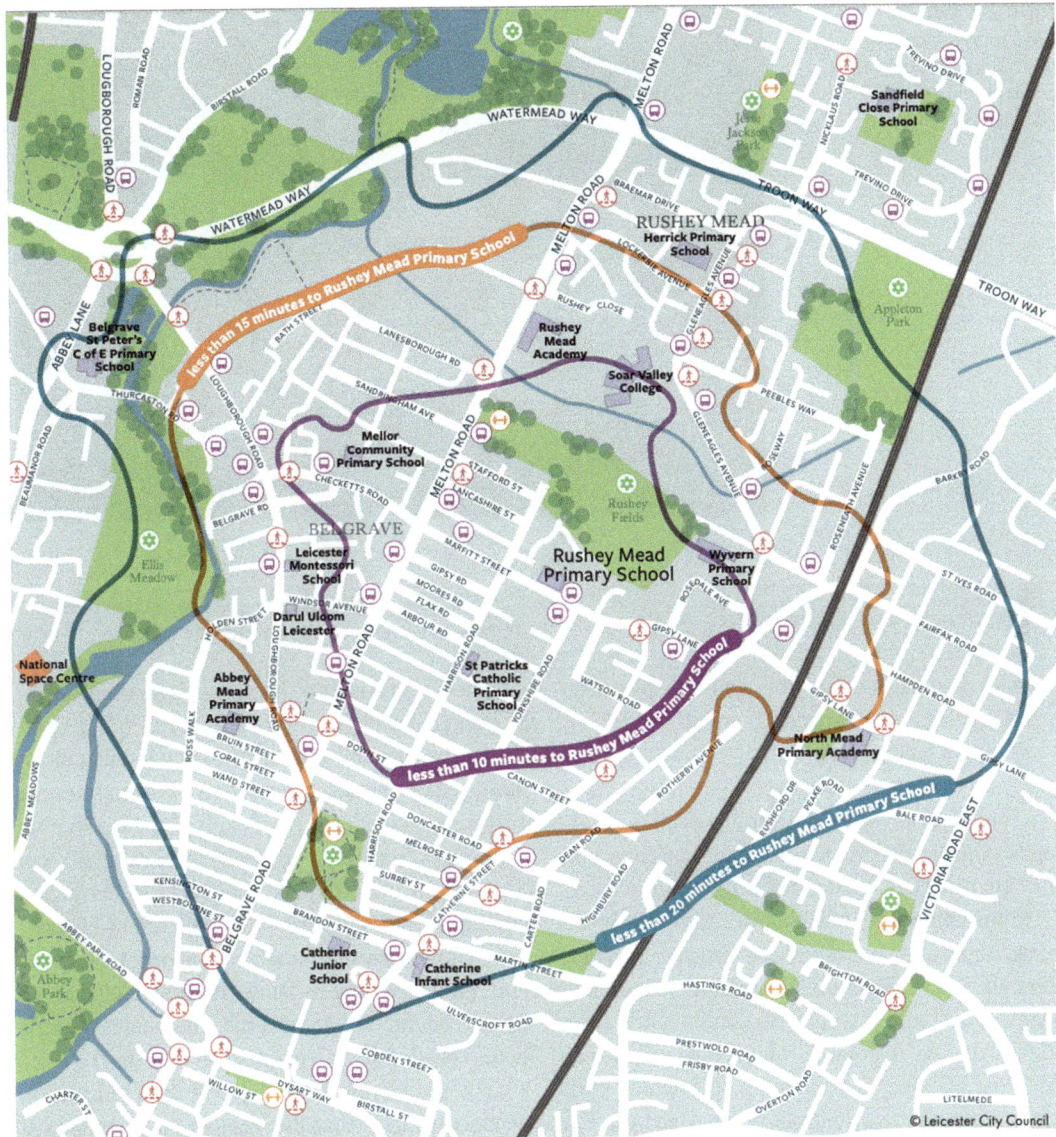

Fig. 6.2 Walking map for a Leicester Primary School

and catchers are hung up at corresponding points around the site. Students revisit their catchers a few weeks later and inspect them for evidence of pollution.

There are clear ESD 2 (critical thinking) elements to such problem-solving activities for students aged 11–16 as evaluating which electric car would be best to buy. ESD 2 is even more clearly present in the activities, 'Behind the headlines: can we work out which are reliable?'. These activities aim to help students sort fact from fic-

tion: thereby directly addressing the 'problem of knowing' (as highlighted in the discussion above about accounts of the Nairobi fire). As part of learning how to work scientifically, students engage critically with sensational news headlines about air pollution. They use two acronyms to critically evaluate the research examples and how they are being presented: SMURF (Sample, Measures, Unbiased Research, Funding) and CASES. The latter asks if the way the evidence is reported is:

Fig. 6.3 Clean Air Clive joins a 'walking bus'. (Photo taken from Leicester City Council with their permission)

- *Contradictory or consistent?*
- *Accurate?*
- *Sourced clearly, accessibly and transparently?*
- *Exaggerated (overstated or embellished)?*
- *Selective (one-sided)?*

This offers a really useful framework for looking at other environmental justice issues and how they are reported, whatever the curriculum area.

In his interview, Ash Sharma said that it has been important for Skips to engage teachers through other concerns: sustainability, the environment and safe travel: 'you couldn't just run a Clean Air Day – nobody would turn up'. Clean Air thus became one element of a wider CPD programme that particularly emphasised the key knowledge that the teachers would then go on to teach children. Action-planning also took place as part of that wider agenda. The offer of sets of complimentary Clean Air Cops books helped attract many schools. Like the Leicester programme, Clean Air Cops has been more about analogue than digital learning. Skips asserts that this is especially true when it comes to activities that parents and children will do together:

With young children, pen and paper still works best … books give you a structure (Ash).

The Clean Air Cops book itself (Sharma, 2018) is engaging and clearly rooted in an ESD 1 paradigm: sharing well-established knowledge with an audience of parents and children, so as to promote particular behaviours (especially for the parents). The Skips approach originally grew out of Ash Sharma's need to explain the world to his young daughter, and this grounded and child-friendly rationale remains the book's hallmark. Children's and parents' right-to-know is absolutely at its heart. The 'air pollution harms your health' page is here reproduced with permission (Fig. 6.4).

These examples of air pollution education especially emphasise the right-to-know agenda at the heart of environmental justice, although some of the strategies aimed at older students also begin to address the more complex 'problem of knowing'. The following examples of educating around the closely connected issue of the climate crisis build further on such strategies, moving in the process from a greater emphasis on ESD 1 and information provision to ESD 2, critical engagement and the testing of ideas.

Fig. 6.4 'Air pollution harms your health': a page from Clean Air Cops. (Photo taken from Skips Educational with their permission)

6.3.2   Issue 2: Teaching About Environmental Justice – The Climate Crisis

The climate crisis throws global fault-and-power lines into particularly sharp relief: especially per-sistent but human-forged lines of international injustice, for example, between sub-Saharan Africa and the Global North:

> Africa is more vulnerable than any other continent to changing weather patterns despite playing barely any part in climate change. (BBC World Service, 2021)

In Africa, temperatures in regions are rising 1.5 times faster than the global average. This in turns is leading to severe droughts, intense heatwave, floods, depletion of natural resources, cyclones and the resultant effect is the displacement of people from their livelihoods. (Adenike, 2021)

The climate crisis has also been powerfully framed as an issue of generational justice.

Now we probably don't even have a future any more. Because that future was sold so that a small amount of people could make unimaginable amounts of money. (Thunberg, 2019)

No-one suffers more from a change in climate than a child. (Fore, 2021)

These claims have an emotional resonance that speaks strongly to young people and their sense of justice, their need-and-right-to-know. It is not accidental that educational change has been one of the demands of youth climate strikers throughout the world (e.g. UK Student Climate Network, 2021).

What might a meaningful educational response to the demand for climate justice look like? How might it go beyond an ESD 1 model of information-giving and individual behaviour change into deeper learning that really embraces criticality, creativity and learner agency? In an attempt to answer these questions that complements what has been said about air pollution, I will adopt a framework of four lenses or 'ópticas' from the network of Catholic Schools in Madrid (Oliveros Palomo, 2016; Ballin, 2016b).

- *The magnifying glass* opens up the issue.
- *3D glasses* invite us to look at the issue from different perspectives.
- *The microscope* subjects it to critical thinking.
- *The telescope* helps us visualise solutions.

With the magnifying glass, learners begin to ask questions about what climate change is and how it works: the greenhouse effect and the basic science and geography of the issue. There are plenty of existing resources that can help with this, which is incidentally the point beyond which many education systems go no further (for some well-referenced and accessible sources, see Gore, 2006; Jackson, 2020; Nelles & Serrer, 2021). Such learning need not be confined to secondary sources: for example, learners might undertake

their own investigations, taking weather measurements or observing, recording and comparing seasonal change.

Using the microscope, learners subject the climate crisis to critical scrutiny and explore why it matters. For example, they evaluate different timelines to zero carbon emissions and consider why some people favour terms like 'net zero'; they look at the differential impacts of climate-related events on the UK and Africa, where questions of justice come sharply into focus; they use a 'mystery' to explore complex chains of cause and effect across the globe: for example, between Leicester in the UK and the floods in Uttarakhand (Table 6.2).

'Mysteries' originate from David Leat's work on thinking skills (Leat, 2001). The example in Table 6.2 is best used as a framework for constructing your own mystery as a teacher, building in real information from local places. The aim of a mystery is for students to work in small groups, sorting varied evidence into ways that make meaningful connections. The mystery offered here as a learning activity is suitable for students aged 9–14. It can be simplified for younger students by removing a few less pivotal statements, although it should not become a closed activity where students create a single 'correct sequence'. Longer statements, including real extracts from source materials, are useful with older students.

Issue or Case Name: The climate crisis: a mystery.

Subjects: Geography, Science, Social Studies.

Materials: See Table 6.2.

Learning Objectives: To critically explore potential connections between students' own lives and a disastrous environmental event elsewhere in the world.

Background of the Topic: Events leading up to the floods which destroyed the Rishiganga Hydroelectric Project in Uttarakhand, 2021.

Description of the Activity: In this particular mystery, students are invited to do three things, using statements from Table 6.2 which have previously been cut up into individual cards:

1. Find out how Anam gets to school each day and why Adrika's family experienced a

Table 6.2 Mystery. Anam's journey to school and Adrika's power cut

There was a power cut while Adrika was watching a Nature programme on TV
The exhaust fumes from diesel cars have been linked to poor air quality and health problems
Anam's father drives an old Volkswagen car that runs on diesel, but he hopes to replace it as soon as he can afford to
Carbon dioxide is the main greenhouse gas, linked to global warming
A person came to school with a puppet to explain why walking to school is a good idea
Diesel is a fossil fuel
Adrika's family love to get together on Sundays to watch wildlife documentaries
After dropping Anam off at school, her father carries on to work
Sometimes, if there is a lot of traffic, Anam's father stops the car outside and leaves the engine running while she gets out
Adrika and her family live in the Chamoli District of Uttarakhand, India
Anam's little brother has bad asthma: the family are thinking of moving to the countryside for his health
Scientists at the University of Geneva have warned that global warming may lead to more frequent avalanches in the Himalayas
Anam goes to school near Abbey Park in Leicester
The emissions from diesel and petrol produce carbon dioxide and other gases
Anam's father drives her to school. They love to listen to story tapes together on the way
On Sunday 7 February 2021, a Himalayan glacier burst in the Indian state of Uttarakhand
Scientists believe that huge ice blocks broke off the glacier due to a temperature rise
The water from the glacier flooded a huge area
The floodwaters destroyed a dam at the Rishiganga Hydroelectric Project, leaving many people missing or dead
When the ice blocks broke, a huge amount of water rushed downhill
Scientists have known for a long time that global warming is causing glaciers to melt
The Hydroelectric Project supplies energy to neighbouring towns and villages
Anam's family love to get together on Sundays to watch wildlife documentaries on television
A busy 'A' road leading to Leicester city centre runs past Abbey Park

power cut one Sunday (are there things here that we recognise?).

2. Organise statements to show some possible connections between these two events (how do they feel about this?).
3. Start thinking about possible solutions to any problems they may have encountered in reading the statements. Some of these might be immediate and local (e.g. travel to school) and others longer-term and bigger in scale (e.g. changing technologies).

Evaluation or Discussion: This should not become a closed activity where students create a single 'correct sequence'.

Using 3D glasses, learners explore the issue 'contrapuntally' (Said, 1993), looking at different perspectives. For example, building on the mystery: if Anam and Adrika were texting each other, what would they say? (Children can play-act this before writing down ideas.) What might these girls then say to their parents? How might the parent respond? (This can be done as 'hot-seating' or with the teacher in role as a parent.) Older students could follow online news reports (e.g. using the web site http://www.headlinespot. com/international/) about what people are saying in different places about the flooding in Uttarakhand (or in the lead-up to a climate summit). Younger children might use a story such as 'The Giant's Embrace' (Big Brum, 2021) to explore different perspectives on needs, wants and care. Another thinking skills activity, 'Talking Graphs', involves students giving voice to data, thereby linking science and mathematics to human stories. What might an Indian hydro plant worker say about the difference between per capita CO_2 emissions in their country and those in the USA? How might an American politician respond? What about an American environmental justice campaigner? Where might their different ideas come from? What interests are involved? (How about our own ideas, our own interests?) What do we think and feel about these different perspectives? What are the common strands that can be built on between these people and the differences to be resolved? This takes learners from surface-level phenomena through to questions about underlying value systems.

And so to *the telescope*. Based on what we now know, *what can we do about it?* The Madrid schools that came up with the 'ópticas' also produced a poster bearing the motto 'Changing the world begins with the imagination' (Oliveros Palomo, 2016). How imaginative can students be about this? Or their teachers? If intergenerational injustice is part of the problem, how far can adults allow themselves to go in empowering learners about these questions, which lie at the heart of their futures? Hart's Ladder of Participation (Table 6.3) is a useful tool for us in this respect: as we move up the rungs, how far can we really enable learner agency?

As well as the geography and science of the issue, students could devise technical solutions through a STEM challenge in design and technology, consider personal action and morality in Personal Social and Health Education and Religious Education and investigate the decision-making processes at climate summits. They might write persuasive texts to send to local delegates and decision-makers. As exemplified by the examples above about air pollution, they should try and – as importantly – test practical changes in their own lives, homes and schools. This is also a chance for students to purposefully share their learning with others. For example, they could create a digital story about the future that they want (Change the Story, 2021). This sort

of social interaction should also be considered as a form of meaningful action. The key thing here is that all the critical and creative thinking (ESD 2) from the magnifying glass, microscope and 3D glasses ópticas is feeding back into and enriching informed action (ESD 1). This action can then be tested and evaluated on the basis of what has been learned: ESD 2 again, 'empowering children not only as learners, but also as confident citizens' (Vare & Scott, 2008). To reiterate a point made earlier, this conceptualises sustainable development itself as a social learning process: Learning *as* Sustainable Development.

In all this, the 'lenses' themselves can serve as tools for students' meta-learning: exploring how they have learned. In reviewing the ways they have learned about the climate crisis, students may well devise further questions for the future. On the way, they will have developed learning strategies that they will be able to apply to other environmental injustices and the crucial question of how to construct a more just, equal and sustainable world.

6.4 Conclusion

This chapter began with two tales of environmental injustice. Such tales raise quintessentially 'wicked' questions about how we understand the world and address these matters as educators. They also raise challenges about how we understand Education for Sustainable Development: proposing that it is not so much as a collection of uncontested knowledge sets to be transmitted from teacher to learner, but rather a form of social, active, agentic and self-critical learning that in itself forms a crucial contribution to social justice and sustainability. This locates environmental justice as an essential educational concern, but one that is often and intrinsically in tension with many established educational and societal norms. I hope that this chapter ends with the reader having a better sense of some of the real-world and pedagogical questions involved in addressing environmental justice issues with children and young people and having become familiar with some practical and adaptable strate-

Table 6.3 Hart's Ladder of Participation (Hart, 1992)

8	Child-initiated, shared decisions with adults	Degrees of participation
7	Child-initiated, but directed by adults	
6	Adult-initiated, shared decisions with children	
5	Children and consulted but informed	
4	Children are told what to do, but informed	
3	Tokenism	Degrees of non-participation
2	Decoration	
1	Manipulation	

gies that will empower learners to realise their right-to-know about the things that impact on their lives, to begin to address the problem of knowing about these things and to begin to work towards actual environmental justice personally, locally and at a global scale.

Acknowledgements With thanks to Danielle Kennell (Leicester City Council), to Lorraine Cookson and Ash Sharma (Skips Educational) and to Tide~ global learning for their assistance and for making material available for use in this chapter.

References

Adenike, O. (2021, February 3). Tweet as @the_eco-feminist. https://twitter.com/the_ecofeminist/status/1356976008633663488. Accessed 22 Feb 2021.

Adichie, C. N. (2009, July). The danger of a single story. *TED Talk*. https://www.ted.com/talks/chimamanda_ngozi_adichie_the_danger_of_a_single_story. Accessed 10 Feb 2021.

Agyeman, J. (2000). Environmental justice. In *From the margins to the mainstream? TCPA tomorrow series, paper 7*. The Town and Country Planning Association.

Alexander, R. (Ed.). (2010). *Children, their world, their education. Final report and recommendations of the Cambridge Primary Review*. Routledge.

Andreotti, V. (2013). *Taking minds to other places. Primary Geography 80*. Geographical Association.

Ballin, B. (2016a). *Getting to know the world*. Cambridge Primary Review Trust. http://cprtrust.org.uk/cprt-blog/getting-to-know-the-world/. Accessed 10 Feb 2021.

Ballin, B. (2016b). Making connections in a changing climate. *Environmental education* (Vol. 114). NAEE-UK. Online at https://naee.org.uk/making-connections-changing-climate/. Accessed 09 Feb 2021.

BBC World Service. (2021, January 29). The comb – Promotional tweet. https://www.bbc.co.uk/sounds/play/p095f76y. Accessed 22 Feb 2021.

Belk, J., et al. (1992). *It's our world too: A local-global approach to environmental education*. Birmingham and Sheffield.

Big Brum. (2021). *The giant's embrace*. Big Brum TIE.

Bland, A. (2021, February 20). Hearths on fire: UK residents incensed by pollution from wood burners. *The Guardian*. https://www.theguardian.com/environment/2021/feb/20/hearths-on-fire-uk-residents-incensed-by-pollution-from-wood-burners. Accessed 22 Feb 2021.

Boardman, B., Bullock, S., & McLaren, D. (1999). *Equity and the environment. Guidelines for socially just government*. Catalyst/Friends of the Earth.

Bruner, J. (1996). *The culture of education*. Harvard University Press.

Change the Story. (2021). Change the story, creating stories across Europe. https://www.changethestory.eu. Accessed 19 May 2021.

Choudhary, S. (2018, March 27). Proceedings of the National Academy of Sciences, reported in *Mint*. https://www.livemint.com/Science/WXUBL8DaEnuHmvruegmljJ/Warmer-winters-leading-to-rise-in-number-of-avalanches-in-we.html. Accessed 08 Feb 2021.

Client Earth. (2021). Air pollution. https://www.clientearth.org/what-we-do/priorities/air-pollution/. Accessed 12 Feb 2021.

Consultative meeting on Environmental Justice and Education. (2001). Unpublished report.

Cowcher, H. (1988). *Rainforest*. Andre Deutsch.

Davies, E., & Scott, G. (2011). *Leicester city's air quality action plan 2011-2016*. Leicester City Council.

Davies, L., Harber, C., & Yamashita, H. (2005). *Key findings from the DFID project, global citizenship: The needs of teachers and learners*. Centre for International Education and Research, University of Birmingham.

DESA (Department of Economic and Social Affairs). (2010). *Rethinking poverty: Report on the world social situation 2010*. United Nations Department of Economic and Social Affairs.

Eco Schools Global. (2021). https://www.ecoschools.global. Accessed 09 Feb 2021.

Fien, J., & White, C. (co-ordinators) (2002–2009). *Teaching and learning for a sustainable future. Multimedia programme*. UNESCO. http://sustainability.edu.au/material/teaching-materials/unesco-teaching-and-learning-sustainable-future/. Accessed 09 Feb 2021.

Fore, H. (2021). Climate change is the other planetary crisis that won't wait. https://www.unicef.org/reimagine/five-opportunities-children-open-letter. Accessed 22 Feb 2021.

Friends of the Earth. (2000). Pollution injustice. Cited in Agyeman, 2000.

Gardiner, B. (2019, February 13). The mother who wants to put air pollution on her daughter's death certificate. *The New York Times*. https://www.nytimes.com/2019/02/13/opinion/ella-kissi-debrah-pollution-london.html. Accessed 17 May 2021.

Gore, A. (2006). *An inconvenient truth*. Bloomsbury Publishing.

Greig, S., Pike, G., & Selby, D. (1987). *Earthrights: Education as if the planet really mattered*. WWF-UK.

Greig, S., Pike, G., & Selby, D. (1989). *Greenprints for changing schools*. WWF-UK/Kogan Page.

Hart, R. A. (1992). *Children's participation – From tokenism to citizenship*. UNICEF International Child Development Centre.

Hickel, J. (2015). Why the new sustainable development goals won't make the world a fairer place. *The Conversation*. https://theconversation.com/why-the-new-sustainable-development-goals-wont-make-the-world-a-fairer-place-46374. Accessed 17 May 2021.

Huckle, J., & Sterling, S. (Eds.). (1996). *Education for sustainability*. Earthscan Publications.

Jackson, T. (2020). *What's the issue? Climate change.* Quarto Publishing.

Khadka, N. S. (2021, February 8). What caused the glacial burst? *BBC News.* https://www.bbc.co.uk/news/world-asia-india-55975743. Accessed 08 Feb 2021.

Laville, S. (2020, December 16). Air pollution a cause in girl's death, coroner rules in landmark case. *The Guardian.* https://www.theguardian.com/environment/2020/dec/16/girls-death-contributed-to-by-air-pollution-coroner-rules-in-landmark-case. Accessed 08 Feb 2021.

Leat, D. (2001). *Thinking through geography* (2nd ed.). Optimus Education.

Leicester Air Quality Education. (2021a). https://schools.leicester.gov.uk/services/environment-health-and-well-being/air-quality-education/. Accessed 10 Feb 2021.

Leicester Air Quality Education. (2021b). Air quality education resources. https://schools.leicester.gov.uk/services/environment-health-and-well-being/air-quality-education/air-quality-education-resources/. Accessed 19 May 2021.

Leicester City Council. (2015). *Healthier air for Leicester - Leicester's air quality action plan (2015-2026).* Leicester City Council.

Leicester Environmental Education Team. (2021). https://schools.leicester.gov.uk/services/environment-health-and-well-being/environmental-education/. Accessed 10 Feb 2021.

Lynas, M. (2004). *High tide – News from a warming world.* Flamingo Books.

Manchester Research Hive. (2018). We are not that hard to reach. https://manchesterresearchhive.wordpress.com/2018/09/13/we-are-not-that-hard-to-reach-involving-seldom-heard-communities-in-research/. Accessed 12 Feb 2021.

McAndrews, J. (2018). *Supporting children in the face of climate change.* https://youtu.be/2bm18_G4n2Y. Accessed 10 Feb 2022.

Nelles, D., & Serrer, C. (2021). *Small gases, big effect.* Penguin Books.

Oborne, P. (2021). *The assault on truth: Boris Johnson, Donald Trump and the emergence of a new moral barbarism.* Simon & Schuster.

Oliveros Palomo, M. E. (2016). *La óptica del aprendizaje global.* FERE-CECA. Available in English as 'Global learning – Lenses on the world' from www.tideglobal-learning.net. Accessed 10 Feb 2021.

Polychroniou, C. J. (2020, November 25). Noam Chomsky: Trump has revealed the extreme fragility of American democracy. *Truthout.* https://truthout.org/articles/noam-chomsky-trump-has-revealed-the-extreme-fragility-of-american-democracy/. Accessed 10 Feb 2021.

Rosane, O. (2020, December 17). 9-year-old Girl's asthma death officially linked to air pollution in unprecedented coroner ruling. *Eco-Watch.* https://www.ecowatch.com/london-girl-death-air-pollution-2649531226.html. Accessed 08 Feb 2021.

Said, E. W. (1993). *Culture and imperialism.* Vintage.

Samarasinghe, N. (2018). Editorial: The colour line. *UNA-UK Magazine 1.* https://una.org.uk/magazine/2018-1. Accessed 08 Feb 2021.

Scott, W., & Vare, P. (2018). *The world we'll leave behind – Grasping the sustainability challenge.* Routledge.

Sharma, A. (2018). *Clean air cops.* Skips Educational. https://www.skipseducational.org/clean-air-cops/. Accessed 10 Feb 2021.

Sheerman, B. (2021, January). WCRAQ already making an impact. *Fleet Vision International.* https://www.flipsnack.com/fleetvision/fleet-vision-international-q1-2021.html?web=1&wdLOR=c944E5FA9-B051-114E-8299-996F0EDD8073. Accessed 15 Feb 2021.

Skips Educational. (2021). https://www.skipseducational.org. Accessed 10 Feb 2021.

Thunberg, G. (2019). Can you hear me? In *No one is too small to make a difference.* Penguin Books.

Tide~ global learning. (2016). *Global learning – Lenses on the world.* Tide~ global learning.

UK Student Climate Network. (2021). Demand 2-teach the future. https://ukscn.org/our-demands/. Accessed 22 Feb 2021.

UN. (2015). Sustainable development goals. https://sdgs.un.org/goals. Accessed 12 Feb 2021.

UNEP. (2012). UNEP establishes new body to advance environmental law, justice and governance. Press release. https://www.unep.org/news-and-stories/press-release/unep-establishes-new-body-advance-environmental-law-justice-and. Accessed 10 Feb 2021.

Vare, P., & Scott, W. (2008). *Education for sustainable development: Two sides and an edge.* Development Education Association.

Wainaina, B. (2005). *How to write about Africa. Granta 92.* Granta.

Walker, G., Fairburn, J., Smith, G., & Mitchell, G. (2003). *Environmental quality and social deprivation.* Environment Agency.

Webster, K., & Johnson, C. (2008). *Sense and sustainability. Educating for a low carbon world.* TerraPreta Publishing with Yorkshire Forward and InterfaceFLOR.

Westminster Commission on Road Air Quality. (2021). http://wcraq.com. Accessed 11 Feb 2021.

Ben Ballin is a consultant in geography, global learning, and sustainability education. He is a fellow of the National Association for Environmental Education – UK, Secretary of the West Midlands Sustainable Schools Network, and a board member of the journal *Primary Geography*. His recently published research includes "Towards an understanding of the contribution of global learning to the well-being and mental health of young people with special educational needs" (2018, with Ann McGuire and Laura Murphy) and "Socially Connected – the displaced teacher and the displaced child" (2021, as part of a European initiative on drama and well-being). Ben is the author of numerous teaching materials and regularly trains teachers throughout the UK.

You Are Part of the Sustainability Picture: Ideas for Implementation of ESD

7

Armağan Ateşkan and Jennie Farber Lane

Abstract

This chapter presents ideas for teachers about how to implement ESD in their classrooms. The essential purpose is to raise awareness of the United Nation's Sustainable Development Goals for all and what could be done as individuals focusing on taking responsibility as a person in this system. Furthermore, it aims to help students develop consequential thinking skills concerning habits of mind, eating ecological food, recycling materials, and exploring water quality. A brief review of the literature discusses the importance of purposefully integrating of education for sustainability into the curriculum. To guide teachers and facilitate implementation and integration of ESD into elementary and middle school level, educators are provided lesson plans; ideas and experiences are provided in this chapter with the support of theoretical foundations and a review of the literature. The approaches to be covered are mainly problem-based, inquiry-based, cooperative, and experiential learning strategies.

Keywords

Implementation · Problem-based learning · Lesson plan · Cooperative learning

7.1 Introduction

From the journal of Armağan Ateşkan
Date: 1 January 2021, Friday
Place: Ankara, Turkey
Time: 13:30
Temperature: 13.1 °C (January average temperature: −0.6 °C)
I have been walking around this lake for 20 years for different purposes: pleasure, fitness, and field trip with pre-service teachers, middle and high school students, and college students. This is an artificial lake made to meet a factory's water needs in the 1980s. Hence, the factory has provided water from the municipality; it has been only used for the campus's irrigation needs. Several basins and streams feed the lake and water from the campus field. While the pond is full of water, the area is approximately 40 square kilometers. The average depth is about 10–12 m., the volume is 400–500 cubic meters, and the deepest point is around 15–16 m. Today, while I was walking around the lake, I took the photographs below in Fig. 7.1 and felt so sad and frustrated. Several questions were raised in mind. How did it happen? Why is the water level so low

A. Ateşkan (✉) · J. F. Lane
Bilkent University, Department of Educational Sciences, Ankara, Türkiye
e-mail: ateskan@bilkent.edu.tr

© The Author(s), under exclusive license to Springer Nature Switzerland AG 2022
G. Karaarslan-Semiz (ed.), *Education for Sustainable Development in Primary and Secondary Schools*, Sustainable Development Goals Series, https://doi.org/10.1007/978-3-031-09112-4_7

Fig. 7.1 The views of the lake. (Photos taken by the author)

at that time of the year? What will happen in the future if the water level continues to decrease? The answer to the last question is simple, water shortage!

While I was thinking about these questions and answers, I remember the lines of Hikmet Birand's [Turkish botanist, writer] in Conversation with Juniper Tree book:

– How much water do you consume? And then how do you consume water? Literally, you are consuming too much water, and I did not know that.

– I am one of the ones that know how to save water. Right now, it is spring, the season of rain, the soil has enough water to supply us. I consume 15-20 kilograms of water daily. When the water level decrease in the soil, I can decrease my consumption of water. ....

– If the water in the soil decreases and then depleted?

– In that case, our roots cannot get the same amount of water that evaporates from the leaves, the water balance is disturbed, and it will be easily observed from our leaves; they will be withered and fade. If the water balance is not fixed, the leaves will get dry, and then the plant will die. (Birand, 2014, pp. 51–52)

It is so discouraging to see that the nature is dying and the world is changing. Let me go back to the questions and have some more: How do you envision the world to be like 30 years from now? What do you want your world to be like? Which of our current practices will contribute to a positive outlook? What should we be teaching our children today to help them create a sustainable tomorrow?

These questions are an important part of ESD. ESD provides people with the knowledge, skills, dispositions, and opportunities to promote a healthy and livable world. It is a holistic and systems-based approach to teaching and learning

(Sterling, 2001; Tilbury et al., 2005) that integrates social justice, economics, and environmental literacy. We begin by examining our current ways of thinking and acting. We then seek to understand and appreciate the interconnections and systems of how our environment works. We are encouraged to become more conscious of our habits, so we become more conscientious of our needs and wants. The ultimate outcome of ESD is to sustain both human and natural communities. The United Nations (UN) identified 17 Sustainable Development Goals (SDGs) to live in a better and sustainable world. These goals are:

Goal 1. End poverty in all its forms everywhere.
Goal 2. End hunger, achieve food security and improved nutrition, and promote sustainable agriculture.
Goal 3. Ensure healthy lives and promote well-being for all at all ages.
Goal 4. Ensure inclusive and equitable quality education and promote lifelong learning opportunities for all.
Goal 5. Achieve gender equality and empower all women and girls.
Goal 6. Ensure availability and sustainable management of water and sanitation for all.
Goal 7. Ensure access to affordable, reliable, sustainable, and modern energy for all.
Goal 8. Promote sustained, inclusive, and sustainable economic growth, full and productive employment, and decent work for all.
Goal 9. Build resilient infrastructure, promote inclusive and sustainable industrialization, and foster innovation.
Goal 10. Reduce inequality within and among countries.
Goal 11. Make cities and human settlements inclusive, safe, resilient, and sustainable.
Goal 12. Ensure sustainable consumption and production patterns.
Goal 13. Take urgent action to combat climate change and its impacts [n 10].
Goal 14. Conserve and sustainably use the oceans, seas, and marine resources for sustainable development.
Goal 15. Protect, restore, and promote sustainable use of terrestrial ecosystems, sustainably manage forests, combat desertification, and halt and reverse land degradation and halt biodiversity loss.
Goal 16. Promote peaceful and inclusive societies for sustainable development, provide access to justice for all, and build effective, accountable, and inclusive institutions at all levels.

Goal 17. Strengthen the means of implementation and revitalize the Global Partnership for Sustainable Development. (UNESCO, 2017)

SDG4 and specifically Target 4.7 highlighted "by 2030 ensure all learners acquire knowledge and skills needed to promote sustainable development, including among others through education for sustainable development and sustainable lifestyles, human rights, gender equality, promotion of a culture of peace and non-violence, global citizenship, and appreciation of cultural diversity and of culture's contribution to sustainable development." The next section discusses how sustainability topics can be included in the curriculum to contribute to the sustainability literacy of today's children.

7.2 ESD Integration

Although more people are appreciating the need for ESD, administrators, policy makers, and curriculum planners need to consider how to ensure it is included in the school curriculum. Teachers can include concepts related to sustainable development in their lesson plans in a variety of ways. They can reference environmental and health issues during classroom discussions. They can use sustainable development to illustrate concepts in their subject area. They can help students analyze problems related to equity and justice. Teachers sprinkle these references throughout their lessons, but sometimes they are blended more thoroughly to the point where their connections are seamless. The term for seamless inclusion of one topic into another has been labeled infusion by some practitioners (Ramsey et al., 1992). Lane (2006) conducted a study to investigate the practice of environmental education (EE) infusion. While her research focused on environmental education, her findings can be applied ESD. Essentially, she concluded there are three ways to include topics related to sustainability in different subject areas. They can be infused, inserted, or integrated. Lane defines these terms as follows:

- Insertion: To add separate activities about the environment into the curriculum
- Infusion: To blend environmental concepts into existing lessons when the opportunity arises
- Integration: To intentionally design lessons to include environmental concepts

Professional environmental educators advocated for infusion with the hope the concepts and topics related to the environment would permeate the curriculum and become more embedded and less easy to remove. The infusion approach may be ideal because it does not require significant changes to the existing curriculum. In essence, because sustainability is so pervasive, it is seamlessly included in the subject area teaching matter. However, its very pervasiveness – one of its key strengths – might also be its primary shortcoming. Teachers may report that they do not teach about sustainable development, when in fact they do; they just do not recognize since it was not a part of their professional development experience (Borg et al., 2012; McKeown, 2013; Redman, 2013).

Teachers may also "insert" or add standalone activities related to sustainability to the teaching plans. Government agencies in the United States produce environmental education activity guides, such as Project Wild (https://www.fishwildlife.org/projectwild), Project WET (https://www.projectwet.org/), and Project Learning Tree (https://www.plt.org/), that are ideal for "adding" a lesson to a unit or possibly replacing one lesson with another. The Turkish Environmental Education Foundation (http://www.turcev.org.tr) and the Regional Environmental Center (REC): Green Pack Project (https://www.rec.org/index.php) include materials that involve syllabuses for teachers' use, games, information documents for students, and visual materials for education for sustainable development. Teachers are informed how to use these documents and how to integrate them into their lessons. The Eco-School project's content assures cooperation between teachers and managers of schools and school and private and public sectors. The problem with this approach is that just as easily as the activities can be added, they can be taken out. Another shortcoming is that rather than permeating a curriculum, sustainability topics might occur in isolated bits and pieces here and there throughout a child's K-12 learning career.

Therefore, when topics related to sustainability are inserted or infused, it is most likely that these concepts are addressed in a hit or miss fashion throughout students' K-12 learning experience. This piecemeal nature of sustainability is compounded because not only might teachers within a single class randomly use activities but some teachers may also choose to exclude education for sustainability altogether. Another shortcoming of infusion and insertion is that there is no assurance that its goals are being met. Therefore, both the insertion and the infusion approaches have the ability to include sustainability concepts in the curriculum, yet both have their shortcomings. Infused concepts can be so diffuse that it is not apparent, and inserted activities do not guarantee a sequential K-12 experience in sustainability. In both cases, they might be lost or omitted from the curriculum without intentional planning to address Sustainable Development Goals.

In her study, Lane (2006) concluded that the ideal approach for ensuring environmental concepts in the curriculum might be integration. She explains that integration is intentionally designing lessons to include concepts from one or more other disciplines. In this case, the goals for sustainable development need to be deliberately included in the curriculum. Lane compares integrated concepts and topics from different disciplines to a piece of rope. Although the topics from different disciplines are intertwined, the individual fibers and braided strands are still discernable. This represents how with integration the elements of the combined disciplines are interdependent but still unique. While a thoroughly infused topic might become lost, a new product is created that highlights key aspects of the product sources with integration. Furthermore, unlike insertion, with integration, the topics are connected and even dependent on each other.

Integration can occur in a single classroom with one teacher pulling together interdisciplinary concepts into a single lesson or unit, or a team of teachers can work together to create an interdisciplinary curriculum. These teachers can teach a class together, or they can work cooperatively in separate classes given that they teach the same student population (i.e., the same group of students travels from one class to another).

All subject areas can promote awareness of Sustainable Development Goals. Students in the social sciences can discuss global and regional environmental issues, language arts students use their writing and communication skills to promote environmental awareness, and participants in STEM apply science, technology, engineering, and mathematics to design projects to support sustainable development. It should be recognized, however, that some concepts and skills are more easily integrated into subject areas than others. For example, research skills may be more suitable for an advanced science or social studies class. Nevertheless, many subject areas can provide foundational skills needed for research; for example, in science classes, students can enhance observing, measuring, and recording skills needed to investigate environmental and societal issues. The interrelationship and interdependency of skills and knowledge related to achieving sustainable development provide further support for the need for purposeful curriculum planning and collaboration among teachers from many disciplines.

7.3   ESD Teaching Strategies

Sustainable development includes hot issues and complicated systems; therefore, teachers need an extensive range of ESD teaching strategies to support students' place-based (Sobel, 2004; Smith, 2007), active, and experiential (Kolb, 1984) learning. Some of the strategies include case studies, role-plays, projects, debates, experiments, lectures, guest speakers, field trips, simulations, computer-based activities, group discussions, and reading activities (Corney, 2006). Students will use their communicative,

critical thinking, creative thinking, collaborative, cooperative, and systems thinking skills in all of the activities designed.

Problem-based learning strategies and especially community-based problem-solving assist students in acting on a local problem that they are concerned with to promote ESD. This helps both students and teachers to use experiential and research-based strategies for a sustainable future via developing goals for citizenship. According to some of the studies, there are nine main steps for community problem-solving: exploring the community-based problem, selecting the problem (the students may use some resources to find out the problem, such as questionnaires, observations, and interviews), evaluating students' skills and developing them, carrying out a research, developing a vision, planning the action, taking an action, evaluating the action, and making alterations (Torrence, 1995).

Inquiry-based learning is a teaching strategy where the learners find out an answer to their questions by following the procedures, formulating hypothesis, and testing them by using scientific methods (Keselman, 2003; Pedaste et al., 2012). These investigations require learner's active participation and taking responsibility of his/her learning. For the benefit of ESD, use of inquiry-based learning may encourage students to explore more and facilitate in-depth understanding (Justice et al., 2009).

7.4   Examples of Implementing ESD in the Classroom

The rest of this chapter includes three sets of learning experiences to implement ESD with students. These experiences are problem-based, inquiry-based, and discussion-based learning strategies. They all involve students working collaboratively. The experiences are based on actual learning experiences used by teachers, the authors, or pre-service teachers.

Each learning experience has been related to one or more of the UN SDGs. Furthermore, learning objectives for these goals, developed by UNESCO (2017), are identified and labelled as

Cognitive Learning Objective (CLO) or Social-Emotional Learning Objective (SELO) or Behavioral Learning Objective (BLO). For example, SDG11.CLO5 means Sustainable Development Goal 11. Cognitive Learning Objective 5.

To integrate these experiences into different subject areas, teachers will need to review the objectives and activities and relate them to their course objectives. Furthermore, teachers will adapt and adjust them based on their students' learning needs and available resources. These experiences are presented here to help educators start thinking about different ways sustainability can be integrated into students' learning experiences.

7.4.1 Exploring Sustainability of the School, the Local Community, and Abroad

This set of learning experiences includes four examples of activities developed by classroom teachers. After a 5-day workshop on integrating education for sustainability in the classroom given by experts in the area, primary and middle school teachers were asked to develop and share activities they developed and implemented with their students.

- *Example Activity 1: Let the Wastes Not Be Garbage*
Teaching strategy: Problem-based learning and cooperative learning
Age: 12–15 years
Subject areas: Science, art, and social studies
SDG learning objectives:

SDG11.CLO5. The learner understands the role of local decision-makers and participatory governance and the importance of representing a sustainable voice in planning and policy for their area.
SDG11.SELO5. The learner is able to feel responsible for the environmental and social impacts of their own individual lifestyle.

SDG11.BLO1. The learner is able to plan, implement, and evaluate community-based sustainability projects.
SDG11.BLO2. The learner is able to participate in and influence decision processes about their community. (UNESCO, 2017)

A questionnaire was implemented to the school's parents and the people living in the community to explore their awareness about the oil waste and how to recycle it. The students at the school get in touch with the municipality to get support for their actions. The posters and presentations were prepared to increase the awareness of the community. The oil wastes were collected and recycled by students, parents, and community members with the municipality's support. The action results were shared with different parties, the action was evaluated, and the next steps were decided.

- *Example Activity 2: The Target Is a Root in Nature*
Teaching strategy: Problem-based learning and cooperative learning
Age: 10–12 years
Subject areas: Science, art, music, and language
SDG learning objectives:

SDG15.CLO1. The learner understands basic ecology with reference to local and global ecosystems, identifying local species and understanding the measure of biodiversity.
SDG15.SELO2. The learner is able to argue for the conservation of biodiversity on multiple grounds including ecosystems services and intrinsic value.
SDG15.SELO3. The learner is able to connect with their local natural areas and feel empathy with nonhuman life on Earth.
SDG15.SELO4. The learner is able to question the dualism of human/nature and realizes that we are a part of nature and not apart from nature. (UNESCO, 2017)

Based on the students' observations, the variety of the plants in their school's garden decreased, and there are plenty of remaining craft papers at art lessons usually thrown away. The students collected the remaining of craft papers; later, they prepared seed balls (mixing seeds with craft papers and water) to plant into their school garden. To increase awareness in the community, they also gave seed balls as gifts to the people around their school. Related to the activities, the students wrote a poem in a language class and sang a song with the poem they wrote about the human is a part of nature. Sustainability was included in science, art, language, and music lessons, where students were actively engaged and taken action.

- *Example Activity 3: Sustainable School, Sustainable Street*
  Teaching strategy: Problem-based learning and cooperative learning
  Age: 11–14 years
  Subject areas: Science and art
  SDG learning objectives:

  SDG6.CLO1. The learner understands water as a fundamental condition of life itself, the importance of water quality and quantity, and the causes, effects, and consequences of water pollution and water scarcity.
  SDG6.SELO3. The learner is able to feel responsible for their water use.
  SDG6.BLO2. The learner is able to contribute to water resources management at the local level.
  SDG6.BLO3. The learner is able to reduce their individual water footprint and to save water practicing their daily habits.
  SDG7.SELO4. The learner is able to clarify personal norms and values related to energy production and usage as well as to reflect and evaluate their own energy usage in terms of efficiency and sufficiency.
  SDG15.CLO1. The learner understands basic ecology with reference to local and global ecosystems, identifying local species and understanding the measure of biodiversity.
  SDG15.SELO3. The learner is able to connect with their local natural areas and feel empathy with nonhuman life on Earth. (UNESCO, 2017)

After observing the school corridors and garden, students decided that, especially in winter, the birds could not find enough food to survive

and there are many waste materials and food in the school that could be recycled. They collected the water bottles used for bird feeding and put the food that they could not consume all. When the birds came to eat the food, they observed the birds and tried to identify their species. Another group of students was disturbed with the water and electricity bills of the school after discovering it. They read the bills not just as money but also as consumption of energy and water. They designed projects to decrease water and electricity, such as awareness activities and warnings of turning off the tap water and electricity. It also helped students to go through their habits at home.

- *Example Activity 4: Environment-Friendly Pen-Pal*
  Teaching strategy: Problem-based learning and cooperative learning
  Age: 10–12 years
  Subject areas: Science and language
  SDG learning objectives:

  SDG15.CLO2. The learner understands the manifold threats posed to biodiversity, including habitat loss, deforestation, fragmentation, overexploitation, and invasive species, and can relate these threats to their local biodiversity.
  SDG15.CLO3. The learner is able to classify the ecosystem services of the local ecosystems including supporting, provisioning, and regulating cultural services and ecosystems services for disaster risk reduction.
  SDG15.SELO3. The learner is able to connect with their local natural areas and feel empathy with nonhuman life on Earth.
  SDG15.BLO1. The learner is able to connect with local groups working toward biodiversity conservation in their area. (UNESCO, 2017)

The students explored that the communication among young generation is not qualified. They decided to run a project via writing mails to peers from other cities explaining their local environmental problems and observations related to these problems. While exchanging the views about different cities, students realized that there are not enough trees in their local area, so they got help from an NGO to plant more trees around their school. The students exchanged their views about use of recycled materials in the school context and carried out similar projects in both cities and shared the results of them.

7.4.2 Concepts of Sustainability: Past, Present, and Future

This learning experience was conducted with fifth grade classes during a guest presentation conducted by the authors at the school for an hour. It is composed of five consecutive tasks. Before conducting the activity, teachers should ask students to learn about lifestyle habits of the past by having a conversation with an elder (such as a grandparent).

Teaching strategy: Problem-based learning and cooperative learning

Age: 10–12 years

Subject areas: Science and art

SDG learning objectives:

SDG7.CLO1. The learner knows about different energy resources – renewable and non-renewable – and their respective advantages and disadvantages including environmental impacts, health issues, usage, safety and energy security, and their share in the energy mix at the local, national, and global level.

SDG11.CLO2. The learner is able to evaluate and compare the sustainability of their and other settlements' systems in meeting their needs particularly in the areas of food, energy, transport, water, safety, waste treatment, inclusion and accessibility, education, integration of green spaces, and disaster risk reduction.

SDG11.CLO4. The learner knows the basic principles of sustainable planning and building and can identify opportunities for making their own area more sustainable and inclusive.

SDG11.SELO5. The learner is able to feel responsible for the environmental and social impacts of their own individual lifestyle.

SDG12.CLO1. The learner understands how individual lifestyle choices influence social, economic, and environmental development.

SDG12.SELO1. The learner is able to communicate the need for sustainable practices in production and consumption.

SDG12.SELO4. The learner is able to envision sustainable lifestyles. (UNESCO, 2017)

Tasks for Exploring Concepts of Sustainability Then, Now, and in the Future

First Task: How Was Life 10 Years Ago?

Students will ask their parents and/or older relative about changes in the lifestyles they have noticed over the past 10 or so years. Students can ask about at least three of the following: food consumption, water management, heating, electricity, communication, transportation, health, education, entertainment, and waste management (this information will be used in the fourth activity).

Second Task: Sustainable House

The teacher will show some images from a sustainable home: collecting rainwater in bowls and using stove and solar panels for heating. The students will be encouraged to think about the term "sustainable." Lead the discussion to something that is short-lived thrown away.

Third Task: Finding New Uses for Discarded Things

The teacher will prepare four or five groups of students, around four students per group, and then provide each group with three items that are normally thrown away. Tell them to think of how each could be used again rather than discarded. Each group picks one item they thought was very interesting and share with the rest of the class how it could be reused.

Fourth Task: Sharing Changes over the Past 10 Years and Envisioning a Future 10 Years from Now

The teacher will ask students to work together to draw how life was like around 10 years ago based on their discussions with elders on one half of a large piece of paper. Later, the teacher asks students to imagine what will be different 10 years from now. Remind students to think about the lifestyle categories (food, travel, communication, etc.). Have students draw their visions on the other half of the paper. Share

(continued)

examples from each category if time allows.

Fifth Task: My Promise to Myself

The teacher will share the SDGs and help students relate these to their categories. Which could they take action on today to help create a better future? Have students write a promise to themselves on a one part of their drawing and tear or cut it out and put it in a safe or secret place. This is their idea for taking action. If time allows and if students wish, they can share their action.

7.4.3 Organic Food, Organic Life

This learning experience consists of two activities to increase students' awareness of locally and organically grown food. This experience was another guest presentation provided by the authors.

Teaching strategy: Discussion-based learning and cooperative learning

Age: 12–14 years

Subject area: Science and art

The learning objectives for achieving SDGs:

SDG12.CLO1. The learner understands how individual lifestyle choices influence social, economic, and environmental development.

SDG12.CLO2. The learner understands production and consumption patterns and value chains and the interrelatedness of production and consumption (supply and demand, toxics, CO2 emissions, waste generation, health, working conditions, poverty, etc.).

SDG12.CLO4. The learner knows about strategies and practices of sustainable production and consumption.

SDG13.CLO3. The learner knows which human activities – on a global, national, local, and individual level – contribute most to climate change.

SDG15.SELO3. The learner is able to connect with their local natural areas and feel empathy with nonhuman life on Earth. (UNESCO, 2017)

- *Activity 1: Organic Farming-Organic Food*

The teacher will start a lesson by showing students pictures from a hobby garden (Fig. 7.2) and facilitate discussion among students related to growing own food and gardening. Organic farming is also included into the discussion, which brings the teacher to the next topic.

Related to the organic farming, the use of fertilizers will be discussed, and then linked to sustainable life and waste management, the worm composting (Fig. 7.3) will be shared with the students.

The definition of organic farming and the figures from the world and local places will be shared with students. After that, the next concept which is related to sustainability is introduced to the students: share of an organic farm, where the people will not be able to have their own organic farm and by paying an annual fee, a box of seasonal vegetables is taken once a week during the growing season (July–November).

- *Activity 2: "More" and "Less" Game*

Students will be put into groups of four or five. They will have flipcharts and board markers. Each group generates a list of outcomes that are "more" and "less" as a result of organic farming. Each group would share one more and one less with the entire audience. The teacher could have a brief discussion of their ideas and wrap up by discussing what this means for their future. The message is that by doing a simple thing like eating organic food, they can have broader outcomes on the local, regional, and global environments and their future. If there is more time, the teacher could discuss what could happen as a result of these things that are "more and less"; for example, if organic farming means more clean water, what does having more clean water mean? More of what? Less of what? Sort of like a ripple effect, when you drop a stone into water, rings radiate out from the one stone. This realization of cause and effect might happen "organically" in their groups anyway. At the end of the activity, besides organic food, the importance of

Fig. 7.2 The hobby garden. (This photo is taken by the author)

Fig. 7.3 Worm compost. (This photo is taken by the author)

eating local food will be reminded as well for sustainable future. The students will be included into the discussion of why we need to think about eating local food.

7.4.4 A Day Out at a Local Lake

This final learning experience is an extensive exploration of a local lake to promote students' observation, experimental, analysis, and synthesis skills. Field trips are valuable to environmental education, but they do require more preparation and different class management skills. Teachers will need to secure equipment and permission for travel. Teachers should conduct pre-field trip lessons and discuss appropriate behaviors and expectations for learning.

There are three activities adapted from a field trip to the lake mentioned at the beginning of this chapter. Pre-service teachers developed the activities and conducted them with local middle school students. The first activity involves general observations and requires no special equipment. The second activity requires some nets and a key to help identify animal species but can be simplified by having students make general observations. The third activity does require special lab and field investigation equipment. These activities are best conducted in small groups of not more than five students per group.

Teaching strategy: Experiential and cooperative learning

Age: 12–14 years

Subject area: Science and art

The learning objectives for achieving SDGs:

SDG6.CLO1. The learner understands water as a fundamental condition of life itself, the importance of water quality and quantity, and the causes, effects, and consequences of water pollution and water scarcity.

SDG6.CLO2. The learner understands that water is part of many different complex global interrelationships and systems.

SDG6.SELO3. The learner is able to feel responsible for their water use.

SDG6.SELO4. The learner is able to see the value in good sanitation and hygiene standards. (UNESCO, 2017)

• *Activity 1: Observations*

The purpose of this activity is to assess water quality by using *general observations*. Ask students to make some general observations about the physical parameters of the lake. Note physical parameters of the pond to look for signs of eutrophication. Consider other ways to monitor the biodiversity of plant and animal life around the pond (e.g., transects, binoculars, insect collection). Look for signs of human impact in and around the pond (e.g., walking trails, litter). Use these observations to decide on the overall water quality. Provide first impressions: sounds, sights, smells, and touch (biotic and abiotic factors, human impact, etc.).

Are there any factors you cannot observe? What are they?

They can practice their mapping skills by charting the shape of the lake on this grid.

Sense:	Abiotic/Biotic/Human Impact?	Observation:

- *Activity 2: Observing and Classifying Pond Invertebrates*

The purpose of this activity is to assess water quality by investigating macroinvertebrates. Macroinvertebrates could be seen by eye without the need of a microscope. Capturing them is easy, and after identification of them, they will give an idea about the pollution of the water.

Students will collect using nets and identify species and dichotomous keys for identification. These keys help identify indicator species to help determine the quality of the water; some species need very clean water, and others can tolerate more adverse conditions. Depending on which species is present indicates the quality of the water.

Materials: Nets, jars, pipette, dichotomous keys, bowl

Methods:

1. Use nets and take samples from the upper surface and bottom of the pond at least three times.
2. Take animals from the net and put them into jar by using pipette.
3. Observe the body structures very carefully (some may not be able to see easily).
4. Try to identify them by using dichotomous key.

Simple drawings	Characteristics	Name
	Case or shell: Legs: Tails: Movement: Wings:	

Discussion:

1. What do the animals living in a stream tell us about the water quality?
2. What can be the pollution sources for the lake? Also what can be done to reduce them to make a healthy lake?
3. Observe the lake in terms of eutrophication. What are the signs of eutrophication?
4. Describe how you imagine the bottom of the lake.
5. Explain the relationship between eutrophication and algal growth.

6. Discuss the causes and consequences of eutrophication.

- *Activity 3: Physiological Properties of the Lake*

The purpose of this activity is to assess water quality through *measuring parameters* such as dissolved oxygen, pH, turbidity, etc. Use data to identify potential sources of pollution.

Salinity: Determines the type of organisms that can survive in the area sampled. The fresh saltwater boundary separates different bodies of water from each other.

Dissolved oxygen (DO): Aquatic animals need oxygen to live. River mixing, temperature, and pollution affect DO levels.

Temperature: Determines the amount of oxygen that can dissolve in water. Determines the rate of photosynthesis of aquatic plants. Determines the sensitivity of organisms to pollution, parasites, and diseases.

pH: Acidity of the water determines what organisms are present.

Total dissolved solids: Silt, sewage, sediment. Used as the "watchdog" of environmental test. Human sources include fertilizers, erosion, and effluent.

Materials: Vernier LabQuest, Vernier probes (pH, temperature, dissolved oxygen concentration, conductivity, salinity), buckets, water depth sampler, sampling plastic jars, distilled water, pipette

Method:

Now, you will make measurements on water samples taken from the lake.

1. Prepare the temperature probe; make sure that it is ready for measurement.
2. Take a water sample from surface of the lake.
3. Measure the temperature of the sample immediately. Record your finding in Table 7.1 (Trial 1) and make two more measurements and record your result.
4. Prepare the dissolved oxygen concentration probe. Measure the dissolved oxygen concentration of the sample.

5. Record your findings in Table 7.1 (Trial 1) and make two more measurements and record your result.
6. Prepare the conductivity probe (measures dissolved solids). Measure the conductivity of the sample.
7. Record your findings in Table 7.1 (Trial 1) and make two more measurements and record your result.
8. Measure the pH of the sample using pH paper strips.
9. Record your findings in Table 7.1 (Trial 1) and make two more measurements and record your result.
10. Prepare the salinity probe. Measure the salinity of the sample.
11. Record your findings in Table 7.1 (Trial 1) and make two more measurements and record your result.
12. Take another sample from another area of the lake (again from the surface). Repeat steps 3–12.
13. Take sample from the bottom of the lake (let your teacher show how). Repeat steps 3–12. Record your findings in Table 7.2.
14. Take another sample from another area of the lake (again form the bottom). Repeat steps 3–12.

Table 7.1 The raw data table for your measurements for surface measurements

	Surface 1			Surface 2		
	1	2	3	1	2	3
Temperature						
Dissolved oxygen						
Conductivity						
pH						
Salinity						

Table 7.2 The raw data table for your measurements for bottom measurements

	Bottom 1			Bottom 2		
	1	2	3	1	2	3
Temperature						
Dissolved oxygen						
Conductivity						
pH						
Salinity						

Results:
  Discussion:
1. Are there any differences between the measurements you got from the bottom and the surface of the lake?
2. Explain the salinity results.
3. Do you think pH, temperature, and oxygen levels are proper for the living organism?

7.5 Conclusion: Each Part in the Sustainability Picture Is Important

This chapter provided implementation ideas to guide teachers for engaging their students in sustainability actions. The chapter began with a memo from one of the author's journal entry showing how serious the climate change is and supported with a famous botanist writers' section from the book. To help address the UN SDGs and their learning objectives, several teaching ideas are provided; these support the integration of problem-based, inquiry-based, and discussion-based activities in elementary and middle school settings. Admittedly, teachers will need to take time to review these activities and relate them to their disciplines. They will need to review the objectives of their subject area and find links to the SDG goals. Ideally, if they can participate in a workshop or course, they can gain experience doing the lessons and better appreciate how they can be implemented in their school settings. They can start with some of the classroom-based, discussion activities and as their experience and confidence grow endeavor to take students to explore their school grounds and on a short field trip to a local pond. Teachers can involve students in discussions about local issues and use the activities in this chapter as a resource for student projects. This chapter provides a beginning to launch ideas for integrating ESD goals into student learning experiences; they provide ideas for teachers and their students to recognize they are all parts of a picture of – a vision for global sustainability. Any action they take to become a positive part of our environmental systems will contribute to a sustainable future.

References

Birand, H. (2014). *Alıç ağacı ile sohbetler. [Conversation with juniper tree]* (5th Ed.) Türkiye İş Bankası Kültür Yayınları.

Borg, C., Gericke, N., Höglund, H. O., & Bergman, E. (2012). The barriers encountered by teachers implementing education for sustainable development: Discipline bound differences and teaching traditions. *Research in Science & Technological Education, 30*(2), 185–207. https://doi.org/10.1080/02635143.2012.699891

Corney, G. (2006). Education for sustainable development: An empirical study of the tensions and challenges faced by geography student teachers. *International Research in Geographical and Environmental Education, 15*(3), 224–240. https://doi.org/10.2167/irgee194.0

Justice, C., Rice, J., Roy, D., Hudspith, B., & Jenkins, H. (2009). Inquiry-based learning in higher education: Administrators' perspectives on integrating inquiry pedagogy into the curriculum. *Higher Education, 58*(6), 841. https://doi.org/10.1007/s10734-009-9228-7

Keselman, A. (2003). Supporting inquiry learning by promoting normative understanding of multivariable causality. *Journal of Research in Science Teaching, 40*(9), 898–921. https://doi.org/10.1002/tea.10115

Kolb, D. A. (1984). *Experiential learning: Experience as the source of learning and development* (Vol. 1). Prentice-Hall.

Lane, J. F. (2006). *Environmental education implementation in Wisconsin: Conceptualizations and practices* (Doctoral dissertation). University of Wisconsin--Madison.

McKeown, R. (2013). Teaching for a brighter more sustainable future. *Kappa Delta Pi Record, 49*(1), 12–20. https://doi.org/10.1080/00228958.2013.759824

Pedaste, M., Mäeots, M., Leijen, Ä., & Sarapuu, S. (2012). Improving students' inquiry skills through reflection and self-regulation scaffolds. *Technology, Instruction, Cognition and Learning, 9*(1–2), 81–95.

Ramsey, J., Hungerford, H., & Volk, T. (1992). Environmental education in the K-12 curriculum: Finding a niche. *Journal of Environmental Education, 23*(2), 35–45.

Redman, E. (2013). Opportunities & challenges for integrating sustainability education into K-12 classrooms. *Journal of Teacher Education for Sustainability, 15*(2), 5–24. https://doi.org/10.2478/jtes-2013-0008

Smith, G. A. (2007). Place-based education: Breaking through the constraining regularities of public school. *Environmental Education Research, 13*(2), 189–207.

Sobel, D. (2004). *Place-based education: Connecting classrooms and communities*. The Orion Society.

Sterling, S. (2001). *Sustainable education: Re-visioning learning and change*. Green Books.

Tilbury, D., Coleman, V., & Garlick, D. (2005). *A national review of environmental education and its contribution to sustainability in Australia: School education*. Australian Government Department of the Environment and Heritage and Australian Research Institute in Education for Sustainability (ARIES).

Torrence, E. P. (1995). *Why fly?* Ablex Publishing Cooperation.

UNESCO. (2017). *Education for sustainable development goals: Learning objectives*. UNESCO.

Armağan Ateşkan is an assistant professor in the Graduate School of Education at Bilkent University. She has an MA in biology teacher education from Bilkent University, and a PhD in computer education and instructional technologies from Middle East Technical University. Dr. Ateskan has been working as an instructor in the Graduate School of Education, Bilkent University, Türkiye, since 2002. In addition, she has worked at a high school in Türkiye as a biology teacher and at an international school in Belgium as an educational technologist. She directed several international and national projects related to IB DP, IB MYP, and environmental education.

Jennie Farber Lane is an associate professor in the Graduate School of Education at Bilkent University. She received her PhD in curriculum and instruction from UW-Madison. Prior to coming to Bilkent, she was the director of the Wisconsin K-12 Energy Education Program (KEEP). Her other work experiences include co-authoring the *Project WET Curriculum and Activity Guide*, which is used throughout the world, teaching public school in New York City and Lewiston, and instructing pre-service teachers in Thailand and at the University of Wisconsin-Stevens Point. Her research areas include environmental, place-based, and sustainability education.

Earth as Self: Healing Our Connection with the Earth Through Education

8

Deniz Dinçel and Birgül Çakır-Yıldırım

Abstract

During the seventeenth century, the Western world witnessed a momentous shift in its perception of the natural world. While earlier societies conceived Earth as a living being, after the scientific revolution, the mechanistic view of nature became a dominant paradigm in science. This shift in our perception of nature promoted anthropocentrism as an ultimate value system, and since then, our planet has been positioned as a reservoir of resources for humanity. This managerial view of the world has also affected environmental education (EE) and education for sustainability (EfS); the role of these disciplines has been increasingly defined to equip students with scientific and technological knowledge to solve problems stemming from the exploitation of resources. This chapter advocates that education with this "resourcist" view of the world is too narrow to solve the ecological crisis we are facing today. We, as humanity, need

another momentous shift to live in harmony with all life forms and within the limits of our planet. To support this shift, as a first step, we can try to take off our anthropocentric glasses by empathizing with all beings in nature. In this chapter, we examine different schools of thought that have influenced the human-nature relationship and how these concepts have affected the EE and EfS programs, and then we provide two activities for educators to build and develop empathy with nature.

Keywords

Anthropocene · Environmental philosophies · Education for sustainability · Empathy with nature

8.1 Introduction

Although humankind has struggled with the limits set by nature since its very existence, the ecological problems have never been as widespread and dangerous as they are today. On the one hand, there is the exponential increase in greenhouse gas emissions; on the other hand, there is enormous pressure exerted through consumerism on the ecological systems. Humans' demand for natural resources has doubled since 1966 (WWF, 2012, 2020). Due to the pressure we are putting on the Earth, the populations of the living species

D. Dinçel (✉)
Istanbul Bilgi University, Department of General Education, Istanbul, Türkiye
e-mail: deniz.dincel@bilgi.edu.tr

B. Çakır-Yıldırım
Ağrı İbrahim Çeçen University, Department of Primary Education, Ağrı, Türkiye
e-mail: bcyildirim@agri.edu.tr

have decreased by 68% in the last 50 years (WWF, 2012, 2020). It takes more than 18 months for the planet to replenish what we consume in a year (WWF, 2012). In other words, we currently use 1.6 planets to sustain human activities. The estimates show that if we continue "business as usual," the equivalent of two planets will be needed in 2030 to meet the annual consumption demands of humans (WWF, 2012). These trends are causing the ecological crisis to deepen and are pushing our planet toward a sixth mass extinction (Kolbert, 2014).

These environmental issues are a worldwide crisis, but some parts of the world are being affected more severely than others. The challenges some disadvantaged countries and cultures have in securing basic needs such as food, water, and health services are exacerbated when their natural resources are destroyed or their climate changes. Every year, 60,000 people, mostly living in developing countries, lose their lives due to the disasters related to climate change, and millions of people are forced to migrate (WHO, 2018). It is expected that 200 million people will have migrated due to climate change by 2050 (Stern et al., 2006). The geologists refer to this period in which humans' influence on the Earth is highest as the Anthropocene, which means "The Human Age" (Crutzen & Stoermer, 2000). The world has now entered a process that is difficult to reverse.

Many efforts have been taken to prevent, mitigate, and ameliorate the disasters caused by environmental crises. Among these is education. Since the 1970s, there has been a rapid increase in the number of environmental education programs, and many different environmental education currents have emerged, such as education about the environment, education for the environment, education in the environment, education for sustainability, and education for sustainable development (Sauve, 2005). The most common practice among these currents is generally "education about the environment" (Dinçel, 2019; Kollmuss & Agyeman, 2002). Various institutions and organizations aim to cultivate environmentally friendly behaviors by sharing practical information about the environment.

Unfortunately, research on pro-environmental behavior shows that early models suggesting a linear relationship between the level of knowledge about environmental problems and environmentally friendly behaviors are not always straightforward (Jensen, 2002; Kollmuss & Agyeman, 2002; Maiteny, 2002). In other words, knowing more about environmental problems does not always lead to environmentally friendly behaviors.

This realization that knowledge alone does not change human behavior raises a crucial debate about how environmental education should be implemented and what its content should be. At the heart of this debate lies a philosophical question about where we position ourselves as human beings in our relationship with nature (Novikau, 2016). This means that environmental issues have ontological and epistemological aspects that need to be considered in addition to just how the issues can be resolved with science and technology. We need to ask fundamental questions such as what we value as human beings, what kind of creatures we are, what kind of life we lead, and what is our place in nature (DesJardins, 2006). Are humans a part of nature, or are they separate from nature and even in some sense superior to it? Is nature a resource to benefit people, or does it have an intrinsic value, independent of its benefits to human beings? The answers to these questions are of immense importance regarding which actions are legitimate in our relationship with nature (Ponting, 2012). However, most of the studies in environmental education mainly focus on the implementation process, while the philosophical and ethical dimensions are relatively ignored.

This chapter's primary purpose is to stimulate a rethinking of the philosophical foundations of the human-nature relationship in EE and EfS and to trigger an inquiry into the potential of EE and EfS to create change and transformation. This rethinking and inquiry begin with an examination of the schools of thought that have influenced the prevailing conceptions of nature. The following section discusses the effects of these conceptions on the environment and EE and EfS programs. The final section shares examples of best practices teachers can do with their students to restore the human-nature relationship.

8.2   The Historical Roots of Prevailing Conceptions of Nature

Jacqueline Russ (2014) begins her book, *L'aventure de La Pensée Européenne [The Adventure of Western Thought]*, by stating that "We are the heirs of two different visions of universes whose seeds were planted thousands of years ago between the Euphrates and Nile" (p. 13). In one vision, human beings, like all other living beings, are a part of nature and subject to its rules. In another vision, humans are superior beings created to dominate nature and transform the planet (Russ, 2014). As it can be understood from the introduction above, we can find the roots of different ideas that dominate the human-nature relationship today in the thoughts and belief systems from thousands of years ago. In antiquity, mind (*logos*) and nature (*phusis*) were two absolutely inseparable elements. Phusis (nature), a Greek term, means a universal power, a growth power immanent to all things, embracing humans and prevails everywhere (Russ, 2014). Nature itself is a divine construct, and all beings, including the gods, arose from it. The gods did not create nature, but rather, they took their power from nature (Gökberk, 2012; Russ, 2014). By the end of antiquity, with the spread of monotheistic religions, a completely different conception of the universe and God emerged. While the gods were part of the physical universe for the Greeks, God is a transcendent being who created reality out of nothing in the monotheistic religions. Thus, God (which is generally conceived masculine) and "His" image on Earth, the human, were considered superior to nature. This "superior" position attributed to people by religion led to the development of a strong anthropocentric worldview in the West and its dominance throughout the Middle Ages (White, 1967).

Medieval Europe underwent a radical change with the Enlightenment Period in the sixteenth and seventeenth centuries and gave birth to a new worldview: the mechanist worldview. The scientific development that lays the foundation for this view is the mathematical physics developed by Galileo and Newton and the philosophical and metaphysical belief systems that Bacon and Descartes built on it (Ünder, 1996). According to Galileo-Newtonian physics, the universe is a mechanical object or machine made up of atoms, consisting of the movements of matter, working like a clock. Number, shape, size, and position are the "primary qualities," which are impossible to separate from objects and expressed in the language of mathematics. Qualities such as taste, smell, color, and sound are subjective, that is, "secondary qualities." While primary qualities exist objectively, secondary qualities are subjective, depending on impressions, feelings, and ideas. Therefore, primary properties can be measured, but secondary properties cannot. A similar distinction is also present in the writings of Descartes. According to Descartes' mind-body dualism, there is matter, which operates according to mechanical principles, occupies space, but does not think, and there is spirit/mind, which does not operate according to mechanical principles, does not occupy any space, but thinks. This thesis of Descartes states that nature is a machine that carries no secrets and can be solved solely by reasoning (Ünder, 1996). Thus, Descartes broke the old animist contract that had dominated the human-nature relationship for centuries and positioned the human as a stranger in nature. The universe is no longer perceived as the work of a divine power. It is a colorless and aimless mass of atoms moving in space (Gökberk, 2012; Ünder, 1996).

The definition of the universe as an enormous machine by Bacon, Descartes, and Newton has profoundly affected humans' relationship with nature in the Western world. With the rising mechanical worldview, nature, which was seen as a goddess in ancient times, was reduced to the sum of mathematical formulas, and the mechanical description of nature became the dominant paradigm in science. Thus, the concept of the ancient Earth, which nourishes and nurtures, has fundamentally changed, and nature has been reconsidered according to the requirements of "utilitarian logic" (O'Sullivan & Taylor, 2004). A new understanding has emerged in order to dominate, control, and rule nature, and the mechanical conception of nature has been a scientific basis for the manipulation and exploitation of Earth (Capra, 2012; DesJardins, 2006; Liftin, 2012; Macy & Brown, 1998; Orr, 1992; Shiva, 2012; Sterling, 2002).

According to O'Sullivan and Taylor (2004), the mechanical worldview not only led to the concept of "instrumental nature" but also caused the emergence of "instrumental value." Ünder (1996) defines "instrumental value" as follows: "Anything that is valuable as an instrument is not considered valuable on its own; it is not a matter of respect in itself. As long as it is useful, it is valuable as a tool. If something is found to take its place, it can be set aside or destroyed" (p. 60). According to the ethical understanding of the mechanistic view, only human beings have a goodness of their own, a value in themselves that is not given, independent of any utility or function for another person or thing. In compliance with this approach, called anthropocentrism, only people are "essentially" valuable; other beings have no intrinsic value: they are tools that people can use for their own good (DesJardins, 2006). This assumption has led to a strong anthropocentric understanding of justice in the West, which leads to the identification of nature with resources and the determination of its value according to the goods it produces for humans. The anthropocentric ethics asks the question of how people will be affected when a decision is to be made about the environment. Environmental rights or wrongs depend on what the consequences are for humans. The view that the human mind is the only source and place of values has led us to reduce the world and natural beings to a tool for our generally materialistic goals, without any social or moral constraints (Garrard, 2016; Plumwood, 2004).

One of the central premises of the Age of Enlightenment was the idea of linear progress or development. In the Age of Enlightenment, inventions in science and technology brought about the idea of the progress of knowledge,

which expresses that humanity is moving toward a better and higher purpose. According to this idea, material progress will be achieved with technology, and the raw materials of this progress will come from nature (Ponting, 2012; Russ, 2014; Şaylan, 2009; Ünder, 1996).

With the advent of capitalism in the sixteenth and seventeenth centuries and the subsequent Industrial Revolution, radical changes occurred in societies' social, economic, and political structures. Capitalism is utterly dissimilar to the prevalent way humans lived and managed resources for millions of years. Hunter-gatherer societies lived with very few possessions, without owning land, and by sharing soil and nutrients with all living things. From the invention of agriculture to the beginning of the nineteenth century, limited trade and transportation facilities required people to live in small, self-contained units (Capra, 2012; Ponting, 2012). However, the Industrial Revolution changed the traditional means of production in European countries and replaced it with batch and mass production. In the late nineteenth century, the readily available oil as a cheap energy source accelerated the spread of capitalism and supported the development focused on unlimited economic growth. Thus, economic growth became the sole target for both developed and developing countries and the most significant indicator of the success of governments (Ponting, 2012). The era of globalization, which started with the invention of the microchip in the 1990s, led to a massive change in the economic production processes and brought along the flexible production organization that allowed the stages of the production chain to disintegrate and spread across the globe. Thus, polluting and destructive technologies banned in industrialized countries are outsourced to the countries where there are no legal restrictions or, even if there are any, they are not applied, while waste removal costs of pollution are inflicted on the host countries (Erdoğan, 2012; Şengül, 2008). These practices have fueled ecological, economic, and social injustice all over the world, accelerated the extinction of natural habitats and species, and created the ecological crisis we face today.

8.3 Education for Sustainability: The Solution or the Problem?

Since the 1970s, when more societies recognized how ecological problems were threatening the future of the entire planet, international partnerships in the field of environmental education also gained momentum. Seminal meetings were held one after another to develop a shared understanding of the aims, objectives, and methods of environmental education. UNESCO organized a workshop on environmental education in Belgrade in 1975. In the Belgrade Charter published at the end of the workshop, the goal of environmental education is stated as follows: "develop a world population that is aware of, and concerned about, the environment and its associated problems, and which has the knowledge, skills, attitudes, motivations, and commitment to work individually and collectively toward solutions of current problems and the prevention of new ones" (UNESCO, 1975, p. 3). Following the Belgrade meeting, UNESCO organized the first "Intergovernmental Conference on Environmental Education" in Tbilisi in 1977 with the participation of delegates from 66 member states. The Tbilisi Declaration published at the end of the conference is accepted as a starting point for environmental education to find a place in the world of education. The ultimate aim of environmental education in the Tbilisi Declaration is also stated as solving environmental problems and preventing environmental destruction (UNESCO, 1977).

Many of the definitions, working objectives, and guiding principles of environmental education in these international documents, like the Belgrade Charter and Tbilisi Conference, which set forth the principles of environmental education and led the development of environmental education curriculum in many countries, are framed by anthropocentric values where Earth is perceived as an object of instrumental value (Berryman & Sauve, 2016; Gough, 1997). There is an implicit assumption that human beings are separate from and superior to the rest of nature (Griffiths & Murray 2017; Bell & Russell, 2000). This is not surprising considering that:

The roots of environmental education are in rational science. This dominant worldview originated in the scientific revolution of the seventeenth century ('the Age of Enlightenment') and is secular, empirical, and mechanistic and characterized by seeing the human species as apart from nature (and thus nature has no intrinsic value). (Gough, 1997, p. 52)

As a reflection of this view, nature is considered as the "source" of sustaining human life and economic growth in most of the environmental education practices, and environmental problems are handled as problems arising from the depletion of natural resources (Berryman & Sauvé, 2016; Hursh et al., 2015; Huckle & Wals, 2015; Jickling & Wals, 2008). Accordingly, mainstream environmental education programs try to solve environmental problems without giving up the modern paradigm and the ideal of unlimited economic growth and generally consider nature as "resources to be managed."

With the neoliberal era that started in the 1980s, there has been a significant change in the definition and scope of environmental education, and it has increasingly begun to be considered within the framework of sustainable development. Notably, in the education section of the "Agenda 21" published at the end of the "Earth Summit - UN Conference on Environment and Development" held in 1992, environmental education was stated as an essential tool of sustainable development, and it was envisaged that children and young people should be trained in the field of environment and sustainable development during their education (UN, 1992). The ideology of sustainable development has dominantly influenced the field of environmental education. So much so that UNESCO has continued its project titled International Environmental Education Programme (IEEP) started in 1975, under the title of "Educating for a Sustainable Future" since 1997. In addition, UNESCO declared 2005–2014 as the Decade of Education for Sustainable Development.

An anthropocentric approach that is dominant in the EE definitions in both the Belgrade Charter and the Tbilisi Declaration did not change in the EfS; on the contrary, it became even more stronger (Berryman & Sauvé, 2016; Hursh et al.,

2015; Huckle & Wals, 2015; Jickling & Wals, 2008). According to ideology of sustainable development:

> The world is positioned as a reservoir of resources for humanity that we need to exploit in a fairer and more sustainable (and sustained) manner. Institutions, including the educational ones in a range of contexts must act accordingly. (Berryman & Sauvé, 2016, p. 106)

Hence, in the context of EfS, the environment is generally considered as a problem to be solved, and EfS is suggested as a part of the solution.

At this point, as educators working in the field of EE and EfS, we should ask ourselves: "What are the central issues of education for sustainability?". "Is the primary purpose of EE and EfS to share objective information about the environment with students, or is it recycling activities?" If that were the case, the ecological crisis we are facing today would have already been overcome. However, despite the EE/EfS practices that have become increasingly widespread in the last 50 years, the ecological crisis is getting deeper each and every day. Of course, it would be wrong to seek the solution to environmental problems only in education, but as educators working in the field of sustainability, it is worthwhile to evaluate our own approaches critically.

Environmental education/education for sustainability is essentially a quest of how human beings should relate to nature and how we should live, and its primary function is to make students critically aware of how they perceive themselves and the world. In this process, we should question where and how we position ourselves as human beings in our relationship with nature, as well as the behavior patterns and lifestyles that cause the ecological crisis. The vast majority of environmental education programs and education for sustainability activities, which have an anthropocentric worldview, turn our relationship with nature into a "monologue" that focuses solely on human's needs and desires. However, as Orr (1992, p. 90) states, "true conversation can occur only if we acknowledge the existence and interests of the other." Therefore, we need to deconstruct the anthropocentric and mechanistic approaches to nature to build another perspective

that is more inclusive, Earth-centric, and open to dialogue with non-human dimensions (Cutter-Mackenzie-Knowles et al., 2020).

So, as educators working with children, how do we support the establishment of genuine dialogue with non-human world? How do we deconstruct our own anthropocentric and mechanistic perspectives? As explained above, the Enlightenment philosophers pushed aside the senses and intuition that humans share with other living beings, exalting the mind and keeping the human being, which they see as the *only being* that possesses mind, separate from and superior to nature. This separation has led to the denial of the fact that life consists of a large network of relationships in which humans also participate. Therefore, the healing of this separation requires both *feeling* ourselves as a part of this network again and learning to think relationally. As educators, one of the most significant steps we can take for children to connect with nature is to open up space for experiences where students can feel themselves as part of nature, even feel as nature itself.

Educational practices for sustainability are becoming increasingly standardized, technical, and context-independent. As a result, "nature has become just another subject, rather than the immanent context for our lives. It is about *them*, separated from us physically, and spiritually, as if nature could be fully understood outside of a connected living context" (Sobel, 1996, p. vi). Therefore, there is a great need for educational practices that will enable the direct experience of nature. We are not beings separate from nature, who pollute or protect it nor who are above it. We are children of nature; we came out of it and are connected with it in every breath we take. All our food comes from nature; our bodies are nourished from the great body of Mother Earth. If we do not provide opportunities for our children to experience that they are "nature," it is unrealistic to expect them to transcend their instrumental view of the non-human world and to feel love, compassion, and responsibility toward nature in the long run. Conservation of nature depends on the essence of the relations of children and young people with nature, how they are connected to nature or not, as well as the institutions working for conservation.

8.4   Different Ways of Relating with Earth

As the discussion above highlights, we need to deconstruct both our and students' anthropocentric and mechanistic perspectives. Toward this aim, as educators, we should provide our students with opportunities to feel as part of nature or feel as nature themselves. To this end, Sobel (1996) advised educators to support their students' feeling of empathy with nature since empathy is the first step in conservation efforts (Sobel, 1996). The studies demonstrated that empathy with nature predicts and leads environmentally friendly behaviors (e.g., Brown et al., 2019; Guergachi et al., 2010; Tam, 2013). In this chapter, we adopted Tam's definition for empathy with nature, that is, "the understanding and sharing of the emotional experience, particularly distress, of the natural world" (Tam, 2013, p. 93).

The two activities we propose in this chapter can support students' ability to develop empathy with nature. One of these activities is based on the activation of imagination, and it is possible to apply it as an indoor activity. The other activity is based on the observation of a specific area over 3 or 4 months and is proposed as an outdoor activity. Both of these activities are aimed at encouraging students to think about their relationship with nature by improving empathy with their surroundings. We believe these experiences will help children connect with nature.

8.4.1   Connecting with Nature Through the Activation of Imagination

Children are exposed to many global environmental issues through media and school. For instance, they hear about the deforestation in rainforests, extinction of species, and melting of the ice caps in the Arctic (Louv, 2005). In the summer of 2021, news of fire which destroyed forests in Europe, the USA, and Africa due to climate change was released. In 2019, a massive wildfire was experienced in Australia, too. Therefore, children are exposed to global news and events that are devastating and shocking. While children are witnessing these global issues, they are generally not aware of the environmental problems in their local area; for this reason, they develop environmental concerns before making a connection with nature in their own neighborhood (Sobel, 1996).

Likewise, Sobel (1996) claimed that before saving endangered species, children should empathize with animals in their neighborhood first. Through this empathy, they can discover the local issues around them. The method suggested by Sobel (1996) is activating imagination, which is one of the best-known methods for developing empathy with animals (Young et al., 2018).

To activate students' imagination, storytelling can be used (Blizard & Schuster, 2007). Stories are personal narratives, and writing a story is a compelling way to share feelings. Therefore, using storytelling in EE and EfS is an excellent way to encourage students to connect with nature (Russell, 2020). For instance, in a climate change context, rather than *guiding* students thinking about polar bears or penguins, which do not live in their neighborhood, they could be asked to write a story on the effect of the climate crisis on endemic species in their village or city. When educators do not have a chance to take students outside, imagination supported by storytelling and role-playing can be effectively used for developing ecocentric worldview (Lithoxoidou et al., 2017). For instance, in her study, Lithoxoidou (2006) used the imagination method by storytelling and role-playing to develop preschool students' empathy for animals. Unlike trying to imagine organisms in remote places, students can actually see and hear the plants and animals that are in stories in their local environments. Therefore, place-based environmental education is an important part of storytelling. Building these connections and relationships can help students understand and empathize with organisms living in their neighborhood. We suggest educators to ask their students to think about how nature is affected by climate change. After discussion, students are encouraged to think of an animal or a plant in their neighborhood. Then, they are asked to write a story on climate change by putting themselves into the animal or plant that they initially selected. In this story, they would feel how the other being experiences climate change and they feel about it. By activating imagination through storytelling and role-playing, we can assist students in expressing their feelings about environmental issues as well as developing empathy for other beings.

8.4.2 Connecting with Nature Through Empathy

Outdoor education practices enable learners to interact with and explore nature. Being in nature in childhood is the basis for building long-term bonds with nature (Chawla & Flanders Cushing, 2007). Today, unfortunately, children's interaction with nature is quite limited, and they are deprived of the benefits of being in nature. When children regularly visit a particular natural area, they broaden their experiences, and this participation helps them develop a sense of place and feel like being a part of this place (Davis & Waite, 2005; Harris, 2017; Murray & O'Brien, 2005; O'Brien & Murray, 2007). It is also known that when children spend time in nature, their well-being, social skills, creativity, and cognitive skills are supported and improved (Louv, 2005; Murray & O'Brien, 2005). Hence, providing outdoor experiences for children is crucial.

Haskell's study (2012) demonstrated that regular visits to a particular natural area strengthen our relationship with nature. In his study, David George Haskell drew a circle with approximately 1 m in diameter in the woods, which he called a mandala. He visited the mandala regularly for 1 year and published his observations in his book, *The Forest Unseen* (Haskell, 2012), in which he states:

> I believe that the forest's ecological stories are all present in a mandala-sized area. Indeed, the truth of the forest may be more clearly and vividly revealed by the contemplation of a small area than

it could be by donning ten-league boots, covering a continent but uncovering little. (p. xii)

Inspired by Haskell (2012), we suggest an outdoor activity, which aims to support primary and middle school students to explore a specific place and their feelings for that place and eventually to develop empathy with nature. In this activity, students select a small area (preferably a circle with 1 m in diameter) in a natural area close to their school or house. They regularly (once or twice a week) visit this place for at least 3 months and take notes. Questions in Table 8.1 can guide students about their observations, and if they wish, they can also add new questions.

While questions from 1 to 5 focus on students' awareness of the mandala and guide them to make a scientific observation, questions from 6 to 8 focus on their feelings about that place. Question 9 focuses on the empathy with non-human nature in the mandala, and this question aims students to develop empathy for non-human beings. Throughout the visits, it is expected that student's empathy would be developed. In addi-

Table 8.1 Examples of questions students can answer during the regular visits to their mandalas

Date:
Time:
Weather:
1. Which species (tree, mushroom, weed, insects, birds, etc.) did you observe today in the mandala?
2. Describe the species in detail Describe the leaf and stem of the tree If you see any animals in the mandala, describe their behavior
3. What do you observe on the ground (the color of ground, the smell, etc.)?
4. Is there any voice you hear around? Describe them
5. Do you have any other observations?
6. How do you feel today while making your observation?
7. Is there anything happened today which made you excited or sad?
8. At the end of your visit, focus on nature, and write your feelings about this place
9. If you were an animal or plant living here, what would be your feelings? Explain them
10. Draw a picture of today reflecting your observations

tion to the questions in Table 8.1, the questions below can also be asked at the end of the visit to help students explore their own sense of place in this activity.

- How do you feel about this place?
- How important is this place for you now? Can you make a comparison between the beginning of this activity and the last day of the activity?
- Throughout the activity, you explored the species in your mandala. Make an investigation if these species have a reflection in your culture. They may have a symbolic meaning, or there may be stories related to them. You can conduct interviews with your parents, grandparents, or people in your community about the local species and discover their memories with them.

At the end of this activity, students can prepare a poster or presentation with their own photos of the mandala to share their learning journey with their classmates and the rest of the school community. In this way, students listen to their peers and explore similarities and differences among mandalas. Furthermore, they would have a chance to share their feelings for the place. As a next step, the instructor can lead a discussion on how to protect and regenerate local habitats with her/his students and make an action plan with them.

8.5   Conclusion

As the discussion in the introduction shows, environmental crisis around the world is getting worse day by day. Urgent action is needed to change this course of events. Everyone has shared responsibilities in this process, from governments to the private sector, from international institutions to non-governmental organizations. As it is seen in the last 3 years, the youth are at the forefront of the groups that deeply feel this need for change. All over the world, children and young people are hitting the streets and protesting to solve the climate crisis and live in a more

just world. That is precisely why schools and teachers are at the heart of this transformation. Thus, as educators working in the field of EfS, we should critically review how the education programs we provide conceive the human-nature relationship and the messages we pass explicitly or implicitly on this subject and re-evaluate whether these messages support the relationship we desire to establish with nature. First and foremost, we should adopt inclusive, Earth-centric perspectives in our educational activities to promote sustainability. One way of doing it is to empathize with the different life forms, to realize the bonds and unity between us, and to find a simpler way of living to leave smaller footprint on Earth, sometimes by putting ourselves in the place of a fox whose forest is regularly shrinking, or a seahorse whose waters are getting acidic, or a heron whose wetland is lost to a construction site.

References

Bell, C. A., & Russell, L. C. (2000). Beyond human beyond words: Anthropocentrism, critical pedagogy, and the poststructuralist turn. *Canadian Journal of Education, 25*(3), 188–203. https://ezp.sub.su.se/login?url=https://www.jstor.org/stable/1585953

Berryman, T., & Sauvé, L. (2016). Ruling relationships in sustainable development and education for sustainable development. *The Journal of Environmental Education, 47*(2), 104–117. https://doi.org/10.1080/00958964.2015.1092934

Blizard, C. R., & Schuster, R. M. (2007). Fostering children's connections to natural places through cultural and natural history storytelling. *Children Youth and Environments, 17*(4), 171–206. https://www.jstor.org/stable/10.7721/chilyoutenvi.17.4.0171#metadata_info_tab_contents

Brown, K., Adger, W. N., Devine-Wright, P., Anderies, J. M., Barr, S., Bousquet, F., ... Quinn, T. (2019). Empathy, place and identity interactions for sustainability. *Global Environmental Change, 56*, 11–17. https://doi.org/10.1016/j.gloenvcha.2019.03.003

Capra, F. (2012). *Batı düşüncesinde dönüm noktası* [The turning point: Science, society and the rising culture] (M. Armağan, Çev.). İnsan Yayınları (1982).

Chawla, L., & Flanders Cushing, D. (2007). Education for strategic environmental behavior. *Environmental Education Research, 13*(4), 437–452. https://doi.org/10.1080/13504620701581539

Crutzen, P. J., & Stoermer, E. F. (2000). The anthropocene. *Global Change Newsletter, 41*, 17–18.

Cutter-Mackenzie-Knowles, A., Brown, L. S., Osborn, M., Blom, M. S., Brown, A., & Wijesinghe, T. (2020). Staying-with the traces: Mapping-making posthuman and indigenist philosophy in environmental education research. *Australian Journal of Environmental Education, 36*, 105–128. https://doi.org/10.1017/aee.2020.31

Davis, B., & Waite, S. (2005). *Forest schools: An evaluation of the opportunities and challenges in early years final report*. University of Plymouth.

DesJardins, J. R. (2006). *Çevre etiği: Çevre felsefesine giriş* [Environmental ethics: An introduction to environmental philosophy] (R. Keleş, Çev.). İmge Yayıncılık (1992).

Dinçel, D. (2019). *Environmental ideologies and environmental education programs of non-governmental organizations in Turkey* (Unpublished doctoral dissertation). Ankara University, Turkey.

Erdoğan, N. (2012). Sancılı Dil, Hadım Edilen Kendilik ve Aşınan Karakter: Beyaz Yakalı İşsizliğe Dair Notlar. In T. Bora, A. Bora, N. Erdoğan, & İ. Üstün (Eds.), *Boşuna Mı Okuduk?* (pp. 75–115). Üçüncü Baskı. İletişim Yayıncılık.

Garrard, G. (2016). *Ekoeleştiri: Ekoloji ve çevre üzerine kültürel tartışmalar* [Ecocriticism] (E. Genç, Çev.). Kolektif Kitap. (2004).

Gökberk, M. (2012). *Felsefe tarihi* [History of philosophy] (23.Baskı). Remzi Kitabevi.

Gough, A. (1997). Founders of environmental education: Narratives of the Australian education movement. *Environmental Education Research, 3*(1), 43–57. https://doi.org/10.1080/1350462970030104

Griffiths, M., & Murray, R. (2017). Love and social justice in learning for sustainability. *Ethics and Education, 12*(1), 39–50. https://doi.org/10.1080/17449642.2016.1272177

Guergachi, A., Ngenyama, O., Magness, V., & Hakim, J. (2010, July). Empathy: A unifying approach to address the dilemma of "environment versus economy". In *Paper presented at International congress on environmental modelling and software*.

Harris, F. (2017). The nature of learning at forest school: practitioners' perspectives. *Education 3-13, 45*(2), 272–291. https://doi.org/10.1080/03004279.2015.1078833

Haskell, D. G. (2012). *The Forest unseen: A year's watch in nature*. Penguin Press.

Huckle, J., & Wals, A. E. J. (2015). The UN decade of education for sustainable development: Business as usual in the end. *Environmental Education Research, 21*(3), 491–505. https://doi.org/10.1080/13504622.2015.1011084

Hursh, D., Henderson, J., & Greenwood, D. (2015). Environmental education in a neoliberal climate. *Environmental Education Research, 21*(3), 299–318. https://doi.org/10.1080/13504622.2015.1018141

Jensen, B. B. (2002). Knowledge, action and pro-environmental behavior. *Environmental Education Research, 8*(3), 325–334. https://doi.org/10.1080/13504620220145474

Jickling, B., & Wals, E. J. A. (2008). Globalization and environmental education: Looking beyond sustainable development. *Journal of Curriculum Studies, 40*(1), 1–21. https://doi.org/10.1080/00220270701684667

Kolbert, E. (2014). *The sixth extinction: An unnatural history*. Henry Holt and Company.

Kollmuss, A., & Agyeman, J. (2002). Mind the gap: Why do people act environmentally and what are the barriers to pro-environmental behavior? *Environmental Education Research, 8*(3), 239–260. https://doi.org/10.1080/13504620220145401

Liftin, K. (2012). Thinking like a planet: Gaian politics and the transformation of the world food system. In P. Dauvergne (Ed.), *Handbook of global environmental politics* (pp. 419–430). Edward Elgar.

Lithoxoidou, L. (2006). *The contribution of an environmental education programme in attitudes and values cultivation at pre-school age* (Unpublished doctoral dissertation). Aristotle University of Thessaloniki.

Lithoxoidou, L. S., Georgopoulos, A. D., Dimitriou, A. T., & Xenitidou, S. C. (2017). Trees have a soul too! Developing empathy and environmental values in early childhood. *International Journal of Early Childhood Environmental Education, 5*(1), 68–88.

Louv, R. (2005). *The last child in the woods: Saving our children from nature deficit order*. Algonquin Books.

Macy, J., & Brown, Y. M. (1998). *Coming back to life: Practices to reconnect our lives and our world*. New Society Publishers.

Maiteny, P. T. (2002). Mind in the gap: Summary of research exploring 'inner' influences on pro-sustainability learning and behavior. *Environmental Education Research, 8*(3), 299–306. https://doi.org/10.1080/13504620220145447

Murray, R., & O'Brien, E. (2005). *Such enthusiasm – A joy to see: An evaluation of Forest School in England*. Report to the Forestry Commission.

Novikau, A. (2016). The evolution of the concept of environmental discourses: Is environmental ideologies a useful concept? *Western Political Science Association 2016 Annual Meeting Paper*. https://ssrn.com/abstract=2754835. Accessed 10 July 2021.

O'Brien, L., & Murray, R. (2007). Forest School and its impacts on young children: Case studies in Britain. *Urban Forestry & Urban Greening, 6*(4), 249–265. https://doi.org/10.1016/j.ufug.2007.03.006

O'Sullivan, E., & Taylor, M. (2004). Glimpses of an ecological consciousness. In E. O'Sullivan & M. Taylor (Eds.), *Learning toward an ecological consciousness* (pp. 5–23). Palgrave Macmillan.

Orr, D. (1992). *Ecological literacy: Education and the transition to postmodern world*. State University of New York.

Plumwood, V. (2004). *Feminizm ve doğaya hükmetmek.* [Feminism and the mastery of nature] (B. Ertür, Çev.). Metis Yayınları (1993).

Ponting, C. (2012). *Dünyanın yeşil tarihi: Çevre ve büyük uygarlıkların çöküşü [A new green history of the world: The environment and the collapse of great civilizations]* (A. Başçı, Çev.). Sabancı Üniversitesi Yayınları (1991).

Russ, J. (2014). *Avrupa Düşüncesinin Serüveni: Antik Çağlardan Günümüze Batı Düşüncesi [L'aventure de la pensée européenne: Une histoire des idées occidentales]*. (Ö. Doğan, Çev.). Doğu-Batı Yayınları (1998).

Russell, J. (2020). Telling better stories: Toward critical, place-based, and multispecies narrative pedagogies in hunting and fishing cultures. *The Journal of Environmental Education, 51*(3), 232–245. https://doi.org/10.1080/00958964.2019.1641064

Sauvé, L. (2005). Currents in environmental education – mapping a complex and evolving pedagogical field, *Canadian Journal of Environmental Education, 10*, 11–37. https://cjee.lakeheadu.ca/article/view/175. Accessed 10 July 2021.

Şaylan, G. (2009). *Postmodernizm*. (4. Baskı). Ankara: İmge Yayıncılık.

Şengül, M. (2008). Türkiye'de Kamu Yönteminde Neoliberal Dönüşümün Çevresel Sonuçları. *Memleket Siyaset Yönetim, 3*(6), 67–87. http://www.msydergi.com/uploads/dergi/153.pdf. Accessed 7 May 2021.

Shiva, V. (2012). *İyilerin Yanında: Çiftçi Haklarına Adanmış Bir Yaşam [Dalle Parte Degli Ultimi]* (Ç. Ekiz, Çev.). Sinek Sekiz Yayınevi (2007).

Sobel, D. (1996). *Beyond Ecophobia: Reclaiming the hearth in nature education*. The Orion Society.

Sterling, S. (2002). A baker's dozen-towards changing our 'loaf'. *The Trumpeter, 18*(1). http://trumpeter.athabascau.ca/index.php/trumpet/article/view/121/130. Accessed 5 May 2021.

Stern, N., Peters, S., Bakhshi, V., Bowen, A., Cameron, C., Catovsky, S., Crane, D., Cruickshank, S., Dietz, S., Edmonson, N., Garbett, S.-L., Hamid, L., Hoffman, G., Ingram, D., Jones, B., Patmore, N., Radcliffe, H., Sathiyarajah, R., Stock, M., ... Zenghelis, D. (2006). *Stern review: The economics of climate change*. HM Treasury.

Tam, K. P. (2013). Dispositional empathy with nature. *Journal of Environmental Psychology, 35*, 92–104. https://doi.org/10.1016/j.jenvp.2013.05.004

UN. (1992). Agenda 21. *Chapter 36*. http://www.un-documents.net/a21-36.htm. Accessed 12 July 2021.

Ünder, H. (1996). *Çevre felsefesi: Etik ve metafizik görüşler* [Environmental philosophy: Ethical and metaphysical views]. Doruk Yayıncılık.

UNESCO. (1975, October 13–22). *The international workshop on environmental education*, Belgrade, Yugoslavia, Final Report. https://unesdoc.unesco.org/images/0001/000177/017772eb.pdf. Accessed 12 July 2021.

UNESCO. (1977, October 14–26). *First intergovernmental conference environmental education, Tbilisi*, Final Report. http://www.gdrc.org/uem/ee/tbilisi.html. Accessed 12 May 2021.

White, L. (1967). The historical roots of our ecological crises. *Science, 155*, 1203–1207.

WHO. (2018). *Climate change and health*. http://www.who.int/mediacentre/factsheets/fs266/en/. Accessed 24 July 2021.

WWF. (2012). *Living planet report 2012: Biodiversity, biocapacity and better choices*. WWF.

WWF. (2020). *Living planet report 2020: Bending the curve of biodiversity loss*. WWF.

Young, A., Khalil, K. A., & Wharton, J. (2018). Empathy for animals: A review of the existing literature. *Curator: The Museum Journal, 61*(2), 327–343. https://doi.org/10.1111/cura.12257

Deniz Dinçel is an independent researcher in environmental education in Türkiye. She holds a Bachelor of Science degree in biological sciences and completed her PhD in environmental education at Ankara University. She has been actively working in the field of environmental education since 2007. Currently Dr. Dinçel gives courses about ecological literacy and sustainability in different universities in Türkiye and also, develops and implements education programs about climate change, sustainability and ecology for teachers and students of all ages.

Birgül Çakır-Yıldırım is Assistant Professor of Primary Education at Agri Ibrahim Cecen University, Türkiye. Dr. Çakır Yıldırım received her PhD from Middle East Technical University, Türkiye, in 2017. She was a visiting scholar at Florida Institute of Technology, USA, in 2016. Her research interests include environmental education, teacher education, and education for sustainability. Dr. Çakır-Yıldırım is currently working on "Change the Story" project, which is funded by the Erasmus + Programme of the European Union.

Nature-Inspired Learning: How Nature Can Teach Us to Be Sustainable?

9

Richard Dawson

Abstract

If humanity is to achieve sustainability, it needs to find a mentor, something with the experience and practice to guide humanity. Nature is that mentor. Nature has evolved over millennia through evolution into a sustainable system using self-organising principles. If humanity can understand and apply these principles wisely, the lofty and perhaps unattainable concepts of education for sustainable development can be dispensed with; all we need to do is live embedded within nature. What this means and how it can be brought about is challenging and requires educators to question the definition of who we are and how we perceive the world, to challenge the mental models through which we understand ourselves. This chapter offers approaches to meet this challenge in the form of understanding the human mind and how we make sense of the world and offers two case studies which have attempted to use nature as model, mentor and measure for a sustainable future.

Keywords

Nature-embedded learning · Biomimicry · Sustainability · Aspiration

R. Dawson (✉)
Wild Awake, Shrewsbury, UK
e-mail: richard@wild-awake.org

9.1 Finding Our Way

In this chapter, we explore how (re)connecting with nature is essential for understanding our dependence on the natural world for survival. In doing so, we also investigate how developing a practice of direct experience within nature is one essential route for understanding ourselves and our perception of our 'self' and how a misperception of 'self' is a root cause of the environmental crisis.

We also explore how (re)connecting with nature is not enough but only a start. We need to reaggregate our experience of nature in such a way that we learn from nature and act 'as nature would'. In this regard, two case studies illustrate approaches which learn from nature. Both offer a more hopeful discourse to address our sustainability crisis, a discourse that addresses the SDGs (Sustainable Development Goals) and at the same time transcends them with far more compelling visions of a sustainable society.

Firstly, we explore how humanity arrived into this state of crisis from a psychological standpoint, how we manufactured a world detached from nature which allowed us to attempt to destroy it. Secondly, we see how the way we view the world is a kind of delusion, a creation of our mind. This is important if we, as educators, are to address the challenges we face, because in doing so we need to address the root causes which lie in the deeply embedded mental metaphors that govern the way we think and act. Thirdly, two case studies are presented of education projects which have attempted to transcend traditional narratives with ESD (education for sustainable development). Neither claim to be perfect, but they do offer valuable lessons for others to stand upon. They are offered as shoulders for those brave enough to take the next step. And finally, conclusions are drawn for the role of education in saving humanity from a crisis of its own making.

9.2 Let's Look Outside: Some Ideas About Nature-Inspired Learning

It's the end of October now and a bit chilly. But let's have a look outside. Have you noticed that the leaves have turned brown and starting to fall down? Did you see the squirrel collect nuts from the beech tree? It makes you wonder. How did the tree grow those nuts and how does the tree know it's time to shed the leaves? The more you look, the more questions start coming. Nature has evolved over 3.8 billion years into model of sustainability. Nature recycles waste efficiently, uses renewable power from the sun, is resilient to sudden changes, is adaptable over time to new conditions and self-regulates through feedback. What if we could use the operating principles found within nature to rethink how we live as humans? To flourish without damaging the natural ecosystems we depend upon for our survival? Nature does not become entangled in a conceptual debate about what is or is not sustainability; it applies the same principles evolved over billions of years into a system that works.

Nature-inspired learning takes us on a journey to discover the principles which makes nature a model of sustainability. It offers an opportunity to explore how these principles can help tackle some of the greatest challenges facing humanity today such as climate change and increasing levels of waste and pollution. It empowers students to apply their new competences to create with real solutions that work. Nature-inspired learning is a uniquely valuable pedagogical teaching practice because of its dramatic potential to engage

students' interests and generate excitement (Stier, 2021) and for its ability to cut through the dogma attached to sustainability debates towards a practical sustainability in action. And it offers the potential to liberate students in their thinking and doing; it does not set limitations on the imagination of the learner.

Let us look at some brief descriptions of nature-inspired approaches to education (Stier, 2021):

- An art teacher exploring shading has students find something living or once-living around the schoolyard to sketch, focus on a detail of it and sketch it at different times of day.
- A teacher exploring the scientific method has students observe natural phenomena outside the classroom over a period of time, preparing questions about features students notice about nature and what functions these features might serve (e.g. Why do squirrels have big, bushy tails? Why tree branches often crooked? What purpose do our toes serve?). Students then choose one question about which to design an experiment and test a hypothesis about a feature's possible functional role.
- A physics class learning about atomic interactions reads research papers about how geckos scale smooth glass using van der Waals forces.
- Students exploring climate change solutions in an afterschool chemistry club make carbon-negative cement out of car exhaust, based on a chemical process corals use to build their stony reefs.
- Students in a maker lab create prototypes of car tailpipes that remove outgoing pollutants, whose design is based on the students' research into how marine sponges filter food out of seawater and other biological strategies for filtering.
- A teacher exploring the material science and structural engineering concepts of stress and strain has students examine a tree in the schoolyard for clues as to how it withstands the passing breeze, despite its massive canopy.
- A teacher exploring the mathematical ideas of volume and mass has students look up from

their desks, textbooks and chalk/white/smart boards and look out the window or go outside to determine how to weigh a cloud passing over the school.

These examples share some common patterns. First, in each instance, a standard academic idea is being explored (hypothesis testing, volume, mass, stress, strain, electrostatic charge, innovation, etc.). That is to say, nature-inspired education does not require new academic content be added to the curricula. Nature-inspired education is not a subject area in-and-of-itself. It does not add to the burden of existing curricula by adding on more content to the school day. It is a pedagogical approach. Instead of adding to the curricular content, existing academic ideas are explored in less abstracted and more meaningful contexts that help engage students' interest and build their contact with and appreciation for the subject under study, as well as with the natural world. In nature-inspired education, already-existing academic ideas can be addressed without lengthening the school day. Standard school topics are simply explored conceptually in connection with the natural world (BioLearn, 2021).

A second thing to notice is that a wide variety of academic ideas and subjects can be approached through nature-inspired education. Many of the academic subjects, such as art, science, math, chemistry, physics and engineering, have been developed historically as a direct result of humankind's observation of natural phenomena. From the night sky and Galileo's ideas of mass and motion to human anatomy and Leonardo da Vinci's artistic realism, the natural world has driven human thought, understanding and creativity. It is thus relatively easy, and completely logical, to teach these subjects and connect their respective academic concepts to the natural world that inspired the development of these subjects in the first place.

A third thing to notice is that the academic ideas and subjects are explored in connection with the engaging context of the natural world. A (re)connecting with the natural world and humanities embedded place within it becomes a critical learning outcome (Baumeister, 2014). Direct

contact with the natural world is also used to explore topics, both in the schoolyard and by leveraging students' experiences to/from school and around their homes. Note that this does not require students go on expensive and rare field trips to natural or semi-natural areas. If direct contact with nature is not possible, it does not even require that students leave the classroom. All academic ideas and subjects explored through a nature-inspired approach can be addressed within the four walls of a classroom (especially with the rich image, video and sound media available to teachers today) (Stier, 2021).

Before moving to exploring practical ways nature-inspired learning has been implemented, we need to understand why it is essential for the survival of humanity, and we need to understand the root causes of how we arrived in our current sustainability crisis.

9.3 How We Forgot About Nature: Our Divided Nature

ESD includes within its remit a reconnection with the natural world, an experience in nature. But is this sufficient to support the major claim of ESD to educate for a sustainable world (e.g. UNESCO (2014) Decade of ESD claim to create a more sustainable future) – to address the SDGs such as sustainable cities and communities (SDG11), responsible consumption and production (SDG12) and affordable and clean energy (SDG7)? An education that does not provide viable and desirable answers to such questions will remain a side line, a sticking plaster at best (Dawson, 2021).

Before exploring the role and purpose of nature-inspired learning, it is pertinent to ask why it is necessary. We need to understand how we arrived into the sustainability crisis and what are the deep-seated cognitive and psychological patterns which continue to perpetuate the crisis now in full sight.

Ask many tribes rooted in the natural world, and they will tell you they have no word for 'nature'. For them, it is not something 'out there' but deeply rooted in their way of being. Why then have the (mainly Western) tribes of the twentieth century become so divorced from nature with increasingly disastrous consequences? (Dawson, 2021) A hint perhaps comes from the subject-object duality of science. Heisenberg in Physics and Philosophy writes 'The Cartesian partition has penetrated deeply into the human mind during the three centuries following Descartes, and it will take a long time for it to be replaced by a really different attitude towards the problem of reality' (Heisenberg, 2000). The Cartesian world-view has become a core metaphor through which we make sense of the world, a sense of how humans fit in. A view that has been reinforced through popular TV documentaries such as Jacob Bronowski's The Ascent of Man (1973) in which he asserts that Man is 'not a figure in the land-scape; he is the shape of the landscape'. In more recent years, this duality has been challenged, for example, by the Santiago Theory of Cognition (Maturana & Varela, 1980), stating that cognition and matter arise as an interplay between phenomena, with neither existing in isolation. This process called 'reciprocal specification' (Weber, 2002) is an act of mutual engendering: 'the world is not an aggregation of things, but rather a symphony of relationships between many participants that are altered by the interaction: a necessarily erotic occurrence' (Weber, 2017). Despite advances in cognitive and quantum science, the dominant mental metaphor is of humans separate from and dominant over nature.

Modern tribes have evolved from a species awed and cowed by nature, and living in balance with their surroundings, to ones who are increasingly dominating and dramatically changing nature. There has been a cognitive shift in the way we perceive the natural world, from being a part of the whole to a disconnected self. We have moved from tribes of nomadic hunter-gatherers in tune and cooperating with the world around them into a species of settled groups largely divorced from a direct sense of connection with nature (Heyes, 2012; Pinker, 2010; Whiten & Erdal, 2012). If you like, we have moved from a time when the principles organising our lives arose from our embedded existence in nature to one where it is replaced with our embedded exis-

tence in the non-natural world of GDP and social status. The core mental metaphors we use to make sense of the world have shifted (Lakoff & Johnson, 2003; Lent, 2017). Clearly, the modern world has provided huge benefits, but are the downsides starting to out-weigh them? Our very conception of self and nature has shifted the way we make meaning of ourselves in relation to the world around us.

To dig deeper into understanding this, we need a model of how we construct reality. One helpful model is presented by Buddhism in the form of the Five Aggregates (Analayo, 2004; Burbea, 2014; Khandha Sutta translated by Thanissaro, 2013); one reason it is useful is that we can test it out for ourselves. The Five Aggregates demonstrate that based on having a body, we interact with the world through our senses (touch, taste, smell, hearing and seeing i.e. the aggregates). The senses create the raw data of our experience. Our consciousness maintains attention with our senses and interprets the data, perceiving what is pleasant or otherwise. Based on this, we create tendencies or habits, certain preferences for what we like and dislike, who we want to spend time with and who we want to avoid. Through this process, we make sense of the world around us. We use concepts to create shared understanding and words to name them. And, as we spin through the model many times each minute, we start to create the patterns and habits that define our self. We become the sort of person who habitually likes (craves) certain experiences and habitually dislikes others. We become a pattern of recurring habits. And of course, there is nothing inherently wrong with this.

From a Buddhist perspective, the model of the Five Aggregates is used through meditation to gain direct insight into the interdependence of all phenomena; nothing exists except in dependence on something else, that an experience of 'something' as solid and substantially existing independent of the conditions which created it is an illusion (or based on ignorance to use the Buddhist terminology). Any sense that the world exists as separate from the self is merely a fabrication of the mind (Burbea, 2015). The Buddhist perspective offers a helpful model, but there is ample secular evidence to support this. Goethe writes 'Nature! We are surrounded by her and embraced by her: powerless to separate ourselves from her, and powerless to penetrate beyond her' (Huxley, 1869).

However, as we have explored above, our patterning and understanding of the world has moved further and further away from our direct experience of it. As our world has become more complex, we have become ever more deeply embedded in the world of concepts. Moving into a world of abstraction where a social media presence has more importance than the presence of a flowing stream. As a result, we no longer 'see' the foundations on which our concepts are built – the natural world and its life supporting services (see Senge et al., 2005). In this way, we have become psychologically disconnected from nature and in doing so eroded these very foundations through actions which act outside of 'natural laws'. No longer connected with our reliance on nature, humanity has freed itself from the necessity to conserve nature resulting in our current planetary crisis.

Scharmer and Kaufer (2013) view this as a crisis of our outdated frames of thought, where nature, work and capital are seen as commodities which we are free to take, sell, use, dump and replace at will. And with the privatisation of commodities, this has led to an ecological divide (overshoot), social divide (inequity and poverty) and spiritual-cultural divide (depression). They argue for a shift from an egosystem to an ecosystem awareness, one where humans are deeply embedded within natural systems, a transformation in the quality of attention from 'I and me' to 'us and all'. A deep (re)engagement with nature has to be an essential component for this shift.

It is this psychological disconnection from the natural world, encouraged through scientific developments such as Cartesian reason, which has led to a dramatic shift in the cognitive metaphors we use to make sense of the world, from metaphors of 'nature as giving parent' (see, e.g. Crow-Apsáalooke nation claims that 'The sky is my Father and these mountains are my Mother') and 'reverent guests of nature' (see Tao Te Ching, 1989) to 'dominion over nature' and 'truly to command the world' (see, e.g. Drucker (1985) 'Human possession and use is what activates the true nobility of any natural object').

These metaphors must matter to educators. If we aim to contribute to a sustainable society for all, engaging in, questioning and changing these dominant and highly destructive metaphors should be a core task in all education work that wants to make a lasting difference.

9.3.1 Our Undivided Nature

Seeking to (re)connect with nature requires, as suggested above, a more detailed map of our psychological disconnection: a map that can provide us with 'routes' to see clearly the web of concepts we weave. Before going further, we need to be clear that concepts are helpful and they enable us to live as human societies, work together and survive. The challenge, however, is that we often reify concepts to the point where we give them independent and solid status. We over-estimate their ability to provide an ultimate truth.

A practice of (re)connecting with nature can provide the start of a shift from a disconnected to a (re)connected self. Study after study has demonstrated that those who choose to spend time in nature feel more connected to something apparently outside of themselves – something bigger, more transcendent and universal (Foster, 2012; Howell et al., 2011; Josipovic et al., 2012). There is a rich and growing list of nature books in which authors describe their contact with and as a part of nature. These range from old Chinese poets:

The birds have vanished into deep skies.
A last cloud drifts away, all idleness.
Inexhaustible, this mountain and I
gaze at each other, it alone remaining.
(Li Po, trans. Hinton, 2007)

…to contemporary writers:

A reverence for life is a reverence for wildness. A reverence for life beyond your control. Something you don't dominate. That is the native habitat of new ideas. Of real humanity – to expose yourself to things beyond your control. And just to ride out the consequences. That is what I seek and want to protect. Elements that are beyond our control.

(Peacock, quoted from MacIver, R. ed (2006, p. 26)

So, simply to look on anything, such as a mountain, with the love that penetrates to its essence, is to widen the domain of being in the vastness of non-being. (Shepherd, 1996, page x)

Fundamental reality is the creative wilderness in which everything interpenetrates, transforms, pushes itself into life, and carries death along with it. (Weber, 2017 page 96)

But is this sufficient? Writers generally describe an experience of connection when in the natural world, and many environmental education programmes aim to achieve the same. But people go back to their towns and cities, and we know when this happens their habits generally revert to type. How can we reaggregate our experiences of nature in a way which makes sense of our daily desires and needs? This is the subject of two case studies explored below.

9.4 Putting Nature-Inspired Learning into Practice

Nature is an integral part of our daily lives; indeed, we depend on it for survival and we are nature. Nature is a source of pleasure and enjoyment and underpins our entire economic model. Whereas nature connection has a strong role to play in (re)connecting ourselves back within nature, it must do more than learn about nature or learn in nature; we must learn from nature. Observed closely, nature offers up key principles as to how nature works to ensure a sustainable system which adapts and flourishes over time. These same principles applied to human endeavours could provide a blueprint for human sustainability embedded within nature.

As outlined above, Scharmer and Kaufer (2013) call for a shift in attention from egosystem to an ecosystem awareness. To change the dominant, and highly destructive, metaphors of our age necessitates a transformation in the quality of attention through which we view the world. To not only see nature through new eyes but to also learn from nature. This comes from both a deep

(re)engagement with nature and applying natural principles to all areas of life. It does not mean a reversion to some past time when humans 'lived off the land', rather viewing nature's principles as the design guide for creating vibrant and flourishing human societies.

The two case studies below come from pan-European projects which attempted to use natural principles to rethink how human society can be 're-integrated' within nature. They both aimed to create learning which truly educated for a sustainable planet. Each attempted to transcend the utilitarian aims of the SDGs towards a more universal way of being through which such functional goals are rendered unnecessary.

In both case studies, there was a deliberate intention to avoid the common framing around 'problem-solving' environmental issues, towards a framing which focuses on igniting a sense of hope and curiosity. A mindset based only on problem-solving places limits on creativity, whereas igniting 'fires which cannot be put out' opens up to a possibility of anything. It becomes about finding solutions based on the aspiration to live the lives we want, rather than giving up things we value based on morally questionable goals.

9.4.1 Case Study 1: Lessons from Nature – Using Insights from Nature to Inspire and Build a Brighter Future

Lessons from Nature was a European-funded project (2009–2013) involving education partners from five countries (Lessons from Nature, 2012). At the time of its inception, the concept of learning from nature within a modern Western education context was largely untried and untested, although it was firmly grounded in industrial ecological thinking (McDonough & Braungart, 2003; Pauli, 2010). The project was posited on the premise that:

We live in an unsustainable world. Pressure on natural resources to drive the European economy is increasing year on year. We cannot continue to meet the needs and aspirations of Europe without significant changes to the way we live and consume. There is a need to promote learning that will change the way we design our economies, businesses and products. LfN (Lessons from nature) takes a unique approach to addressing this need. Nature is inherently sustainable: it recycles waste efficiently, uses renewable power from the sun, is resilient to sudden changes, adaptable over time to new conditions, and self regulates through feedback. A truly green economy and society will share the same characteristics. Such approaches are being pioneered by businesses e.g. Interface, Nike, Wal-Mart, evidencing a real need for this approach to learning. (written by the author)

Lessons from Nature encouraged young people (aged 12–16 years) to reflect on what the future might look like and what models can provide hope for a bright and prosperous future. The Lessons from Nature (LfN) learning programme explored these questions and how nature can be a mentor. It revealed nature as a source of information and inspiration for creative thinking and redesign. Nature does all these things far better, far more effectively and far more beautifully than humans do. By learning the Lessons from Nature, we can redesign our societies to be more like nature, to be abundant, beautiful and free from pollutants.

9.4.1.1 A Different Approach to Learning

An example will serve to highlight the different approach that Lessons from Nature took. A traditional freshwater ecology lesson sees students in a local stream catching macroinvertebrates, measuring temperature and dissolved oxygen and plotting data into their notebooks. By the end of their time at the stream, they have come to understand the parameters of a healthy stream. And they return home, moving on to the next part of their curriculum. For some, perhaps only a few, the experience will have moved them to think more deeply about the natural world, but for many it is just a day out and another 'thing' they need to learn.

But what are we really learning from nature? Names of plants and animals…the names were created by humans. Relationships in a freshwater ecosystem? Learning we almost certainly never use. What if nature could teach us about health

care, running economies on limited resources, providing homes that self-cool and heat? Because nature does all these things far better, far more effectively and far more beautifully than humans do. Now there is a lesson worth going into a stream for.

Learners need to make sense of their environmental experiences within the context of their own lives. The way Lessons from Nature (LfN) worked was simple: we wanted young people to be inspired and excited about their future, the possibilities it holds and the role they will play in it. LfN expressed the belief that the future can and should be bright for all young people. This does not mean the future will be the same as today; change is the only constant. Communities will need to learn how to redesign themselves to combat serious environmental and social issues, and business will need to learn how to develop economics in a natural resource-constrained world. LfN aimed to provide the tools for a hopeful future.

LfN was not mechanistic. It did not 'preach' to young people, telling them what they can and cannot do. Rather, it attempted to apply the principles of nature to whatever choices young people wished to take. Want a fast car? How can nature help us design such a car? It acknowledged that we all have our own dreams for the future and of course as we grow, they change and develop. How can we keep those dreams alive? The answer, according to LfN, was by learning and applying the same principles that nature has used for over 3.8 billion years to design human systems that are abundant, beautiful and resilient. Four key principles were drawn from the fields of ecology and circular design (Pauli, 2010; McDonough & Braungart, 2003):

- *Waste Equals Food*: In nature, everything is cycled so what looks like waste is actually food for the next cycle. For example, dead tree leaves decompose to become food for insects. This insight can be applied to turn current linear human production systems into closed-loop systems in which waste is eliminated.
- *Multiple Benefits*: In nature, organisms have multiple benefits; they do not simply have one

purpose. The goal of a tree is to reproduce to provide the next generation, and in doing so, it also provides food for insects, shelter for animals and nutrients for the soil from their decomposing leaves, turns carbon dioxide into oxygen and helps regulate temperature and rainfall.

- *Run on Solar Income*: Nature runs on renewable energy; it does not use more energy than it can produce itself. Nature does not create energy sources that pollute the atmosphere and designs its processes to work efficiently.
- *Diversity Gives Strength*: Nature relies on a large variety of species, systems and organisms that allow it to withstand external shocks. Diversification effectively reduces risk.

Returning to the stream ecology field trip we started with, how could we weave in the principles of nature? Students could be tasked with finding out how waste is dealt with in a stream; they would discover that detritivores consume decomposing plant and animal parts, reducing the risk of eutrophication and contributing to the nitrogen cycle. *Waste equals food.*

Students could be asked what happens in a healthy stream if one species, say mayfly nymphs, dies out. They would discover that other species move it to fill the niche and perform a similar function. But in an unhealthy stream, if the mayfly nymphs die out, then their function is not replaced, and the whole system suffers (fish would lose a food source). *Diversity equals strength.*

How does the stream sit within the wider ecosystem? Perhaps it provides water, limits flooding, creates a variety of habitats for other species, is an attractive place for a picnic and is a place to de-stress. In other words, it does not provide a single function but has *multiple benefits.*

And how is the stream 'powered'? What is the primary energy source? How is this energy converted into a form animals (including humans) can use? It is, of course, *run on solar income.*

In this way, a traditional stream ecology trip introduces key principles about how nature works. Now take these four principles and apply them to a town, perhaps the one you live in now.

Can we build factories in which waste from one provides the raw materials for another…a circular economy? How can we encourage a diversity of thinking to ensure our politicians get the best advice possible before taking decisions, for example, through participatory budgeting? Are we able to create town centre spaces that serve a multiple of functions and benefit a wide range of people?

If we can learn the lessons of nature, then we can start to redesign human spaces and systems to function in the same way. And we know that nature has done this sustainably for billions of years – nature can be our teacher.

9.4.1.2 An Inquiry Process

ESD often starts by stating or asking the student to understand the problem. This can lead to a sense of disempowerment – it is just about solving problems rather than seeing opportunity. LfN reframed this by asking what sort of future young people would like to live in and how this preferred future can be made reality. This challenges traditional assumptions about how things are made, economies managed and lives lived. Learning can move from individuals feeling guilty or doing less harm and delaying a point of crisis towards rethinking the future with their aspirations to the fore. This presents a more hopeful and realistic way for young people to achieve the sort of future they want. After all, it will be their future.

Within LfN, a four-step inquiry model emerged: inspiring the learner, helping them discover how nature works, understanding how nature's principles can be applied to human systems and finally applying these new insights to their own lives (see Fig. 9.1). This model was uniquely created by the project partners, but other inquiry models support the efficacy of the approach (e.g. Morgan et al., 2007).

The model presents opportunities to learn through first-hand experience helping to inspire discovery and foster real understanding in relation to insights from nature that can be applied to the modern world. It encourages students to develop a wide range of competencies for the jobs of the future. These include creativity, criti-

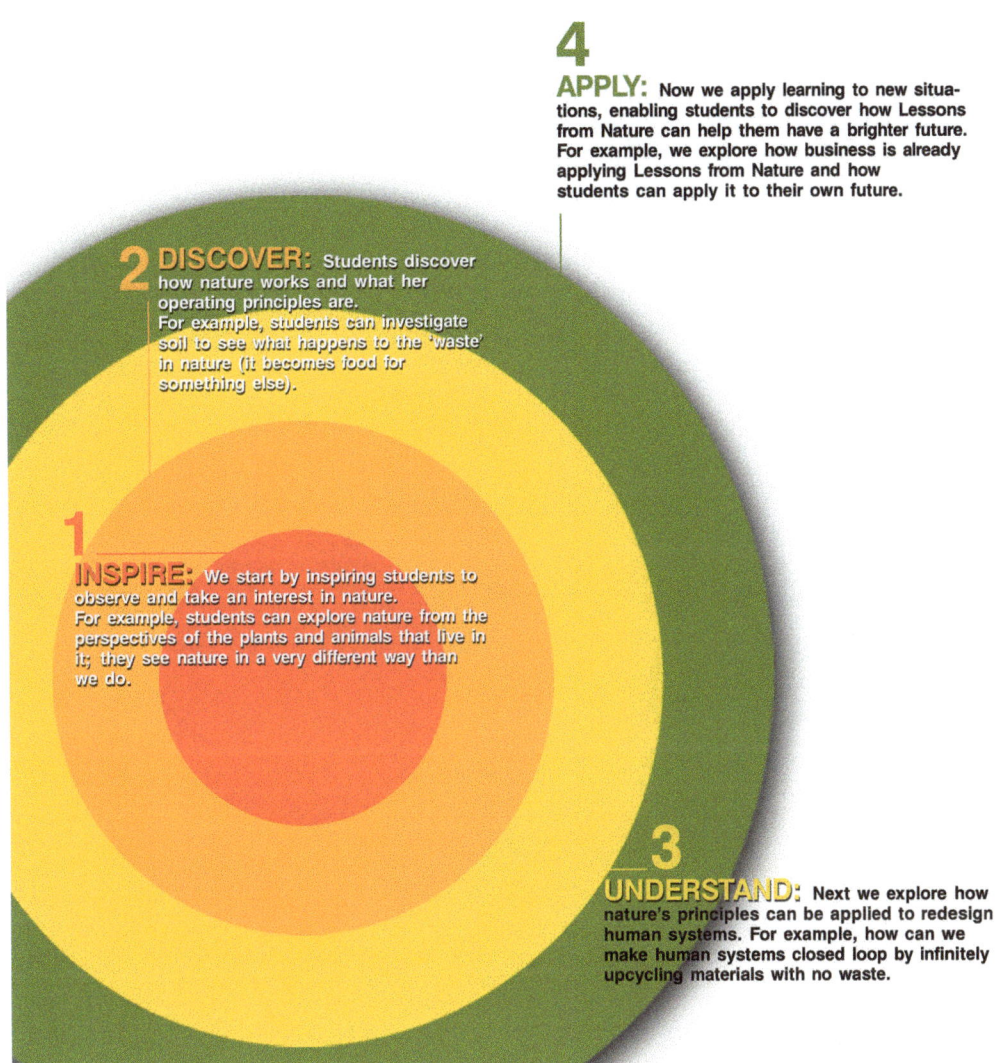

4

APPLY: Now we apply learning to new situations, enabling students to discover how Lessons from Nature can help them have a brighter future. For example, we explore how business is already applying Lessons from Nature and how students can apply it to their own future.

2 DISCOVER: Students discover how nature works and what her operating principles are. For example, students can investigate soil to see what happens to the 'waste' in nature (it becomes food for something else).

1 INSPIRE: We start by inspiring students to observe and take an interest in nature. For example, students can explore nature from the perspectives of the plants and animals that live in it; they see nature in a very different way than we do.

3 UNDERSTAND: Next we explore how nature's principles can be applied to redesign human systems. For example, how can we make human systems closed loop by infinitely upcycling materials with no waste.

Fig. 9.1 Lessons from nature four-step inquiry model. (Source: Lessons from Nature, 2012)

cal thinking, evaluation and reflection, sharing and entrepreneurship.

This simple model can be applied again and again and again. By learning how nature works, we can find solutions that continue today, tomorrow and always. These solutions are not limited to the few; nature's solutions can provide opportunities for all regardless of wealth or location.

9.4.1.3 Fancy a Coffee?

As I write this, I am drinking my favourite cup of coffee. Think for a few minutes, and we will see

that my cup of coffee has very few benefits. Sure, it is a nice drink for me, but what about all the waste produced in processing the coffee? In fact, only 0.2% of the coffee tree reaches my cup. Surely, can we do better? If we learn from nature, then yes.

The 99.8% that is traditionally seen as waste can be used. The coffee pulp can be used as a low-grade fertiliser or used as a substrate to grow mushrooms. It can be fed to goats (for milk or meat) which themselves produce very-high-grade fertiliser or even used to produce thread for producing t-shirts. You can add the coffee plant cuttings to a bio-digester to produce bio-gas. So, what looks like a single-use product can provide multiple benefits for the farmer by applying the principle of waste equals food. The next time coffee prices fall, the farmer has several sources of alternative income and can provide all his/her cooking gas for free. A great economic and ecological model (Fig. 9.2).

Since 2013, LfN has continued to inspire projects across Europe and beyond. The external evaluation confirmed (Wageningen University and Research Centre, 2013):

> Many scholars argue that the complexity of sustainable development requires a system transition to radically change our way of designing economy and society. Participation in this transition requires new competencies. The modules and learning framework from Lessons from Nature offer students a new perspective to current sustainability challenges that stimulates them to explore new solutions for design questions, inspired by nature. The evaluation shows that after their participation in Lessons from Nature, students feel that we as humans can and should learn from nature.

Fig. 9.2 A new story for coffee. (Source: Lessons from Nature, 2012)

Furthermore, the students and teachers are inspired by the opportunities looking at nature offers for concrete design challenges.

Lessons from Nature is a complex concept. Lessons from Nature helps students and teachers to explore new approaches to sustainability challenges and develop competencies to address the complexity of sustainable development.

9.4.2 Case Study 2: Biomimicry – Nature as Design

BioLearn was a 3-year European-funded project which introduced biomimicry as a pedagogical approach to learning from nature. It created learning resources for students aged 12–16 and piloted them in schools in five European countries. The case study below refers only to the project worked carried out in the United Kingdom (BioLearn, 2021).

Biomimicry (meaning 'to copy life') takes us on a journey to discover the principles which makes nature a model for sustainability (see Textbox 9.1, Benyus 1997): a model which achieves dynamic balance, sustains the whole and provides the conditions for survival. Biomimicry offers a design process to apply these principles to address human challenge: to seek solutions which are sustainable and enable humans to thrive within the natural systems we are dependent upon.

Biomimicry has been defined as 'learning from and then emulating natural forms, processes and ecosystems to create more sustainable designs' (Biomimicry 3.8, 2016). It does this based on three intertwined values (Dawson & Winks, 2020):

1. Emulate – observe nature closely, and we can see how organisms use a vast array of strategies to provision their needs; these strategies are being emulated by many companies in product design. It occurs when we tackle human problems through the inspiration of nature and minimise our impact on the Earth.

2. Reconnect – learning from nature requires deep curiosity and observation, reconnecting to the natural world at a level beyond mere utilitarianism. It is about regaining the recog-

---

Textbox 9.1 Biomimicry Principles

Nature runs on sunlight.
Nature uses only the energy it needs.
Nature fits form to function.
Nature recycles everything.
Nature rewards cooperation.
Nature banks on diversity.
Nature demands local expertise.
Nature curbs excesses from within.
Nature taps the power of limits.
  After Janine Benyus

Textbox 9.2 Key Biomimicry Terms

Function: In biomimicry, a function refers to an organism's adaptation which helps it survive and thrive. For example, the purpose of bear fur is to keep warm; in technical terms, its function is to conserve heat (insulation). Often, 'designs' in nature have more than one function. A leaf can photosynthesise (convert energy from the sun into sugar), and it can distribute water (through its veins). Human products also have functions; a kettle has the functions to both contain water and heat water (modify its physical state). In brief, a function is 'what it does'. (BioLearn, 2021)

Strategy: Organisms meet functional needs through biological strategies. This is a characteristic, mechanism or process which performs the function for them. In the bear example, fur is the strategy for delivering insulation. In a kettle, electrical energy is transferred into physical heat which modifies the temperature of water. In brief, a strategy is 'how it does it'. (BioLearn, 2021)

nition that we are a part of nature and the relationship between humans and nature is essential for our survival.

3. Ethos – understanding how we are an integral part of nature, and how nature brings about balance and harmony, we choose how to apply biomimicry thinking with an ethos of care and empathy with all life. It signals the intention to proceed only in ways which work alongside natural balance.

When these values are taken together, biomimicry offers a different way of seeing nature and supports a shift in view from learning about nature to learning from nature. It explicitly places the natural world as a source of solutions to human challenge, providing a moral and practical reason for the conservation of the natural world. It facilitates a deeper looking at the natural world, through which a sense of sacredness can emerge (Dawson & Winks, 2020).

As the three values briefly laid out above suggest, biomimicry is not simply taking ideas from nature to create better products to serve human needs. It necessitates a deep observation of how nature works, the inter-relationships between organisms and their environment and an innate sensing of nature's cycles. In this way, biomimicry offers a method to rethink our relationship with nature, rediscovering our place within the natural world, finding balance, harmony and renewal. In this sense, biomimicry can be far more than a neat engineering solution; it can be used to engage learners deeply with the natural world. In this way, biomimicry can itself become a 'natural pedagogy' rather than simply a tool (Dawson & Winks, 2021). As an approach based on natural principles, it avoids the often abstract ideas and concepts of ESD. It retains an elegant simplicity relying on direct experience and observation of nature as key tools. Biomimicry asks what challenge do you want to solve and invites learners to look into nature for the solutions. Nature is the source of learning and innovation.

Biomimicry offers multiple points into the natural world. It can be entered from a deep ecology point of wholeness and connectivity, broadening out into seeing how nature works and applying nature's principles to address human needs. Or it can be entered from the perspective of an engineering challenge, exploring nature to see how organisms have solved similar challenges, or even stripped back to look at the properties of materials and structures and how they can be applied in a variety of situations. In this sense, biomimicry can appeal to a wide range of people regardless of their current views of nature (Dawson & Winks, 2020).

As a result, biomimicry can help us see beyond the usual events and patterns of our daily lives. It can suspend our usual way of seeing nature and offers a new mental model to reshape our relationship as a part of nature. Biomimicry provides inspiration to go beyond simply copying nature and presents learners with opportunities to enrich and broaden their learning beyond facts and into a new relationship with the natural world.

9.5 Variation and Classification: A Different Approach to Traditional Learning

The following sections take a narrative journey through rethinking learning with a biomimicry approach. They are intended to inspire educators to reflect on how biomimicry can enrich learning rather than be taken as 'finished' learning resources.

Take, for example, a traditional unit in biology on Variation and Classification (see Table 9.1). This might start with students identifying the differences between living and non-living things (Movement, Respiration, Sensitivity, Growth, Reproduction, Excretion, Nutrition – MRS GREN). It could then move on to understanding how different species vary and that these variations are inherited or caused by environmental factors; a dichotomous key might then be used to sort organisms into groups based on certain structures or behaviours. How would a biomimicry approach deliver this?

Any tree in the school grounds can provide a focus for using dichotomous keys, identifying different trees or species of invertebrates found on trees. However, thinking beyond simple iden-

Table 9.1 A biomimicry approach within curriculum specification

Specification	School resources	Biomimicry extensions
Identify differences between living and non-living things (*M*ovement, *R*espiration, *S*ensitivity, *G*rowth, *R*eproduction, *E*xcretion, *N*utrition – MRS GREN) Understand that these variations may be inherited or caused by environmental differences Observe that members of a species have features in common Use keys to sort organisms into groups by common features Appreciate that there are different ways of classifying living things	Compare the characteristics of living things – go outside if fine Key resource: Identifying organisms DVD Life on Earth	Build on use of dichotomous key with trees – consider answers to questions: what characteristics provide strength/energy/transport of water, etc.? *How do trees inspire designers?* How do trees MRS GREN? How do non-living things MRS GREN? Can non-living things mimic MRS GREN? Provide a set of images of inventions/designs inspired by trees. Ask students what inspired each of these designs?

tification, an apparently instrumental approach to learning about trees can yield a host of questions driven by the curiosity of students. For example, while thinking about classification, students might consider:

- What characteristics give this tree strength?
- What characteristics provide protection?
- What characteristics provide energy?
- How do these characteristics provide functions which enable the tree to thrive?
- How can we use features to identify and classify the tree?

Take this a step further by focusing on one function of trees: provide strength, transport liquids and harness energy. Ask how the tree delivers these functions and what strategies does it use? Consider how these strategies might be applied within human design and what opportunities are there to learn from trees?

9.5.1 Biomimicry Within Design

Within Design and Technology subjects, students might be required to demonstrate the necessary knowledge, understanding and skills required to undertake iterative design processes and to place this in a real-world context.

Provide students with a sycamore or maple seed (or any winged seed). Ask them to spend 5 min looking at their seed and analysing its movement and structure in as much detail as possible using these points as a guide:

- Examine the seed structure in detail.
- Throw it in the air (sensibly!).
- Look at how it flies – what allows it to move like this?
- As you are doing this, try and think about what the seed has these features.

Now set a mini-design challenge:

Mini-design Challenge: We are now going to use the seed to help use solve a design challenge. The purpose of this is to help you to consider how looking to nature's designs can help to prompt us to think and see differently – applying the ingenuity of the seed to a human design problem!

How can the design and function of the seed be copied to help people who are at risk of flooding?

The purpose of this mini-design challenge is to get students to consider how looking to nature's designs can help to prompt us to think and see differently – applying the ingenuity of the seed to a human design problem. Students draw their design and label it with functions as observed in the previous activity making use of the nine principles of biomimicry already given out to assess their design. This approach might lead students to ideas about generating energy, similar to an Archimedes screw, or perhaps ways to slow down the flow of water to minimise flooding risk.

9.5.1.1 A Natural Source of Innovation

'Nature is full of solutions looks for problems to solve'. This quote by Christopher Viney (2000) from Heriot-Watt University sums up the biomimicry view of nature. Trees offer wonderful lessons in building structures which provide strength using minimal materials. Also, trees grow by capturing carbon from the air through photosynthesis rather than mining minerals from the ground as we do. Companies such as Solidia are exploring how making cement can actually remove carbon from the atmosphere rather than release it. Can we start to envision buildings as net absorbers rather than producers of carbon?

Taking a study of trees further, we can observe the characteristic curvature of the base of older trees. Why might the tree have grown in this way? In strong winds, the forces generated are channelled to the base where tensile stress is greatest. This particular curvature is very effective in reducing the concentration of stress, and you will find in at other points in a tree where there are points of join. Throughout its life, a tree adjusts its growth to better distribute mechanical stress (we see this when we look at a cross section of felled trees).

Another mechanism some tree species use is by interlocking root systems with different trees. This has been found to reduce tree damage during hurricanes, and architects are now exploring whether buildings can be made hurricane resistant using a series of interlocking 'roots' between buildings.

Taken to the next step, we might then consider how a tree might be able to inform smarter and more efficient human design – perhaps such asymmetrical growth might inform the way buildings are designed – to increase strength on the side of the prevailing wind while reducing the need for materials on the other. Another related point of learning might come from watching trees

sway in the breeze. Is the strength we associate with long-lived and tall specimens related to their immovability or their flexibility?

Compare how a tree addresses the challenge of tensile stress compared with human-built products. Figure 9.3 illustrates how right angles concentrate stress and lead to weakness and fracture. This lesson was tragically learnt on ships with portholes designed in a square pattern. These designs created weak points which weakened the ship's integral strength. In high-stress environments such as airplane windows, you will now only find cut-out corners. Trees already learn this lesson as can be seen in figure 9.4 – there are no right angles (Figs. 9.3 and 9.4).

A final example involves photosynthesis which both extracts carbon from the air for growth and converts solar to chemical energy (D'Augustino, 2015). The function of chloroplasts is to capture photons of light, and this is being mimicked using dyes, called dye-sensitised solar cells (Stier, 2014). Because they use dyes, they do not rely on purified silicon and can be made from recycled plastic. Additionally, leaves have tiny wrinkles and folds which allow for maximum light absorption (Kim et al., 2012). Typically, solar panels are flat which means they absorb little of the longest wavelengths. It turns out that the wrinkles and folds in a leaf absorb these far more effectively. Dye-sensitive solar cells made from recycled plastic could both remove silicon and enable surface structure to

Fig. 9.3 'Cracked pavement' by Erin Mallinson is licensed under CC BY-NC-SA 2.0

Fig. 9.4 'File:Keeler Oak Tree – distance photo, May 2013.jpg' by Msact at English Wikipedia is licensed under CC BY-SA 3.0

mimic leaves, leading to a potential 600% increase in light absorption.

By attending closely to nature, we are able to learn the forms, patterns and processes nature applies to be sustainable. We are able to discover a chain of insight and imagination which begins with identification and ends with creative application of natural principles to solve human challenges. The above examples illustrate the ways in which learning about nature can also become learning from nature and furthermore can offer real-world opportunities to enhance problem-solving. While on the face of it biomimicry is inherently knowledge based, it also offers important avenues for creative exploration of the natural world. Taking the step from considering the function of natural phenomena to examining the strategies nature takes to achieve such functions deepens student knowledge and indeed enhances wonder and appreciation for nature. Yet, one of the most important benefits of a biomimicry approach to learning is the application of these strategies into complex human challenges. For this to occur, the learning has to take a creative direction, and students must be able to play with their ideas – adapting, changing and learning to accept failure as an aspect of evolution and improvement rather than an end in itself.

From this point of view, biomimicry has much to offer – from deepening knowledge of the natural world to divergent thinking and problem-solving rooted in exploration and creativity.

9.6 Concluding Thoughts: Nature-Inspired Integral Education

Much has been written about education for sustainable development, and much has been achieved in the name of education for sustainable development. It is clear, however, that the result of education for sustainable development is not a sustainable planet. Perhaps education for sustainable development is the problem? Attempts to codify and rationalise education for sustainable development has resulted in more documents, meetings and textbooks. Perhaps ESD needs to be a sub-sect of nature-inspired learning, rather than the other way round. ESD has become another concept created by humanity to explain a reality it does not fully and cannot fully understand – the complexities of planet Earth. Only with deep attention to how nature (the Earth) functions, and its principles, can humanity hope to live sustainably. Only with the humility to accept we are not in charge can humanity learn how to survive and flourish. Without this deep appreciation and sense of service towards nature, sustainability will remain just a policy maker's dream. We do not need ESD; we need nature – we are nature. ESD only have value insofar as it supports humanity to live within nature's boundaries.

The ideas and examples outlined above offer some starting points, ways in which nature-inspired learning can become a part of schooling. They offer a way in, connecting deeply through our senses directly to the world around us, moving away from concepts towards experience. To seeing nature as a source of inspiration for human learning and innovation. This is a connection with the world as it really is, as it really operates and as we are really integrated within it. And it provides meaning to why we need to educate and innovate as a part of nature rather than separate from it. It also offers a key link between how nature works and the aspirations of modern Western societies (not forgetting that the majority of the world also needs more to reach an acceptable standard of living). Biomimicry, in particular, is demonstrating that business can successfully innovate and thrive based on natural principles and that innovation and enterprise can deliver a better future (Smith et al., 2015).

Nature can teach us to live sustainably on planet Earth. Will we choose to listen?

References

Analayo. (2004). *Satipatthana: The direct path to realization*. Windhorse Publications.

Baumeister, D. (2014). Biomimicry resources handbook: A seed bank of best practices. *Montana: Biomimicry, 3*, 8.

Benyus, J. (1997). *Biomimicry-innovation inspired by nature*. Harper Collins Publishers.

BioLearn. (2021). *BioLearn inspired by nature*. https://biolearn.eu/united-kingdom/. Accessed 10 May 2021

Biomimicry 30.8. (2016). *What is biomimicry?* https://biomimicry.net/what-is-biomimicry/. Accessed 12 May 2021

Bronowski, J. (1973). *The ascent of man*. TV series created for BBC in a 13-part series. www.bbc.com/historyofthebbc/anniversaries/may/the-ascent-of-man. Accessed 10 May 2021.

Burbea, R. (2014). *Seeing that frees: Meditations on emptiness and dependent arising*. Hermes Amara Publications.

Burbea, R. (2015). The Buddha and the sacred earth. *Resurgence and Ecologist Magazine, 288*, 34–37.

D'Augustino, A. (2015). *Where do trees get their mass from?* Michigan State University. www.canr.msu.edu/news/where_do_trees_get_their_mass_from. Accessed 12 May 2021.

Dawson, R. (2021). *Our divided nature*. Nature United blog. www.natureunited.org.uk/blog/divided-nature/. Accessed 22 July 2022.

Dawson, R., & Winks, L. (2020). *Exploring biomimicry*. National association for environmental education blog. www.naee.org.uk/exploring-biomimicry/. Accessed 9 May 2021.

Dawson, R., & Winks, L. (2021). Biomimicry – A nature-based approach to designing sustainable futures. *School Science Review, 102*(381), 43–47.

Drucker, P. (1985). *Innovation and entrepreneurship*. Routledge.

Foster, I. (2012). *Wilderness, a spiritual antidote to the everyday: A phenomenology of spiritual experiences on the Boundary Waters Canoe Area Wilderness*. University of Montana. https://www.etd.lib.umt.edu/these/available/etd-06262012-124555/unrestricted/Foster.pdf. Accessed 15 May 2021.

Heisenberg, W. (2000). *Physics and philosophy: The revolution in modern science*. Penguin Books Ltd.

Heyes, C. (2012). New thinking: The evolution of human cognition. *Philosophical Transactions of the Royal Society B: Biological Sciences, 36*, 2091–2096.

Hinton, D. (2007). *Mountain home: The wilderness poetry of ancient China*. Anvil Press Poetry Ltd.

Howell, A. J., Dopko, R. L., Passmore, H., & Buro, K. (2011). Nature connectedness: Associations with well-being and mindfulness. *Personality and Individual Differences, 15*(2), 166–171.

Huxley, T. H. (1869, November 4). Goethe: Aphorisms on nature. *Nature, 1*(1), 9–11. www.nature.com/nature/about/first/aphorisms.html. Accessed 11 May 2021.

Josipovic, I., Dinstein, J., Weber, J., & Heeger, D. J. (2012). Influence of meditation on the anti-correlated networks in the brain. *Frontiers in Human Neuroscience, 5*, 183. https://doi.org/10.3389/fnhum.2011.00183

Khandha Sutta: Aggregates (SN 22.48), translated from the Pali by Thanissaro Bhikkhu. (2013, November 30). *Access to Insight (BCBS Edition)*. http://www.access-toinsight.org/tipitaka/sn/sn22/sn22.048.than.html. Accessed 26 May 2021.

Kim, J., Kim, P., Pegard, N., Oh, S., Kagan, C., Fleischer, J., Stone, H., & Loo, Y. (2012, May). Wrinkles and deep folds as photonic structures in photovoltaics. Published in *Nature Photonics, 6*, 327–332.

Lakoff, G., & Johnson, M. (2003). *Metaphors we live by*. University of Chicago.

Lent, J. (2017). *The patterning instinct*. Prometheus Books.

Lessons from Nature. (2012). https://www.lessonsfromnature.org. Accessed 10 May 2021.

MacIver, R. (2006). *Meditations on nature, meditations on silence*. North Atlantic Books.

Maturana, R., & Varela, F. (1980). *Autopoiesis and cognition. The realization of the living*. Reidel.

McDonough, W., & Braungart, M. (2003). *Cradle to cradle remaking the way we make things*. North Point Press.

Morgan, J., Williamson, B., Lee, T., & Facer, K. (2007). *Enquiring minds*. Futurelab.

Pauli, G. (2010). *The blue economy: 10 years, 100 innovations, 100 million jobs*. Paradigm Publications.

Pinker, S. (2010). The cognitive niche: Coevolution of intelligence, sociality and language. *Proceedings of the National Academy of Sciences, 107*, 8993–8999.

Scharmer, O., & Kaufer, K. (2013). *Leading from the emerging future: From ego-system to eco-system economics*. Berrett-Koelher Publishers Inc.

Senge, P., Scharmer, O., Jaworski, J., & Flowers, B. (2005). *Presence: Exploring profound change in people, organisations and society*. Nicholas Brealey Publishing.

Shepherd, N. (1996). *The Grampian Quartet*. Canongate Books Ltd..

Smith, C., Bernett, A., Hanson, E., & Garvin, A. (2015). *Tapping into nature*. Terrapin Bright Green LLC.

Stier, S. (2014). *Engineering design inspired by nature. The center for learning with nature*. www.learning-withnature.org. Accessed 25 May 2021.

Stier, S. (2021). *Introduction to nature-based design*. BioLearn. www.biolearn.eu/wp-content/uploads/2019/10/background-article-.pdf. Accessed 10 July 2021.

The Tao Te Ching: A new translation with commentary, trans. Ellen M Chen. (1989). Paragon House.

UNESCO (2014). *Shaping the future we want. UN decade education for sustainable development (2005-2014) Final Report*. Paris: UNESCO.

Viney, C quoted in Snail slime 'could mend bones.' (2000, August 19). *BBC new article by Robert Aitken*. http://news.bbc.co.uk/1/hi/health/900869.stm. Accessed 25 May 2021.

Wageningen University and Research Centre (2013). Lessons from Nature. Final Evaluation Report. http://www.lessonsfromnature.org/images/project_files/External%20evaluation%20report.pdf Accessed 10 Sept. 2022.

Weber, A. (2002). The 'Surplus of Meaning'. Biosemiotic aspects in Francisco J. Varela's philosophy of cognition. *Cybernetics & Human Knowing, 9*(2), 11–29.

Weber, A. (2017). *Matter & desire: An erotic ecology*. Chelsea Green Publishing.

Whiten, A., & Erdal, D. (2012). The human socio-cognitive niche and its evolutionary origins. *Philosophical Transactions of the Royal Society: Biological Sciences, 367*, 2119–2129.

Richard Dawson has developed a reputation for bringing fresh insights to education and learning over 25 years. His focus is on helping teachers, schools, and organizations to improve the quality of their learning, create learning for a sustainable future, and enhance their capacity to deliver learning effectively with lasting benefits. He has worked in over 30 countries with NGOs, government, business, and civil society organizations. He is the director of Wild Awake, a not-for-profit social enterprise whose purpose is to develop and provide learning which inspires change towards a more sustainable planet, and support people to live healthy and happy lives which respect natural limits.

Inspirational Outdoor Education: Learning for Sustainability with the Educational Polygon for Self-Sufficiency in Dole

10

Ana Vovk, Janja Lužnik, and Danijel Davidović

Abstract

The Educational Polygon for Self-Sufficiency in Dole (In Slovene, the expression polygon refers to an area intended for learning or training. In this case, the visitors learn about self-sufficiency and see it in practice. The full name of the place in Slovene is Educational Polygon for Self-Sufficiency in Dole "Učni poligon za samooskrbo Dole"; we shortly refer it as "Polygon Dole.") has been established as an outdoor learning environment in the municipality of Poljčane, 40 km south of Maribor. In 2010, we began implementing outdoor education, so we needed an open learning environment where both young people and adults could gain experience in integrated self-sufficiency. Integrated self-sufficiency includes ecosystem food production, water collection and reuse, circular bio-waste management, renewable energy sources, wild insect care, and sustainable construction. Polygon Dole demonstrates the advantages of this type of education, which results in increased interest in self-sufficiency; greater awareness of natural laws; care for natural resources like energy, water, and biodiversity; and for personal development. Common goals include personal growth and building a community integrated with place consciousness and environmental goals. Outdoor environments are extremely important for young people who do not yet have life experience, to distinguish theoretical knowledge from practical knowledge and to connect the importance of multiple knowledge types for life. In this chapter, we provide insight into the activities that young people perform to strengthen thinking, develop manual skills, delve into ecosystem laws, and understand life on Earth. The emphasis is on holistic self-sufficiency, both material and spiritual, which develops a person into a vital, mentally connected, and responsible citizen, which is a great need in today's society.

Keywords

Polygon Dole · Holistic self-sufficiency · Outdoor education environment

A. Vovk (✉) · D. Davidović
Department of Geography, International Center for Ecoremediation, Faculty of Arts, University of Maribor, Maribor, Slovenia
e-mail: ana.vovk@um.si

J. Lužnik
Project Office, Faculty of Arts, University of Maribor, Maribor, Slovenia

10.1 Introduction

Outdoor learning in natural areas can be an enrichment for children, enabling them to learn beyond the borders of their classroom, and has the potential to strengthen primary schools' educational practice directly and indirectly (Rickinson et al., 2004; Blair, 2009; Wistoft, 2013; Goodall, 2016). For effective education about and comprehension of sustainability, it is necessary to develop potential in students by motivating them, arousing curiosity to gain new knowledge, and enabling them to enter the educational process on their own. The academic literature suggests that early childhood nature experiences are particularly important for developing affective connections with the natural environment (Ernst & Theimer, 2011; Kahn & Kellert, 2002; Raudsepp, 2005; Wells & Lekies, 2006). Based on decades of practice, we find that natural environments, classrooms in nature, or outdoor learning environments are extremely suitable for an effective and meaningful education (Vovk, 2015, 2019). Moreover, teachers find that they have lack of time for in-depth work in nature; they often focus on data, numbers, and facts and less on the process of acquiring knowledge. The results are reflected in a decline in the creativity of both teachers and students, which is a major obstacle to lifelong learning (Oberbillig et al., 2014), according to which everyone is responsible for their personal development. In outdoor learning environments, teachers focus on strengthening students' deficient contact with nature, which affects the decline of genuine relationships and behavior. Many examples of education in nature show that students connect readily with nature if they have opportunities for practical experience (Beery, 2014). The transformation of educational approaches by strengthening the role of nature and connections to it has shown the potential for enhancing the acquisition of lasting knowledge. There is an increasing level of commitment to nature and its recognition, the importance of green systems in urban environments, and complex thinking about the importance and future of connections with the environment and nature (Root et al., 2017).

This paper focuses on explaining the concept of an outdoor learning environment and showing the opportunities that teachers have for implementation of practical education. Outdoor learning environments are upgraded classrooms in nature, as they also include experiences, personal development, and behaviors that are not necessarily tied to the school curriculum (lifelong knowledge).

In the case of the Polygon Dole presented below, we have shown the advantages of outdoor education, which results in increased interest in self-sufficiency; greater awareness of natural laws; care for natural resources like energy, water, and biodiversity; and for their personal development. Traditional outdoor experiential activities, such as hiking, orienteering, bird watching, and environment education, have been commonly understood by many to be the domains of white people with greater means to access leisure experiences (Rose & Paisley, 2012). Outdoor learning environments are extremely important for young people who do not yet have life experience, to distinguish theoretical knowledge from practical knowledge and to connect the importance of multiple knowledge types for life. Outdoor learning is a combination of formal and non-formal education, with the help of modern technology and appropriate motivational spaces (learning environments), a combination which increases access to knowledge and education, while equipping learners properly with information (Košir & Habe, 2015). Thus, students can learn anywhere and anytime, not just in school. Open education should not be confused with excursions, since it is a new form of education that upgrades students' existing knowledge with additional information, personal experience of processes, and the possibility to create the result themselves.

In the scientific literature, learning environments are defined in connection with the achievement of learning goals. Thus, in the paper "Analysis of the factors of the learning environment based on the model of the hierarchy of needs of Abraham Maslow," Košir and Habe (2015) analyze factors in the learning environment from the motivation point of view. Based on a review of many studies, they have derived recommendations that enable the teacher to effectively promote the holistic development of students. In the *Journal of Elementary Education*, the authors in the article "Reverse learning and teaching as an opportunity for innovative and flexible implementation of learning forms in higher education" emphasized the importance of ensuring quality in higher education. They represent the relationship between direct and indirect forms of work through the approach of reverse learning and teaching, where they emphasize the importance of student self-regulation of learning, mental activity, achieving higher learning outcomes, and promoting a higher level of understanding of knowledge and skills (Plešec et al., 2020). However, we know from our practice that students cannot acquire skills if they are not involved in appropriate learning environments. They also cannot achieve higher goals if they do not have the opportunity to think differently.

Education in nature covers several competences: relationship skills, responsible decision-making, self-awareness, social awareness, and self-management. A promising practice to promote positive social and emotional learning is school garden programming (Ambrose et al., 2020). Since the concept of an open learning environment is relatively new, it is associated with learning paths, classrooms in nature, landscape elements (forest, meadow, stream), and other practices, including sports and recreation. We are using open learning environment with the same meaning as outdoor learning environment.

The term outdoor learning environment is therefore understood as contact with nature and the possibility of perceiving sounds, smells, vibrations, and other sensations from the outside. After reviewing the situation in the field of open learning environments, we find in the monograph *Slovenia: Educational Region* (Vovk, 2015) that Slovenia has invested considerably in equipping learning paths and various natural centers. These centers of nature and learning paths have been set up by various associations and municipalities, but they do not have the expertise to connect these arrangements with education systems.

In the following part, we present the Polygon Dole, an outdoor learning environment in the municipality of Poljčane, which enables the realization of all physical needs as well as the highest needs of being and the close relationship between nature and human. The emphasis is on holistic self-sufficiency, both material and spiritual, which develops a person into a vital, mentally connected, and responsible citizen, which is vital in today's society.

10.2 Polygon Dole Is an Outdoor Learning Environment for All Generations

In 2010, on a 1.5 ha extensive meadow in a small valley in the hamlet of Dole near Poljčane, we developed an open learning environment for holistic self-sufficiency, with an emphasis on the material and spiritual levels, which enables different generations their personal development. Ten years ago was the beginning of the financial crisis, and it became increasingly clear that we needed a holistic view of life, that is, food, work, and relationships between us and overall well-being and health. Above all, we needed a new way of thinking, since the material world collapsed for many when they lost their property because of various investments and many also lost their jobs. These needs encouraged us to develop an open learning environment and offer help to those looking for new paths of development, for themselves and others. Because it is necessary to first meet material needs, we have developed a mini eco-village with all the arrangements that enable self-sufficient living, including water, energy, arable land, sustainable buildings, and supporting educational facilities for new knowledge development. In the following, we present the elements of the outdoor learning environment at Polygon Dole for achieving various educational and developmental goals (Table 10.1 and Fig. 10.1):

There are also many smaller elements on the Polygon Dole, all of which are intrinsically connected to nature. In that way, all these practices represent an irreplaceable source of information,

Table 10.1 The elements of the outdoor learning environment at Polygon Dole

Plants	Traditional plants: local traditional species of trees, herbs, and vegetables, adapted to local natural conditions, for food, biomass, and energy
	Special plants: plants not indigenous to the area but not harmful to the natural ecosystem such as miscanthus, paulownia, Chinese yam, sweet wormwood, sedges, reeds, and chicory
	Phytoremediation plants: special plants for cleaning and prevention of soil and water degradation
Production	Food: raised, mound, and keyhole beds, also forest and a vertical garden for most effective self-sufficiency
	Energy: solar panels for electricity generation, greenhouses for heat retention, and a solar water heater
	Water: well, pond, rainwater collectors, mulches, rain gardens, retention trenches for water collection and treatment
	Compost: treebog toilets for organic matter decomposition and compost bins for biomass reuse
Facilities	Sustainable buildings: yurt and hobbit house made of natural materials for living and storing food, greenhouses, and a hayrack for food production and processing, dryer for phytopharmaceutical preparations
	Research facilities: laboratory for soil and water testing, weather station for meteorological data
Pedagogical tools	Experiential infrastructure: pedagogical profile, geological wall, pond, wildlife feeders, sand bioaccumulation filters
	Educational boards: additional information about herbs, pollinators, clouds, microorganism
Biodiversity areas	Green systems: natural meadows, shrubs, trees, vegetation belts
	Blue systems: ponds, puddles, water reservoirs
	Artificial habitats: for insects and other beneficial organisms to support biodiversity

insight, feelings, experiences, and spiritual and personal growth and are therefore essential for achieving holistic development for all generations.

10.3 Practical Examples for Education in Nature

In the following, we list some starting points that teachers can adapt to a specific age, level of previous experience, motivation, and skills. The starting points can help teachers to plan individual research assignments and teamwork. The listed activities have all been tested in practice at the Polygon Dole.

Activity 1: Listen to Life in the School Garden

We walk quietly around the garden, between the flowerbeds, and listen to birds, crickets, moving branches, and our steps. Do we always hear these sounds?

Result: Students find that the sounds are diverse, in origin, duration, and intensity. Let them recognize the source of these sounds and infer the meaning of the living beings that create these sounds (Fig. 10.2).

Activity 2: Observe Life in the School Garden

A more detailed exploration into the ground or under its surface and under the branches of plants reveals the richness of life in the garden. Prepare a table and fill it in with the specimens we have seen and how many of each (e.g., ladybug 3 times, black snail 1 time, and so on). Students can follow life in the garden in groups, in pairs, or individually. We take care to avoid touching the animals, just sketching or photographing them, instead. We present all the observed animals to the whole group and think about their habitats.

Result: Students identify rarer animal species and their role in the ecosystem (Fig. 10.3).

Activity 3: Colors, Smells, and Sounds in the School Garden

When identifying plants and animals in the garden, we use a color method that requires observation and comparison (Molek et al., 2010). Herbs are also distinguished by their smell, so the combination of color and smell is a reliable method for recognizing plants. Birds are also distinguished by color and sound, as are various beetles and insects.

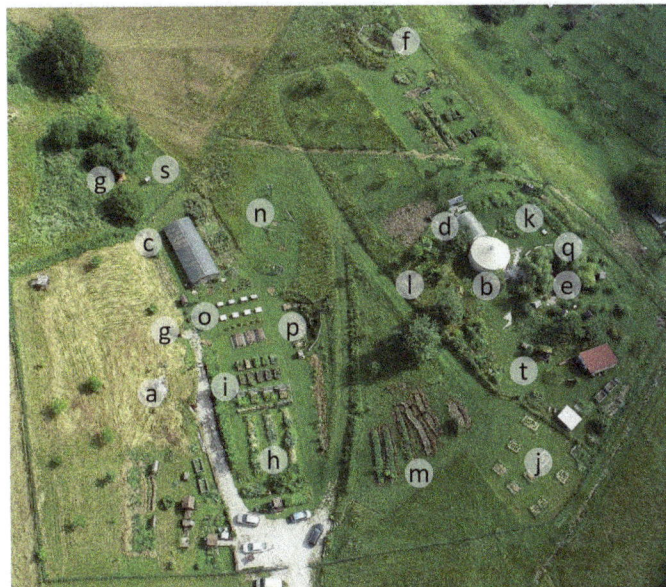

a. Hobbit house
b. Yurt
c. Greenhouse
d. Winter garden
e. Pond
f. Water retainer
g. Compost toilet
h. Hugel beds with mulch
i. Raised beds
j. Beds on pallets
k. Spiral beds
l. Forest garden
m. Terraces
n. Natural playground
o. Learning boards
p. Educational soil profile
q. Benches from recycled plastic
r. Pet house
s. Wildlife feeder
t. Insect house

Fig. 10.1 The elements of the outdoor learning environment at Polygon Dole. (Photo taken by the authors)

Result: Students become aware of the heterogeneity of garden life and develop an affinity for certain colors, smells, and sounds, which they also mimic (Fig. 10.4).

Activity 4: How Warm Is It? And How Much Rain Falls in the School Garden?

Students can collect temperature data using a thermometer. The data about warmth can be connected to everyday life. They can also make simple containers from packaging to measure precipitation in mm. The amount of precipitation is measured in millimeters: for example, 10 mm of precipitation falls on 1 m^2, which equals 10 l. To make it easier for students to understand these units, it is recommended that they measure the amount of precipitation themselves and then infer the importance of water for plants and animals.

Result: Students become interested in official weather data and understand weather forecasts and climate zones more clearly (Fig. 10.5).

Activity 5: Which Soils Do We Have in the Garden?

Soil knowledge is important for choosing specific plants for planting. Most plants need light soil to grow, which is moderately moist, friable, and darker in color because of the presence of humus. The soil on open surfaces often compacts

Fig. 10.2 The garden is full of diverse sounds (photo taken by the authors)

Fig. 10.3 Garden animals approach us when they feel we want to get to know them (photo taken by the authors)

and becomes completely hard. Plants cannot thrive in such a habitat.

Result: Students observe the color, moisture, and fragility of the soil and distinguish between favorable soil for plants and less favorable (Fig. 10.6).

Activity 6: Who Lives Under the Ground?

Soil, earth, dirt, or ground (all terms can be interchangeable) is home to many organisms. Even if we observe the surface of the ground carefully, we can see ants, spiders, and centipedes. But if we dig up the soil, there we can see even more animals.

Result: Students gain awareness about the importance of preserving the soil. By observing, counting, comparing, and touching the organisms in the soil, they become aware that the soil is alive (Fig. 10.7).

Activity 7: Soil Seed Balls

If we have clay soil in the garden, we can make balls and add grass, flower, or vegetable seeds. Clay retains water for a long time, so the seeds will be moist and ready to germinate. Seed balls can be placed in planting beds, or they can be thrown on the lawn to sow flowering grass.

Fig. 10.4 Many shades of color tell us about processes in nature (photo taken by the authors)

Fig. 10.5 Temperature data gathering is a simple method for motivation (photo taken by the authors)

Fig. 10.6 Prolonged dry periods completely dry out the soil (photo taken by the authors)

Fig. 10.7 The richness of life in the soil is indescribable. We do not see most living things with the naked eye (photo of the display information board taken by the authors)

Result: Students recognize the seeds of plants and learn that the seeds are the smallest fragment of the plant and that we need to store them dry so that they do not begin to germinate (Fig. 10.8).

Activity 8: Let's Make a Pond in the Garden
Ponds are a habitat for aquatic animals that make an important contribution to maintaining the ecosystem balance in the garden. Therefore, every garden should have a pond. Of particular importance are frogs and toads, which eat the larvae of various beetles and worms that would otherwise eat our crops. Students can make a puddle themselves by digging a 30-cm-deep pit and placing foil in it. It is then filled with water, at least one

bucket of which should be from a natural pond or stream to start life developing.

Result: A puddle is a living ecosystem and students become aware of this. Aquatic and terrestrial ecosystems can be explained in the garden as students experience life in both (Fig. 10.9).

Activity 9: Let's Make a Garden Bed of Willow Branches
We always want extra flower or vegetable beds in the garden, especially if the plants are growing well. The beds must be as sustainable as possible, so we use willow branches found in the vicinity. Then we add soil to the bed and plant herbs and vegetables.

Fig. 10.8 Seed balls for seed storage (photo taken by the authors)

Fig. 10.9 A pond can greatly enrich the garden, as animals are always heard in it (photo taken by the authors)

Result: Students acquire bed design skills and learn the importance of wood in self-sufficiency (Fig. 10.10).

Activity 10: Let's Make Clay Pots

We collect seeds from the garden, so the students will make clay containers for seed storage. Clay can also be collected in nature; it must be as clean as possible. Students form medium-sized clay pots and dry them in the shade. The outdoor learning environment can be designed and used to enrich perceptual development through sensory and motor stimulation.

Result: Students learn how to design a container and develop fine-motor skills (Fig. 10.11).

Activity 11: Filling the Raised Beds with Biomass and Soil

To design the garden and obtain arable land, we need larger garden beds, including raised beds.

The wooden framework and its protection are prepared for us at school, and we can do everything else with the students. Students learn about all stages of bed design. Then students place twigs on the bottom of the bed, biomass (leaves, hay) above them, and fertile soil on top. The bed is then sown or planted and protected with mulch.

Result: Students can observe the whole process of growing local herbs and vegetables (Fig. 10.12).

Activity 12: Prepare Your Seedlings

The greatest joy is to observe the seeds becoming seedlings. Seeds can be collected by us or obtained at exchange communities. Certain plants should be sown early (peas, beans) and

Fig. 10.10 Round beds made of willow branches enrich the school garden (photo taken by the authors)

Fig. 10.11 Making clay pots helps develop manual skills (photo taken by the authors)

others later. Fill the seed pot halfway with soil, and then add up to three seeds and then more soil.

Result: Students recognize the whole process from seed to seedling and the conditions for seed growth. They learn to care for seeds and later seedlings (Fig. 10.13).

Activity 13: Caring for Wild Pollinators

It is crucial for our garden that the plants have a healthy environment and that the site is suitable for plant requirements. Wild pollinators are important for seed germination, especially wild bees and bumblebees. If the garden is in an urban environment, it is important to set up a home for wild pollinators, like a hotel for insects. We

assemble it from materials of varied color and composition.

Result: Students express their creativity, recognize varied materials, and understand the importance of wild pollinators for self-sufficiency (Fig. 10.14).

Activity 14: Caring for Birds in the Garden

Birds are of great importance in regulating the balance in the ecosystem, so they are an indicator of environmental health. In the garden, we prepare straw for their nests by stuffing it into standing cups, so the birds can take it to their nests. Birds that do not migrate need help in the winter.

Fig. 10.12 The beds are made directly on the grass (photo taken by the authors)

Fig. 10.13 Seedlings are protected with boxes against the weather (photo taken by the authors)

Result: Students learn about birds and their living conditions (Fig. 10.15).

Activity 15: Make a Mixture for Brewing Tea
Herbs are very suitable for teas, assuming the garden is in a place safe from external influences such as traffic and random walkers. Therefore, when preparing the garden, we must pay atten-tion to protection from traffic and other sources of pollution. Almost all herbs can be used for tea. Tea is made from fresh or dried herbs. Teas can also be mixed into blends to emphasize desirable flavors. Otherwise, dry the leaves of the plants and store them in a glass jar.

Result: Students develop an interest in plants and get used to drinking tea (Fig. 10.16).

Fig. 10.14 Materials for insect hotels are found mostly in the garden or at home (photo taken by the authors)

Fig. 10.15 The straw pots are emptied in just a few days, as birds quickly carry straw to their nests (photo taken by the authors)

Fig. 10.16 Chamomile prevents inflammation and cheers us up with their flowers and scent (photo taken by the authors)

10.4   Holistic Effects of Outdoor Learning Environments

Green schoolyards and other natural areas such as forests, parks, woodlands, and gardens afford a meaningful context for childhood education, as they provide children with numerous opportunities for both informal and formal learning experiences (Auer, 2008; Ballantyne & Packer, 2009; Dyment & Reid, 2005; Sahrakhiz, 2018). The scholarly literature proposes that outdoor learning engaging with the natural environment provides opportunities to improve academic achievement and social emotional intelligence for students; thus, the features of the landscape can influence the acquisition of awareness in the natural environment, while enhancing training for sustainability and various green systems,

learning in a number of subject areas such as science, and learning by smelling, feeling, and tasting (Riemer, 2003). Living in outdoor learning environments impacts human life from the basic physiological level to the higher needs of human beings. Students can learn how to work together to take responsibility for tasks such as planting by group learning. Access to nature should help to develop self-confidence, responsibility, a sense of pride, cooperation, problem-solving skills in the classroom, and relationships between students and teachers (Bowker & Tearle, 2004). Everyone has physiological needs, which one satisfies in diverse ways, and how one copes with them affects a person's vitality, well-being, and appearance. Security needs are related to the way of life and local regional conditions; people regulate these through their ways of living and connecting with others. The need for affiliation is expressed through networking in societies, associations, and organizations, which is supported by offering diverse learning environments that connect people with similar interests. The need for respect and self-esteem arises from the relationships between people and social communities and is also reflected in open learning environments based on respect for the individual (Košir & Habe, 2015). The deficit needs we have mentioned so far can be met in open learning environments, as they are based on a strong connection between nature and humans. Outdoor learning environments also enable self-realization, self-development, and meeting the needs of one's being (Plešec et al., 2020). Most importantly, however, we achieve transcendence through open learning environments, that is, transcending the observable world. Too often we become immersed in routines and habits and do not see beyond our limitations; open learning environments allow us to transition and personally grow, which is above the needs of everyday life. Since the condition for achieving transcendence is a willingness to open one's perceptions, this is exactly what open learning environments offer. This is especially important if we are making major changes when we feel we should transcend our present state, when we have outgrown our current world, and when we want to move forward from limitations to another reality. It is important to realize that open learning environments allow us to perceive what is around us all the time, but which for various reasons, we do not see or perceive. In discussions of transcendence, the fundamental question is how to be or how to design and live such openness (understanding) that it will lead both our being and other beings to authenticity and to self-ownership (Klun, 2015).

Since the world is becoming more complex because biodiversity is rapidly diminishing, there are an increasing number of hungry people, people feel alienated, and social problems are becoming more serious, learning environments offer a timely form through which to achieve observation and response to opportunity.

10.5 Conclusion

In this chapter, we have shown the importance of open learning environments, which differ from learning paths, and learning tables along paths in that outdoor education have several dimensions. This alone is often not enough to provoke thought and develop deeper levels of learning, as much information is overlooked (longer texts on cultural heritage, descriptions of phenomena), because people, unfortunately, do not read lengthy texts. Although we have many different learning paths and learning tables in Slovenia (perhaps too many), we do not recognize their effects on people's thinking. Therefore, we find that the open learning environment offers a specific connection between humans and nature, where not much arrangement is needed, but the perception of space must be emphasized.

At the Polygon Dole, we have developed an integrated approach, which promotes people's vitality (their life energy, will to live) and leads them to transcendence, which is related to the ability to understand diverse approaches, thus contributing to happiness, personal development, and the well-being of our planet Earth.

References

Ambrose, G., Das, K., Fan, Y., & Ramaswami, A. (2020). Is gardening associated with greater happiness of urban residents? A multi-activity, dynamic f in the Twin-Cities region, USA. *Landscape and Urban Planning, 198*, 103776. https://www.sciencedirect.com/science/article/pii/S0169204619307297

Auer, M. R. (2008). Sensory perception, rationalism and outdoor environmental education. *International Research in Geographical and Environmental Education, 17*, 6–12. https://doi.org/10.2167/irgee225.0. https://www.tandfonline.com/doi/abs/10.2167/irgee225.0

Ballantyne, R., & Packer, J. (2009). Introducing a fifth pedagogy: Experience-based strategies for facilitating learning in natural environments. *Environmental Education Research, 15*, 243–262. https://doi.org/10.1080/13504620802711282. https://www.tandfonline.com/doi/abs/10.1080/13504620802711282

Beery, T. (2014). People in nature: Relational discourse for outdoor educators. *Research in Outdoor Education, 12*, 1–14. https://doi.org/10.1353/roe.2014.0001. https://www.researchgate.net/publication/286453791_People_in_Nature_Relational_Discourse_for_Outdoor_Educators

Blair, D. (2009). The child in the garden: An evaluative review of the benefits of school gardening. *The Journal of Environmental Education Research, 40*, 15–38. https://doi.org/10.3200/JOEE.40.2.15-38. https://www.tandfonline.com/doi/abs/10.3200/JOEE.40.2.15-38

Bowker, R., & Tearle, P. (2004). Gardening as a learning environment: A study of children's perceptions and understanding of school gardens as part of an international project. *Learning Environment Research, 10*, 83–100. https://eric.ed.gov/?id=EJ812601

Dyment, J. E., & Reid, A. (2005). Breaking new ground? Reflections on greening school grounds as sites of ecological, pedagogical, and social transformation. *Canadian Journal of Environmental Education, 10*, 286–301. https://files.eric.ed.gov/fulltext/EJ881791.pdf

Ernst, J., & Theimer, S. (2011). Evaluating the effects of environmental education programming on connectedness to nature. *Environmental Education Research, 17*(5), 577–598. https://eric.ed.gov/?id=EJ812601

Goodall, J. S. (2016). Technology and school–home communication. *International Journal of Pedagogies and Learning, 11*, 118–131. https://doi.org/10.1080/22040552.2016.1227252. https://www.tandfonline.com/doi/abs/10.1080/22040552.2016.1227252

Kahn, P. H., & Kellert, S. R. (2002). *Children and nature: Psychological, sociocultural, and evolutionary investigations*. MIT Press. https://mitpress.mit.edu/books/children-and-nature

Klun, B. (2015). *Ontološka diferenca in transcendence [Ontological difference and transcendence]*. http://kud-logos.si/2015/ontoloska-diferenca/. Assessed 10 Feb 2022.

Košir, K., & Habe, K. (2015). Analiza dejavnikov učnega okolja na osnovi modela hierarhije potreb Abrahama Maslowa [Analysis of learning environment factors based on the hierarchy model needs of Abraham Maslow]. *Revija za Elementarno Izobraževanje, 3*, 21–30. Pedagoška fakulteta Maribor. https://journals.um.si/index.php/education/article/view/471

Molek, I., Golob, L., & Franken, G. (2010). Barve, barvna metrika in barvno upravljanje [Colors, color metrics and color management]. Konzorcij šolskih centrov Slovenije.

Oberbillig, D., Randle, D., Middendorf, G., & Cardelús, C. (2014). Outdoor learning in formal ecological education: Looking to the future. *Frontiers in Ecology and the Environment, 12*(7), 419–420. http://www.jstor.org/stable/43187839. Accessed 29 Jan 2021

Plešec, G. R., Valenčič, Z. M., & Kalin, J. (2020). Obrnjeno učenje in poučevanje kot priložnost za inovativno in prožno izvajanje učnih oblik v visokošolskem izobraževanju. [Flipped learning and teaching as an opportunity for innovative and flexible implementation of learning forms in higher education]. *Revija za Elementarno Izobraževanje, 13*, 51–80. Pedagoška fakulteta Maribor.

Raudsepp, M. (2005). Emotional connection to nature: Its socio-psychological correlates and associations with pro-environmental attitudes and behavior. In B. Martens & A. Keul (Eds.), *Designing social innovation: Planning, building, evaluating* (pp. 83–91). Hogrefe Publishing. https://psycnet.apa.org/record/2005-10271-010

Rickinson, M., Dillon, J., Teamey, K., Choi, M. Y., & Benefield, P. (2004). *A review of research on outdoor learning*. https://www.informalscience.org/sites/default/files/Review%20of%20research%20on%20outdoor%20learning.pdf. Assessed 10 Feb 2022.

Riemer, M. J. (2003). Integrating emotional intelligence into engineering education. *World Transactions on Engineering and Technology Education, 2*(2), 189. https://www.researchgate.net/publication/299566618_Integrating_emotional_intelligence_into_engineering_education

Root, E., Snow, K., Belalcazar, C., & Callary, B. (2017). Playing naturally: A case study of schoolyard naturalization in Cape Breton. *Research in Outdoor Education, 15*, 1–20. https://doi.org/10.1353/roe.2017.0001. https://www.researchgate.net/publication/322928580_Playing_Naturally_A_Case_Study_of_Schoolyard_Naturalization_in_Cape_Breton

Rose, J., & Paisley, K. (2012). White privilege in experiential education: A critical reflection. *Leisure Sciences: An Interdisciplinary Journal, 34*, 136–154. https://www.tandfonline.com/doi/abs/10.1080/01490400.2012.652505

Sahrakhiz, S. (2018). The 'outdoor school' as a school improvement process: Empirical results from the

perspective of teachers in Germany. *Education 3–13, 46*(7), 825–837. https://doi.org/10.1080/03004279.2017.1371202

Vovk, A. (2015). *Prepoznavnost Slovenije z učnimi regijami*. [Recognizability of Slovenia with learning regions]. Nazarje: GEAart, 2015. 106 str., ilustr. ISBN 978-961-93683-8-1.

Vovk, K. A. (2019). The art of outdoor learning. *International Journal of Inspiration, Resilience & Youth Economy, 3*(1), 1–9. https://doi.org/10.18576/ijye/030102

Wells, N. M., & Lekies, K. S. (2006). Nature and the life course: Pathways from childhood nature experiences to adult environmentalism. *Children Youth and Environments, 16*(1), 1–24. https://www.researchgate.net/publication/252512760_Nature_and_the_Life_Course_Pathways_from_Childhood_Nature_Experiences_to_Adult_Environmentalism1

Wistoft, K. (2013). The desire to learn as a kind of love: Gardening, cooking, and passion in outdoor education. *Journal Adventure Education and Outdoor Learning, 13*, 125–141. https://doi.org/10.1080/14729679.2012.738011

Ana Vovk   holds two PhDs, one in physical geography and one in environmental protection. She works as a full professor in the Department for Geography and the Faculty of Agriculture and Life Sciences. She founded the Educational Polygon for Self-Sufficiency Dole, which was awarded the national prize for the second-best open learning environment in Slovenia. She cooperates with most European countries and with Asia and Africa. Prof. Vovk is very actively involved in the development of several municipalities and educational regions, as well as in many councils at the ministries for agriculture and education. She is author of many books, handbooks, and articles in Slovene and other languages.

Janja Lužnik   is a university graduate engineer of landscape architecture. She graduated in landscape ecology. Within the framework of national and international development and research projects, she focuses on spatial planning, regional and sustainable development, nature protection, preservation of the cultural landscape, and the establishment of open learning environments for all generations. She deepens her knowledge at various national and international seminars. She is a member of the European Heritage Interpretation Association.

Danijel Davidovič   has an MA in geography and philosophy. He is currently employed as a researcher at the International Center for Ecoremediation, University of Maribor, where they deal with projects in sustainable regional development and environmental education. He also works in the Department of Geography as an assistant, where he teaches tutorial seminars in physical and economic geography. As part of his doctoral studies, he deals with the use of geographic information systems GIS and satellite imagery to study the landscape. In his spare time, he enjoys drone photography.

Implementation of Education for Sustainable Development Through a Whole School Approach

11

Niklas Gericke

Abstract

The whole school approach to Education for Sustainable Development (ESD) embraces a holistic and participatory educational philosophy that aims to enhance the potential of the school environment to function as an authentic and meaningful learning place. There is rich diversity in ways this approach plays out in different places, both within countries and globally. However, a common guiding principle is the integration of three lines of action: environmental management ("greening") of the school, establishment of ongoing partnerships with the broader local community to address issues of social-environmental sustainability, and incorporation of sustainability in the curriculum. Hence, a whole school approach demands involvement of all parts of schools and stakeholders in the society to expose students to real sustainability issues. In this chapter, I describe how the idea of whole school approaches has developed in the literature by presenting different models and present a school organization model that can be used to guide implementation of ESD in a whole school approach. Aspects of school organization that facilitate implementation of ESD from a whole school approach are highlighted, with a focus on school leadership.

Keywords

Education for sustainable development · School leadership · School organization · Sustainability · Whole school approach

11.1 The Role of Whole School Approaches in ESD

The term *whole school approach* was originally used in research about schools' ability to incorporate perspectives of health, well-being, and anti-bullying (Wyn et al., 2000). The goal has been to engage all parts of the school in common efforts at all levels to improve targeted characteristics, for example, the well-being of staff and pupils. This has been found to be a successful approach in health education, according to several studies (e.g., Rowe & Stewart, 2009; Rowe et al., 2007). The underlying idea has also been subsequently adopted in Education for Sustainable Development (ESD) and sustainability education and developed especially in Australia and New Zealand (Eames et al., 2010; Ferreira et al., 2006; Henderson & Tilbury, 2004). The concept includes involvement of all parts of a school and expert stakeholders in society (Henderson & Tilbury, 2004), to expose students

N. Gericke (✉)
Department of Environmental and Life Sciences,
Karlstad University, Karlstad, Sweden
e-mail: niklas.gericke@kau.se

to authentic problems in the wider society in order to transform the school itself into an agent of change in a sustainable direction (Mogensen & Schnack, 2010). In that sense, whole school approaches can be seen as a continuation of a progressive teaching tradition (Mathar, 2015), although the aim has become specifically directed toward sustainability, i.e., to transform schools, including all their actors, into agents of change toward a more sustainable world.

ESD and sustainability education can be seen as two different ideas, as sustainability education only addresses the sustainability aspect and ignores the development aspect of ESD. In that sense, sustainability education can be described as a more eco-centric concept than ESD, which to a greater extent recognizes the needs of humans, the anthropocentric perspective (Gericke et al., 2020b). Nevertheless, in this chapter I ignore these differences and treat the concepts as synonymous, because a whole school approach by definition cannot ignore the human aspect. Instead, I address how ESD or sustainability education can be implemented according to a whole school approach in order to improve the conditions both for humans and the environment, although I mainly use the ESD concept.

McKeown and Hopkins (2007) identify four levels that are important for implementing ESD in a school—the *disciplinary, whole school, educational system*, and *international policy* levels. At the disciplinary level, they point out that traditional disciplines such as language, mathematics, and science form the core school subjects that are regularly tested and reported. Due to the importance assigned to them, it is very important to implement ESD in these high-stake school subjects in order to prioritize it strongly in education at large. In addition, they claim that other school subjects, such as civics, history, geography, art, and physical education, form a second tier of school subjects that should also address ESD. ESD is also closely related to the disciplinary content of some subjects, such as biology, geography, and civics, which could provide valuable foundations for catalyzing the incorporation of ESD in other school subjects (Sund & Gericke, 2020). However, ESD calls for transdisciplinary

approaches that integrate, rather than separating, content knowledge and skills from different disciplines, so it is important to engage teachers of subjects in ESD-oriented efforts (Sund et al., 2020).

Due to the requirement for a transdisciplinary perspective, McKeown and Hopkins (2007) identify a need to adopt a whole school approach for ESD. Hence, this is their next important level to address when implementing ESD. The whole school approach recognizes that more than knowledge and information are required to induce the behavioral shifts needed in schools and society for effective reorientation toward sustainability (Henderson & Tilbury, 2004). From a holistic perspective, including environmental, social, and economic aspects, according to McKeown and Hopkins (2007), a whole school approach should involve efforts to embed sustainability beyond disciplines by incorporating compatible practices in all activities and everyday aspects, such as travel, purchase, and physical surroundings. Hence, according to this view, whole school approaches should include not only the teaching and learning in a school but also all the other activities and practices outside the classroom. This element of whole school approaches is often referred to as "greening," through which all staff and students of the school should become involved. The moral is that the school should live in the same way as it is tutoring its students (Henderson & Tilbury, 2004) or as formulated by McKeown and Hopkins (2007, p. 22): "In the whole-school approach, the curriculum, programmes, practices, and policies of an educational institution are engaged to contribute to building a more sustainable future. In this approach, sustainability is lived as well as taught." To accomplish this ambitious goal, the importance of ESD must also be recognized at higher levels of the society.

To support ESD efforts at the disciplinary and whole school levels, McKeown and Hopkins (2007) emphasize that it must also be institutionalized at the national educational level, i.e., in the official steering documents that regulate activities in schools. At the national level, an appropriate legislative framework is needed to create good policies, build administrative support, enable provision of sufficient resources, support school leadership, and assess success in terms of learning outcomes. Without support at this level, it will be difficult to establish a whole school approach because officials at this level write curricula and are responsible (under governmental direction) for activities such as teacher certification and provision of in-service development that support, or hinder, the implementation process (McKeown & Hopkins, 2007).

The way ESD is addressed differs substantially among countries, but in Sweden (the source of empirical findings reported in this chapter) the national curriculum for the 9-year compulsory school states that teachers of all school subjects are responsible for teaching and promoting sustainable development. Moreover, 1 of 16 overall aims concerns sustainable development (Education, 2011, p. 8):

> It is the responsibility of the school that all individual students can observe and analyze the interactions between people in their surroundings from the perspective of sustainable development.

However, ESD and sustainability are not mentioned at all the syllabi of various school subjects and only emphasized in syllabi of a few subjects such as biology and geography.

The last level of importance identified by McKeown and Hopkins (2007) for incorporation of an ESD perspective is international policy and agreement, to support its inclusion at national level. Currently this is strongly manifested in the work of UNESCO and the global action plan for ESD, a roadmap to generate and scale up ESD action at all levels and in all areas of education and learning (UNESCO, 2017). Thus, I would argue that it is already incorporated well at this level.

To conclude, a whole school approach may be crucial for successful implementation of ESD in schools, at a level between steering documents and the teaching and learning activities. However, it is clearly important to determine the key reasons for adopting such an approach. Hence, this is the objective of the next section.

11.2 Why Adopt a Whole School Approach to ESD?

As previously pointed out, a whole school approach to ESD can be defined as the involvement of a whole school community in efforts to promote sustainable development, in addition to orientation of the teaching, learning, and curriculum toward its promotion (Breiting & Meyer, 2015; Gough, 2005; Henderson & Tilbury, 2004; Gough, 2005; Mathar, 2015; McKeown & Hopkins, 2007). Moreover, the references show there seems to be consensus about this definition. The reason for engaging the whole school, and not just the teachers and students, is that ESD and sustainability education call for action. Specifically, there is a need to develop competence to act in a sustainable manner (Mogensen & Schnack, 2010; Sass et al., 2020, 2021) rather than merely developing knowledge, which is the most common goal of education. Research and practice have shown that higher levels of knowledge do not necessarily bring about a change in people's way of life (e.g., Kollmuss & Agyeman, 2002). Therefore, the environmental and sustainability education practice and research community has long taken inspiration from the progressive teaching and learning movement (Mathar, 2015), which dates all the way back to Jean Jacques Rousseau in the eighteenth century and the call for a "return to nature" (Sarabhai, 2007). The underlying idea is that the learner needs to experience nature or the environment in order to fully understand it.

In progressive teaching, affective dimensions of learning such as emotions, attitudes, and values are often given equal importance to the cognitive knowledge dimension. This kind of teaching, or rather learning, is referred to as transformative education (Mezirow, 2009). Transformative education is supposed to "foster deep engagement with and reflection on our taken-for-granted ways of viewing the world, resulting in fundamental shifts in how we see and understand ourselves and our relationship with the world" according to a leading journal in the field (https://journals.sagepub.com/home/jtd). Hence, by empowering individual learners, and transforming their ways of understanding themselves in relation to the world, it is believed that they will be prepared to become agents of change in their societies and the world, as a consequence of that education. Therefore, issues related to participation, power, and democracy are important aspects of ESD (Shallcross & Robinson, 2008), and deployment of transformative teaching strategies that develop learners' action competence for sustainability is regarded as crucial for sustainable development (Parra et al., 2020).

This is because sustainable development often poses complex, or "wicked," problems (Rittel, 1973). The complexity stems from the multidimensional character of sustainability issues, which can span multiple levels (from individual to global), and disciplines as they require integrated consideration of environmental, cultural, and economic dimensions (Parra et al., 2020), as illustrated in Fig. 11.1. A wicked problem is one that is difficult or possibly even impossible to solve in a completely satisfactory manner, because any solution will involve trade-offs between environmental, social, and economic pros and cons. There is often no right or wrong answer, and choices must be made that favor

some groups and goals more than others, based on values as well as knowledge. For example, constructing a new mine may impair local biodiversity, but boost local employment and meet needs for valuable metals required to produce batteries in order to reduce "greenhouse gas" emissions driving climate change. In such cases there are no simple solutions, and some values must be prioritized at the expense of others when reaching a decision. Many sustainability issues have such "wicked" complexity and must be solved based on well-developed action competence for sustainability rather than knowledge alone. Thus, there is a clear need for transformative teaching according to the ESD discourse.

However, the transformative aim of ESD, in which the learner and world are viewed as corresponding vessels transforming each other, has proven very difficult to achieve in reality. An initiative called the green school movement has been launched globally in efforts to accomplish it (Gough et al., 2020). However, effects of implementation efforts involving whole school accreditation in the green school movement have detected small effects and usually manifested in improved levels of knowledge rather than action

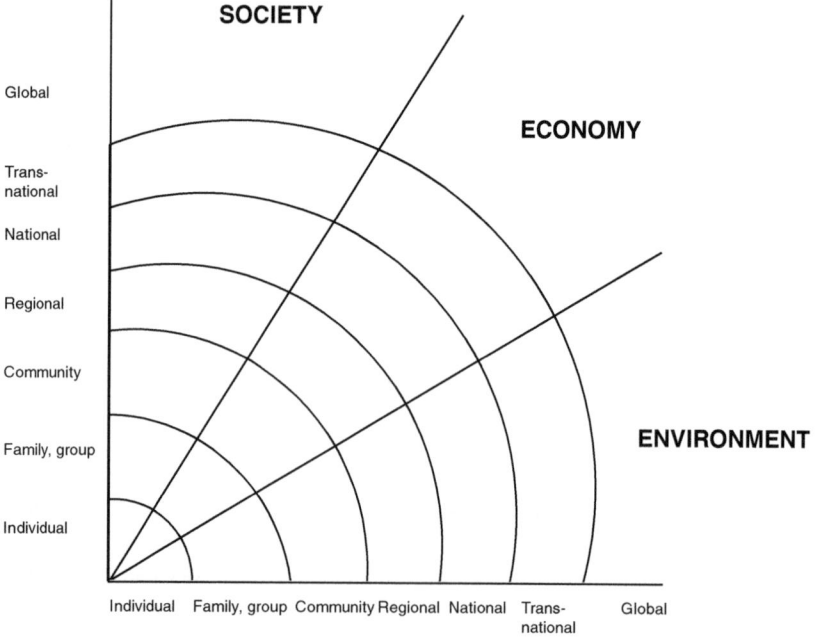

Fig. 11.1 Dimensions and levels of sustainability. (Modified by Gericke from Mathar (2015))

competence. This is exemplified by studies of efforts in the Czech Republic and Slovak Republic (Cincera & Krajhanzl, 2013), Flanders in Belgium (Boeve-de Pauw & Van Petegem, 2013), Indonesia (Riastini et al., 2019), Israel (Goldman et al., 2017), Portugal (Spinola, 2015), Sweden (Berglund et al., 2014; Olsson et al., 2016), Taiwan (Berglund et al., 2020; Olsson et al., 2019), and the USA (Warner & Elser, 2015). It has been shown that transformative ESD affects students' self-reported behavior (Boeve-de Pauw et al., 2015), but difficulties seem to lie in features of schools' organization and functions that hinder schools implementing ESD in a transformative manner (Gericke et al., 2020a). Thus, there is a clear need to identify and counter the difficulties in implementation of transformative ESD.

It is here that the whole school approach comes into play. It seems to be very difficult to implement truly transformative ESD teaching through support solely at the disciplinary level, or the educational and national level, as outlined by McKeown and Hopkins (2007). Instead, previous research on ESD implementation suggests that the whole school needs to be engaged if truly transformative ESD is to be enacted in any school.

11.3 How Can a Whole School Approach Be Described?

In the preceding sections, I discussed the role and reasons for using whole school approaches. However, very little has been written about how whole school approaches could be arranged in practice and as pointed out by Shallcross and Robinson (2008, p. 312): "What is often missing from discussions about sustainability education, participation, and whole school approaches is any consideration of design." Hence, from a practical perspective, there is a clear need to determine key features of a whole school approach, and how a school interested in adopting one should be organized, and how the whole school should be integrated. Shallcross (2005) and Henderson and Tilbury (2004) have provided a

basic model of a whole school approach with five specific features: *formal curriculum, research and evaluation, institutional practice, social and organizational aspects,* and *community links* (Fig. 11.2). First, regarding the *curriculum* it is important to establish participatory learning and cross-curricular teaching in the school. Second, regarding *research and evaluation*, it is important to initiate and maintain continuous monitoring, with reflection and evaluation of the staff and students in combination with practitioner research. Third, *institutional practices* should include greening activities of the school and its surroundings, reducing the school's ecological footprint, and most importantly ensuring alignment with the sustainability ideas of the formal curriculum. Fourth, in terms of *social and organizational aspects*, sustainability should be at the heart of all school practices that reflect the formal curriculum, involve activities that promote participation of all of the school, and be fully embedded in the continuous professional development of teachers, support staff, and other stakeholders. These four aspects together highlight the core importance of the school *community's links* with the outer society, which should include partnerships with external groups that foster alignment with key ideas of the formal curriculum, as illustrated in Fig. 11.2.

By securing these five aspects, Shallcross and Robinson (2008) argue that a whole school approach will create interaction between the learner and the environment (outside and inside the school) in intellectual, material, spatial, social, and emotional senses. In this way action competence can be developed as part of the subjectification process of schooling, i.e., as a development of the student's self, instead of being acquired as a social norm in the form of internalized behavior, as part of the socialization process of schooling. Developing action competence for sustainability, instead of following specific norms, will be more advantageous in the long run according to transformative ESD theory, because a competence can be moderated and adopted according to the context or demands of a specific situation, unlike a specific behavior. To develop action competence, it is essential to know about

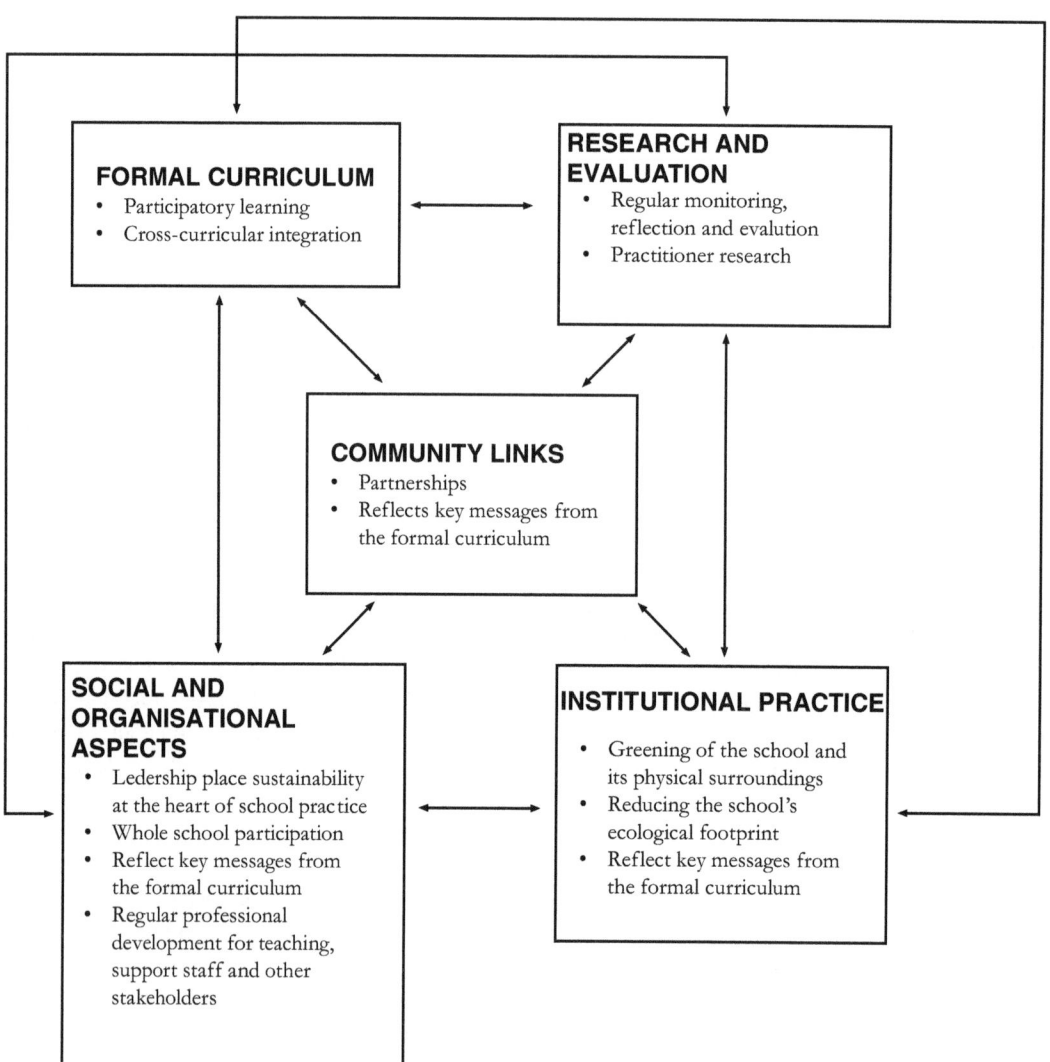

Fig. 11.2 A general model of a whole school approach and its central features. (Modified by Gericke from Shallcross and Robinson (2008))

different ways to act, i.e., action possibilities. Self-efficacy, or confidence in one's own ability to influence sustainability when taking action, is also required (as people with strong self-efficacy are more likely to engage in pro-environmental and sustainability actions) together with willingness to engage and make a difference in sustainability issues (Olsson et al., 2020). If action competence for sustainability is developed by a learner, she/he will be better equipped to deal with complex and wicked problems. However, such learning demands enactment in both affec-

tive and cognitive dimensions, which is the aim of transformative education and whole school approaches.

The model of Shallcross and Robinson (2008) highlights the centrality of links with the community to establish authentic learning with relevance for the society in which the students live and thus enable transformative learning. This builds on theory developed by Lave and Wenger (1991) to build communities of practice that include stakeholders both within and outside the school. These communities should establish joint actions to address real-world sustainability problems. In that way the students can be engaged from cognitive and affective perspectives in ways that might lead to double transformation, i.e., the students changing the world and the students being changed by the world. However, it is very demanding and difficult to achieve such joint action in a whole school approach for three reasons.

First, it is very difficult to engage the whole school in an endeavor that is not clearly prescribed in formal steering documents. Second, there may be a need for the school community to agree that sustainability should be embraced as a common vision and curriculum (Mogren, 2019). Third, when consensus regarding the whole school approach has been reached, links with the community must be established, which has proven difficulty for schools because they are traditionally inward-looking institutions (Mogren & Gericke, 2017a, 2017b).

One way to overcome these difficulties suggested by Rowe and colleagues, in the context of health education (Rowe & Stewart, 2009; Rowe et al., 2007), is to introduce the idea of *school connectedness* through whole school approaches. School connectedness is described as a sense of belonging to the school environment and can be defined as the cohesiveness between diverse groups in the school community, including students, families, school staff, and the wider community (Rowe et al., 2007). In short, school connectedness refers to the social relationships within school communities. It is recognized as an important characteristic of a whole school approach and should be strengthened by positive reinforcement in multiple contexts, such as family and community settings. Therefore, it can be described as an ecological perspective that takes into account the quality of connections among multiple groups in the school community. If successful, the school connectedness should be characterized by strong social bonds, featuring high levels of interpersonal trust and norms of reciprocity, also known as social capital (Rowe et al., 2007).

Rowe and colleagues identify two mechanisms that they consider to be important to achieve school connectedness in whole school approaches: *inclusive processes* and *supportive structures*. The processes of the whole school should be characterized by inclusiveness, involving the full range of members that make up a community, including members outside the school. The participation of community members should also be active and characterized by democratic equal "power" relations and partnerships among the members. Structural elements such as school policies, school organizational arrangements, the physical environment, and teaching and learning approaches should also reflect the values of participation, democracy, and inclusiveness and promote processes based on these values (Rowe et al., 2007).

Rowe et al. (2007) also propose a model based on the identified processes and structures, with three levels (the community context, broad school environment, and classroom environment). According to the model, all of these should be connected by numerous transversal social relationships (school connectedness) characterized by the core concepts of equal partnerships, participation, and value of diversity (Fig. 11.3). The cited authors highlight the importance of creating consistency between school policy and the organization of the broad school environment and the class policy, class organization, and physical environment of the classroom. The school connectedness model of Rowe et al. (2007) incorporates many of the same main features as the model by Shallcross and Robinson (2008) such as participation, institutional practices, organizational structures, policies, and curriculum. Most importantly, both models identify links to the wider community outside the school as of crucial importance and perhaps the most important aspect of the whole school approach.

The model of Rowe et al. (2007) also expands and contributes important dimensions of whole school approaches that are not included in the model of Shallcross and Robinson (2008). First, by using the concept of school connectedness, it highlights the importance of social relations between individuals within and outside the school. Without social relationships, it is impossible to build common visions and cultures in the

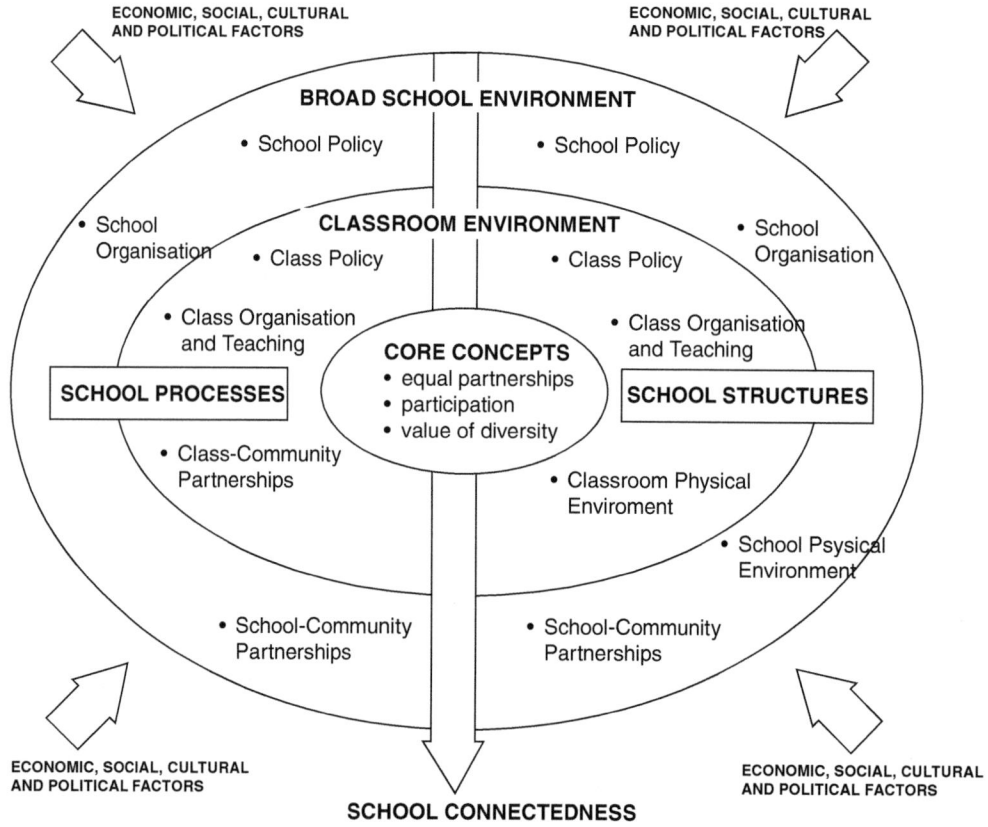

Fig. 11.3 A model of school connectedness in relation to a whole school approach. (Modified by Gericke from Rowe et al. (2007))

school, thereby creating common norms, which is important for establishment of a whole school approach. Second, it points out the different hierarchical levels of the school organization that are important to recognize and understand in order to implement a whole school approach. From a school organization perspective of whole school approaches, it is important to recognize that schools can be viewed as multilayered, with at least three layers, consisting of school leadership (with administrative support), a teacher collegium, and students (Mogren, 2019). These groups partly create separate communities within the school, although school leaders and teachers regularly meet, and teachers and students meet daily. However, these meetings often take place under specific school conditions, with expectations that communication will be subject to implicit norms.

For example, in the most important arena for communication, the classroom, recontextualized knowledge from the outer world is presented, and there are specific expectations about what the teacher and students are supposed to do. Inter alia, there are clear expectations about how they should interact, what is to be learnt, etc. according to what has been called the didactical contract by Brousseau (1997). Thus, most social interactions in the school follow norms or patterns that might be far from those that participating actors would freely choose in an authentic situation outside school. Therefore, it might be difficult to enact transformative teaching, connecting authentic experiences of students, teachers, and school leaders in the school although the social interactions might be plentiful. Moreover, in the didactical contract, contact with the surrounding society, creating links to the community, as proposed for whole school approaches, is difficult because it challenges the existing traditions and organizational structures of most schools (Gericke et al., 2020a). The whole school model of Rowe et al. (2007) visualizes the importance of these aspects, unlike the model of Shallcross and Robinson (2008). However, neither gives any suggestions for strategies to overcome these hurdles in implementation of ESD via whole school approaches, which are considered in the next section.

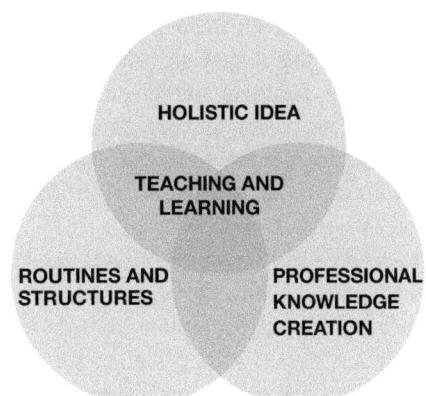

Fig. 11.4 Visualization of the Scherp model of the four dimensions of a school organization. (Modified by Gericke from Mogren et al. (2019))

11.4 A Model for Facilitating Implementation of ESD via a Whole School Approach

It has been theoretically suggested that the school organization is one of the most crucial facilitators (or hindrances) of ESD implementation in practice (Scott, 2013). As discussed in the previous sections of this chapter, different models of whole school approaches have integrated some aspects of the school organization in efforts to illuminate factors that facilitate achievement of the ultimate goal (i.e., a double transformation through which the schools become agents of change of the world and the schools become changed through interaction with the world). However, as also shown in this chapter, implementing ESD via whole school approaches has proved difficult, and there are few successful examples (published at least). Therefore, the main idea here is to align the whole school approach to the school organization and its modus operandi, rather than imposing conflicting changes from the outside. For this, we first need to know how a school organization normally works, which is described in a previously developed theoretical model based on empirical research (Scherp, 2013). This theoretical model has been described and empirically validated for whole school approaches (Mogren et al., 2019) and incorporates four dimensions describing how a school organization functions in relation to its surroundings (Fig. 11.4).

The Holism dimension or holistic idea refers to the degree that schools have an articulated, holistic vision of their aims concerning student outcomes and the pedagogic methods and perspectives that should be applied to realize the vision. The construct encompasses the pattern of common understandings about learning and teaching among school's teachers and leaders. It can be used to gauge the degree that educational priorities are aligned with an articulated school vision, priorities of the local community and the school itself. If there is no common holistic vision in a school organization, every issue in the school will be handled in a vacuum, rather than in a manner that leads the organization in a clear, desirable direction. Instead it will meander, go astray, and respond to the latest fad. Elements of the holistic dimension should also meet demands in curricular and other national steering documents (Mogren et al., 2019). To establish a whole school approach for ESD, clearly the holistic ideas should be sustainable development.

The routines and structures dimension concerns the degree to which stability and security are maintained, and teaching is protected from disturbances, through routines, scheduling, locations, and teachers' working units. It incorporates arrangements for student healthcare, staff services, and evaluation. In the everyday activities it encompasses, the underlying logic is that any disturbance is related to a lack of fit between individuals and structures, rather than the structures per se. Consequently, the processes and structures of this dimension are conservative in nature and follow a top-down leadership logic. There is high quality in this dimension when the routines and structures facilitate implementation of the school's holistic vision and are adjusted to new learnings through interactions with factors associated with the other dimensions of school organization (Mogren et al., 2019).

Professional knowledge creation is the dimension that accommodates the organization in relation to changes in the community outside school. Disturbances in everyday pedagogical practice are indications that the general educational (and particularly teaching) arrangements do not match changes in the outer world or community and thus should be adapted. This may involve changes in the curriculum or local demands on the school, such as demands for sustainability or greening. Professional knowledge creation is important for developing new understanding of learning and teaching, so it is strongly linked to critical reflection in education and understanding different values that underpin different views of reality. This requires application of a bottom-up logic of leadership based on experience from praxis. High quality in the professional knowledge creation dimension includes willingness to change routines and structures, e.g., through promoting peer-learning and adjusting the teaching according to societal needs. The underlying idea for an effective school organization is that suggestions for improvement may be evaluated within the professional knowledge creation domain and subsequently incorporated in the organization's routines and structures (Mogren et al., 2019).

The teaching and learning dimension refers to the classroom level of school organization. Ideally, teachers and students should create learning platforms together, handling situations in the learning interaction in a manner that promotes both the students' learning of new knowledge and the teacher's teaching ability. The teacher has a key role in structuring environments and situations for students to learn and is helped or hindered in this by factors stemming from the other organizational dimensions (*Holism, routines and structures*, and *professional knowledge creation*). If there is coherence between the other three dimensions, they will jointly support the practical

classroom activities, thereby enhancing student outcomes. For example, if the scheduling (part of the routines and structures domain) is flexible in a school that holistically embraces ESD (for instance), a teacher may let students participate in a debate arranged by a local TV station regarding a topical issue such as exploitation of a local lake. Thus, factors associated with these three dimensions would promote quality in the *teaching and learning* dimension. Alternatively, in a school with the opposite attributes, poor knowledge creation and rigid scheduling would prevent such interest being raised, and any such out of school activities taking place, in alignment with the lack of holism, resulting in poorer teaching and student outcomes (Mogren et al., 2019).

As can be seen in Fig. 11.4, if the other three dimensions (*holistic idea, routines and structures*, and *professional knowledge creation*) are aligned, they can reinforce the quality of teaching and learning, thereby leading (if the holistic idea is ESD and sustainability) to development of action competence for sustainability among the students. However, if the dimensions of the school organization are disjointed, they will impair attempts to make elements of the school work synergistically, and it will be difficult to work with ESD or sustainability according to a whole school approach.

In order to implement ESD or sustainability education according to a whole school approach, it is first necessary to ensure that the holistic idea is ESD or sustainability. Moreover, the holistic idea needs to be negotiated and agreed at all levels of the school, including school leaders, teachers, other staff, and students, as well as stakeholders outside the school with whom the school forms partnerships and community links (Fig. 11.2). For this, relevant elements of the knowledge creation dimension of the school organization, which channels new ideas and influences from outside school, such as ESD and sustainability, into the organization, must be activated. The agreed holistic idea of ESD and sustainability must also be clearly manifested in relevant documents (such as those describing school policy, class policy, and community partnerships) and integrated as a vivid and lived experience in the processes and cultures of the school organization and class organization and, of course, in the teaching and learning (Figs. 11.3 and 11.4). If all this is accomplished, ESD or sustainability will have been successfully institutionalized in the school organization (Fullan, 2001), and I would say that this is a precondition for work according to a whole school approach.

However, it is very difficult to achieve true institutionalization, and it requires integration of ESD and sustainability into the routines and structure dimension of the school organization. In that way, by including and aligning ESD and sustainability to the conservative structures and routines of the organization, sustainability will always be at the top of the agenda. As can be seen from the Scherp model, in order to work with ESD or sustainability according to a whole school approach, it is important to establish a balance with the logics of the organization that are linked to the outside community. It is then important to strive to change (in response to those links) relevant elements of the professional knowledge creation dimension, as well as elements of the routines and structures dimension that maintains the organization's stability and focus. When these dimensions are in tune and connected to the same agreed holistic idea, it is possible for a school to work according to a whole school approach with the potential to achieve a double transformation, i.e., the students changing the world and the students becoming changed by the world.

This idea is contrary to what has often been proposed in ESD and sustainability literature, where the focus has been on transgressing borders between the school and society rather than working in alignment with them (Lotz-Sisitka et al., 2015). Whole school approaches have also been advocated using the same rhetoric as exemplified by this quotation: "What we are advocating is a participative democracy that transfers power to people, and not necessarily a representative democracy" (Shallcross & Robinson, 2008, p. 309). Based on empirical studies, I would instead argue for the suggested approach to work in alignment with the school organization, a position also supported in other recent studies (Verhelst et al., 2020).

11.5 A Strategy to Implement a Whole School Approach

The previous section describes a model of the school organization to implement a whole school approach for ESD. However, as shown in the literature, accomplishing whole school approaches has empirically proven difficulty. In this final, section of the chapter, I describe strategies for accomplishing the goal of implementing ESD through whole school approaches, based on findings from empirical studies.

Studies of the top Swedish schools striving to adopt whole school approaches have identified several ESD quality criteria that guide the schools' work according to their leaders (Mogren & Gericke, 2017a). Clustering of the schools' practices according to ESD implementation strategies, based on these criteria, identified four main principles: *collaborative interaction and school development, student-centered education, cooperation with local society,* and *proactive leadership and continuity* (Mogren & Gericke, 2017b). Interestingly, closer examination of the results revealed that the schools focused on either the internal part of the organization, i.e., the dimension of routines and structures, or the external community links, i.e., professional knowledge creation. The main problem in implementing whole school approaches for the participating schools was in combining the internal aspects with the community links. However, one school was able to institutionalize the ESD perspectives within its routines and structures after first introducing ESD via the knowledge creation process. The results indicated that students at this school played an active role in planning their education and learning, primarily in a way that was linked to societal needs. The school had a flexible scheduling system that was used as a resource for planning transformative activities that supported students' collaboration with society (Mogren & Gericke, 2017b). This, I would argue, could be the very essence of a school working according to a whole school approach. Let us now consider the road this school took to reach that position.

A major aspect of this school's historical development to practice a whole school approach revealed by the study was that it had undergone something of a transition from proactive leadership established by its previous school leader toward a more student-centered education-oriented style of leadership by the current school leader (Mogren & Gericke, 2017b). This transition seems to have influenced the development of the school's practices over time. The whole school unit eventually cooperated with society in a more inclusive manner than previously. In contrast, in the other participating schools in the cited study, collaborations with society were primarily treated as the school leader's responsibility (Mogren & Gericke, 2017b). Thus, the study suggests that an excessively strong leadership focus might hinder truly participatory school development that engages the whole school. However, by gradually transitioning from a proactive leadership focus toward a student learning focus, this school managed to use external qualities to support development of its internal qualities. This suggests a potentially useful two-step strategy for establishing whole school approaches: first, starting with a proactive leadership in the ESD implementation process in which the school leader is primarily responsible for managing societal input into the school organization and, second, allowing gradual transition to a student-centered school organization in which the students, and their teachers at the classroom level, are empowered to become actors that influence and are influenced by society. This integration of internal and external processes will probably be required to achieve a truly transformative ESD whole school approach. Finally, to institutionalize the ESD perspective in the organization, it is important to use anchor functions of the routines and structures dimension, and collegial work methods and groups must be constituted within the organization to organize ESD in the teaching and learning domain (Mogren & Gericke, 2019).

To conclude, if a school aims to implement ESD or sustainability through a whole school approach, empirical evidence indicates that the process should focus on school organization qualities that integrate internal and external processes and structures. A whole school approach should take into consideration internal qualities such as teaching and learning, long-term planning aspects, and external qualities including school operations such as integrated governance or stakeholder and community involvement. The internal learning processes of the school and the external demands of society can then synergistically contribute to transformative education, thereby making the learning process more meaningful. In this way a whole school approach with the potential to equip students with action competence for sustainability could be implemented that is characterized by a process of rethinking systemic dysfunction of the society, as called for in the ESD discourse (Lotz-Sisitka et al., 2015).

References

Berglund, T., Gericke, N., & Chang Rundgren, S. N. (2014). The implementation of education for sustainable development in Sweden: Investigating the sustainability consciousness among upper secondary students. *Research in Science & Technological Education, 32*(3), 318–339. https://doi.org/10.1080/0 2635143.2014.944493

Berglund, T., Gericke, N., Boeve-de Pauw, J., Olsson, D., & Chang, T. (2020). A cross-cultural comparative study of sustainability consciousness between students in Taiwan and Sweden. *Environment, Development and Sustainability, 22*(7), 6287–6313. https://doi.org/10.1007/s10668-019-00478-2

Boeve-de Pauw, J., Gericke, N., Olsson, D., & Berglund, T. (2015). The effectiveness of education for sustainable development. *Sustainability, 7*(11), 15693–15717. https://doi.org/10.3390/su71115693

Boeve-de Pauw, J., & Van Petegem, P. (2013). The effect of eco-schools on children's environmental values and behaviour. *Journal of Biological Education, 47*(2), 96–103. https://doi.org/10.1080/00219266.2013.764 342

Breiting, S., & Meyer, M. (2015). Quality criteria for ESD schools: Engaging whole schools in education for sustainable development. In R. Jucker & R. Mathar (Eds.), *Schooling for sustainable development in Europe* (pp. 31–46). Springer.

Brousseau, G. (1997). *Theory of didactical situations in mathematics* (trans: Balacheff N, Cooper M, Sutherland R, Warfield V). Kluwer.

Cincera, J., & Krajhanzl, J. (2013). Eco-schools: What factors influence pupils' action competence for pro-environmental behaviour? *Journal of Cleaner Production, 61*, 117–121. https://doi.org/10.1016/j.jclepro.2013.06.030

Eames, C., Barker, M., Wilson-Hill, F., & Law, B. (2010). *Investigating the relationship between whole-school approaches to education for sustainability and student learning. A summary.* New Zealand Council for Educational Research.

Education, T. S. N. A. f. (2011). *Curriculum for the compulsory school, preschool class and school-age educare (revised 2018).* https://www.skolverket.se/getFile?file=3984. Accessed 15 Mar 2021.

Ferreira, J., Ryan, L., & Tilbury, D. (2006). *Whole-school approaches to sustainability: A review of models for professional development in pre-service teacher education.* Australian government department of the environment and heritage and the Australian Research Institute in Education for Sustainability (ARIES).

Fullan, M. (2001). *The new meaning of educational change.* Routledge.

Gericke, N., Huang, L., Knippels, M. C., Christodoulou, A., Van Dam, F., & Gasparovic, S. (2020a). Environmental citizenship in secondary formal education: The importance of curriculum and subject teachers. In A. Hadjichambis et al. (Eds.), *Conceptualizing environmental citizenship for 21st century education* (pp. 193–212). Springer.

Gericke, N., Manni, A., & Stagell, U. (2020b). The green school movement in Sweden – Past, present and future. In A. Gough, J. C. Lee, & E. P. K. Tsang (Eds.), *Green schools globally: Stories of impact on education for sustainable development* (pp. 309–332). Springer.

Goldman, D., Pe'er, S., & Yavetz, B. (2017). Environmental literacy of youth movement members–is environmentalism a component of their social activism? *Environmental Education Research, 23*(4), 486–514. https://doi.org/10.1080/13504622.2015.1108390

Gough, A. (2005). Sustainable schools: Renovating educational processes. *Applied Environmental Education and Communication, 4*(4), 339–351. https://doi.org/10.1080/15330150500302205

Gough, A., Lee, J. C., & Tsang, E. P. K. (Eds.). (2020). *Green schools globally: Stories of impact on education for sustainable development.* Springer.

Henderson, K., & Tilbury, D. (2004). *Whole school approaches to sustainability: An international review of sustainable school programs.* Report prepared by Macquarie University for the Australian Government Department of the Environment and Heritage. Sydney.

Kollmuss, A., & Agyeman, J. (2002). Mind the gap: Why do people act environmentally and what are the barriers to pro-environmental behavior? *Environmental Education Research, 8*(3), 239–260.

Lave, J., & Wenger, E. (1991). *Situated learning: Legitimate peripheral participation.* Cambridge University Press.

Lotz-Sisitka, H., Wals, A. E., Kronlid, D., McGarry, D., & D. (2015). Transformative, transgressive social learning: Rethinking higher education pedagogy in times of systemic global dysfunction. *Current Opinion in Environmental Sustainability, 16*(17), 73–80. https://doi.org/10.1016/j.cosust.2015.07.018

Mathar, R. (2015). A whole school approach to sustainable development: Elements of education for sustainable development and students' competencies for sustainable development. In R. Jucker & R. Mathar (Eds.), *Schooling for sustainable development in Europe* (pp. 15–30). Springer.

McKeown, R., & Hopkins, C. (2007). Moving beyond the EE and ESD disciplinary debate in formal education. *Journal of Education for Sustainable Development, 1*(1), 17–26. https://doi.org/10.1177/097340820700100107

Mezirow, J. (2009). An overview on transformative learning. In K. Illeris (Ed.), *Contemporary theories of learning: Learning theorists in their own words* (pp. 90–105). Routledge.

Mogensen, F., & Schnack, K. (2010). The action competence approach and the 'new' discourses of education for sustainable development, competence and quality criteria. *Environmental Education Research, 16*(1), 59–74. https://doi.org/10.1080/13504620903504032

Mogren, A. (2019). *Guiding principles of transformative education for sustainable development in local school organizations: Investigating whole school approaches through a school improvement lens* (Doctoral thesis). Karlstad University Studies.

Mogren, A., & Gericke, N. (2019). School leaders' experiences of implementing education for sustainable development—Anchoring the transformative perspective. *Sustainability, 11*(12), 3343. https://doi.org/10.3390/su11123343

Mogren, A., & Gericke, N. (2017a). ESD implementation at the school organization level, part 2 – Investigating the transformative perspective in school leaders' quality strategies at ESD schools. *Environmental Education Research, 23*(7), 993–1014. https://doi.org/10.1080/13504622.2016.1226266

Mogren, A., & Gericke, N. (2017b). ESD implementation at the school organization level, part 1 – Investigating the quality criteria guiding school leaders' work at recognized ESD schools. *Environmental Education Research, 23*(7), 972–992. https://doi.org/10.1080/13504622.2016.1226265

Mogren, A., Gericke, N., & Scherp, H.-Å. (2019). Whole school approaches to education for sustainable development: A model that links to school improvement. *Environmental Education Research, 25*(4), 508–531. https://doi.org/10.1080/13504622.2018.1455074

Olsson, D., Gericke, N., Pauw, B.-d., Berglund, T., & Chang, T. (2019). Green schools in Taiwan – Effects on student sustainability consciousness. *Global Environmental Change, 54*(1), 184–194. https://doi.org/10.1016/j.gloenvcha.2018.11.011

Olsson, D., Gericke, N., & Chang Rundgren, S. N. (2016). The effect of implementation of education for sustainable development in Swedish compulsory schools-assessing pupils' sustainability consciousness. *Environmental Education Research, 22*(2), 176–202. https://doi.org/10.1080/13504622.2015.1005057

Olssen, D., Gericke, N., Sass, W., & Boeve-de Pauw, J. (2020). Self-perceived action competence for sustainability: The theoretical grounding and empirical validation of a novel research instrument. *Environmental*

Education Research, 26(5), 742–760. https://doi.org/1 0.1080/13504622.2020.1736991

Parra, G., Hansmann, R., Hadjichambis, A., Goldman, D., Paraskeva-Hadjichambi, D., Sund, P., Sund, L., Gericke, N., & Conti, D. (2020). Education for environmental citizenship and education for sustainability. In A. Hadjichambis et al. (Eds.), *Conceptualizing environmental citizenship for 21st century education* (pp. 149–160). Springer.

Riastini, P. N., Wati, C. S., Prodjosantoso, A. K., & Suryadarma, I. G. P. (2019). Is there any difference in waste consciousness between national eco-schools and others? *International Journal of Instruction, 12*(4), 513–528. https://doi.org/10.29333/iji.2019.12433a

Rittel, H. (1973). Dilemmas in a general theory of planning. *Policy Sciences, 4*(2), 155–169. https://doi.org/10.1007/BF01405730

Rowe, F., & Stewart, D. (2009). Promoting connectedness through whole-school approaches: A qualitative study. *Health Education, 109*(5), 396–413. https://doi.org/10.1108/09654280910984816

Rowe, F., Stewart, D., & Patterson, C. (2007). Promoting whole school connectedness through whole school approaches. *Health Education, 107*(6), 524–542. https://doi.org/10.1108/09654280710827920

Sarabhai, K. V. (2007). *Tbilisi to Ahmedabad, the journey of environmental education.* Centre for Environmental Education (CEE).

Sass, W., Boeve-de Pauw, J., Olsson, D., Gericke, N., De Maeyer, S., & van Petegem, P. (2020). Redefining action competence: The case of sustainable development. *The Journal of Environmental Education, 51*(4), 292–305. https://doi.org/10.1080/00958964.2020.1765132

Sass, W., Boeve-de Pauw, J., De Meyer, S., Gericke, N., & van Petegem, P. (2021). Actions for sustainable development through young students' eyes. *Environmental Education Research, 27*(2), 234–253. https://doi.org/1 0.1080/13504622.2020.1842331

Scherp, H. (2013). *Lärandebaserad skolutveckling. Lärglädjens förutsättningar, förverkligande och resultat* [Learning based school development. The joy of learning, Conditions, Implementation and Results. Studentlitteratur.

Scott, W. (2013). Developing the sustainable school: Thinking the issues through. *Curriculum Journal, 24*(2), 181–205. https://doi.org/10.1080/09585176.2013.781375

Shallcross, T. (2005). Whole school approaches to education for sustainable development through school-focused professional development (The SEEPS project). *Education for a Sustainable Future.* NSW: Macquarie University. http://kpe-kardits.kar.sch.gr/Aiforia/international_review2.pdf. Accessed 21 March 2021.

Shallcross, T., & Robinson, J. (2008). Sustainability education, whole school approaches, and communities of action. In A. Reid et al. (Eds.), *Participation and learning* (pp. 299–320). Springer.

Spinola, H. (2015). Environmental literacy comparison between students taught in eco-schools and ordinary schools in the Madeira Island region of Portugal. *Science Education International, 26*(3), 392–413.

Sund, P., & Gericke, N. (2020). Teaching contributions from secondary school subject areas to education for sustainable development – A comparative study of science, social science and language teachers. *Environmental Education Research, 26*(6), 772–794. https://doi.org/10.1080/13504622.2020.1754341

Sund, P., Gericke, N., & Bladh, G. (2020). Educational content in cross-curricular ESE teaching and a model to discern teacher's teaching traditions. *Journal of Education for Sustainable Development, 14*(1), 78–97. https://doi.org/10.1177/0973408220930706

UNESCO. (2017). *Global action programme on education for sustainable development.* UNESCO.

Verhelst, D., Vanhoof, J., Boeve-de Pauw, J., & Van Petegem, P. (2020). Building a conceptual framework for an ESD-effective school organization. *The Journal of Environmental Education, 51*(6), 400–415. https://doi.org/10.1080/00958964.2020.1797615

Warner, B. P., & Elser, M. (2015). How do sustainable schools integrate sustainability education? An assessment of certified sustainable K–12 schools in the United States. *The Journal of Environmental Education, 46*(1), 1–22. https://doi.org/10.1080/00958964.2014.953020

Wyn, J., Cahill, H., Holdsworth, R., Rowling, L., & Carson, S. (2000). Mind matters, a whole-school approach promoting mental health and wellbeing. *Australian and New Zealand Journal of Psychiatry, 34*(4), 594–601. https://doi.org/10.1080/j.1440-1614.2000.00748.x

Niklas Gericke is Professor of Science Education and director of the research center SMEER (Science, Mathematics and Engineering Education Research) at Karlstad University in Sweden, and visiting professor at NTNU in Trondheim, Norway. His main research interests are biology education and sustainability education. His area of research includes education for sustainable development from both conceptual as well as implementation perspectives. Much of his work relates to how disciplinary knowledge is transformed into school knowledge and what impact these transformations have on teachers' work and students' understanding. Niklas Gericke http://orcid.org/0000-0001-8735-2102

Building Primary Schools as a Model of Sustainable Communities: Hints for Teachers

12

Elvan Şahin

Abstract

Children seem to be victims of deterioration in ecological systems but might be also actors having roles in shaping the future of our planet when equipped with the necessary skills, worldviews, and values. Education for Sustainable Development (ESD) has been viewed as an approach that may enable future generations engage in pro-sustainable actions within their local environment. This chapter touches upon sustainability practices carried out in Green Schools and Eco-Schools Programme utilizing 'whole school approach'. Learning outcomes including knowledge, values, value orientations and motivation as precursors of pro-environmental behaviours have been considered while revealing the impacts of these certification programmes. Primary schools that declare their usage of whole school approach succeed in increasing knowledge of children but fail to cultivate non-cognitive constructs. Educators should have in-depth insight into factors acting as barriers and facilitators while transforming primary schools into sustainable communities. For instance, the degree of participation in actions, interaction with nature, and local community involvement are among the issues that we need to evaluate if we intend to observe lifelong impacts of ESD initiations.

Keywords

Education for sustainable development in primary schools · Whole school approach · Climate change education

12.1 Introduction

As an impact of human actions, our planet has entered the Anthropocene where highly interconnected global challenges such as deforestation, water scarcity, and climate change threaten survival of every living thing. Many calls declare a need for an urgent shift in human practices in order to enable more sustainable lifestyles (Corcoran et al., 2017). Children have appeared to be victims of deterioration in ecological systems but might undertake important roles in shaping their own future when equipped with the necessary knowledge, skills, worldviews, and values (Hickman & Riemer, 2016). Education for Sustainable Development has been viewed as a contemporary educational approach that may enable future generations engage in pro-sustainable actions within their local environment (Hofman, 2015). This chapter covers

E. Şahin (✉)
Department of Mathematics and Science Education, Middle East Technical University, Ankara, Türkiye
e-mail: selvan@metu.edu.tr

current reflections and implications on ESD schools utilizing 'whole school approach' which has taken attentions of practitioners starting from the 1990s. As proposed by McKeown and Hopkins (2007), this approach does not only touch upon multidisciplinary nature of ESD but embrace diverse disciplines to model some practices for a more sustainable future in all school activities from transportation to energy and water usage. ESD schools pursuing 'whole school approach' have responsibilities to build up some opportunities for children to learn sustainability and to experience sustainability. Community engagement, parent involvement, planning, monitoring, and evaluation as well as curriculum are incorporated in the teaching and learning process. This chapter intends to draw some implications on ESD practices for primary school teachers using whole school approach while delving into climate change education as one of the 'adjectival' educations (UNESCO, 2012). To be more specific, firstly, it is aimed to present the goals of two initiatives using whole school approach, namely, Green Schools and Eco-Schools and their connection with theoretical models of pro-environmental behaviours. Secondly, learning experiences that students of these two ESD initiatives engaged in are exemplified for ESD practitioners considering the context of climate change education. Lastly, some empirical evidences and their implications from EE and ESD literature utilizing the theoretical models are provided to portray the effects of Green Schools and Eco-Schools in various countries.

12.2   Education for Sustainable Development in Primary Schools Utilizing Whole School Approach

Whole school approach has been widely accepted by various initiatives following the targets of ESD in primary schools. Green Schools and Eco-Schools are depicted among these school-based initiatives implemented in different regions of the globe (Dzerefos, 2020). The Green Schools

Alliance with a head office in the USA connects more than 10,000 schools across 91 countries in order to empower school communities in their transformation for a sustainable future (Green Schools Alliance, 2021). Whole school sustainability positioned in the core of the alliance incorporates innovative and sustainable attempts into their educational programs, physical place, and organizational culture. Students in these sustainable schools are viewed as 'environmental stewards' donated with problem-solving and critical thinking skills through hands-on and outdoor learning opportunities. Accordingly, stewards of our planet should have their significant life experiences in such a school that will inspire them and cultivate their sense of responsibility and awareness. As a community member of the Green Schools, teachers, students, and administrators have an opportunity to join free online platforms that offer great potential for widespread use on a global scale.

The Eco-Schools project has self-reported assessment systems indicating improvements in quality. With respect to the number of schools actively engaged in the programs, Eco-Schools connecting about 56,000 schools across 70 countries appears as the most widely implemented initiative serving for ESD (Foundation for Environmental Education [FEE], 2021). The Eco-Schools Programme is a global model for sustainability practices which was developed in 1992 considering the needs claimed at the United Nations Conference on Environment and Development. This sustainable system seeks to 'encourage young people to engage in their environment by allowing them the opportunity to actively protect it' (FEE, 2021). To be more specific, Eco-Schools cultivates young people's motivations and attitudes to make a difference in sustainable and environmentally responsible behaviours while carrying these impacts to local communities beyond the school gates. Stated by Boeve-De Pauw and van Petegem (2013), Eco-Schools employs a multidisciplinary and participatory approach to create a sustainable community while linking local issues with global challenges such as waste reduction and climate change.

12.3 Precursors of Pro-sustainable Behaviours as Learning Outcomes in the Context of ESD?

Paths might differ from each other in ESD initiatives, but the ultimate goal pursued is common. These programmes aim to build up sustainable communities while leading a transformation in young people's actions with a lifelong impact. At this point, it is a matter of what we, as sustainability practitioners, should touch upon to achieve this transformation for pro-sustainable behaviours. We will focus on knowledge, motivation, and values regarded as the most commonly cited learning outcomes of ESD while slightly delving into theoretical models of pro-environmental behaviours. These theoretical models depict how knowledge, motivation, and values act as precursors of behaviours that lead educators to consider these constructs as learning outcomes in the context of ESD.

According to Kollmuss and Agyeman (2002), the most primitive assumption that the more knowledge you have the more enlightened you act directs communication campaigns and plans managed by environmental non-governmental organizations (NGO). It seems that governmental agencies and various companies use this claim, for instance, Profilo company runs a campaign titled 'Turkey is Collecting its Energy' to develop women's awareness of energy conservation. Similarly, CEVKO as an environmental NGO in Turkey and Migros, a grocery chain, prepared a recycling campaign for awareness and consciousness of public. This linear and simplest model proposed that informing people about environmental issues would lead to more favourable attitudes towards the environment and which in turn assumed to result in more pro-environmental behaviours (Robottom & Hart, 1995), but in reality changing behaviours is not so simple (Kollmuss & Agyeman, 2002). Although more recent theoretical models do not take account of knowledge as the direct determinant of behaviours or of attitudes, ESD practitioners and teachers still claim that when their students know more about the environment, they will take an action for sustainability (Anyolo et al., 2018). It is noteworthy to say that educating young people on knowledge of environmental issues is significant whereas it is just surprising to expect such a direct impact of knowledge on attitudes and behaviours.

An individual's motivation to act for the environment based on the Self-Determination Theory was introduced to the field by Deci and Ryan (1985). These pioneers represented a framework on motives with different degrees of self-determination which may result in behaviours. The motives were categorized as amotivation, controlled motivation, and autonomous motivation. Autonomous motivation, the most self-determined category, reflects an inherent interest and intrinsic reasons to perform the specific behaviour. The behaviours associated with autonomous motivation seem to be consistent and long-lasting, and individuals could show resistance to change them when they come across with a challenge. If people do not experience any driving force to demonstrate a specific behaviour, they tend to be amotivated. Amotivation placed at the opposite end of the spectrum is regarded as a barrier to engage in an action. Controlled motivation, on the other hand, represents an individual's goal set by external or internal pressures. Regarding these motives that span the continuum of self-determination, De Young (1996) noted the different levels of engagement potentials in a behaviour. To be more specific, environmental practitioners expect individuals with self-determined motivations to take the necessary actions for the environment although it is complex and uncomfortable to perform these actions (Green-Demers et al., 1997).

Values were previously defined by Schwartz (1992) as a favourable, trans-situational target which serves as a guiding principle in human's life. They are indicated as significant agents since being relatively stable over time, but more than that values are evaluated in terms of their influences on beliefs, attitudes, norms, and behaviours (Stern, 2000). Stern and colleagues underlined the role of values in his Value-Belief-Norm Theory suggesting that values affect an individual's environmental worldviews, which in turn influences beliefs about adverse consequences of engagement in an action and beliefs regarding perceived responsibility to eliminate these adverse consequences (Stern et al., 1999). These beliefs also collectively activate personal norms which in turn lead to sustainable behaviours. In this respect, due to being a precursor of behaviours, environmental values have been widely emphasized as a target in ESD and EE literature. What makes 'values' so important in primary education in the context of EE and ESD goals is also based on suggestions of Dietz et al. (2005). These researchers emphasized that 'values' are formed early in life and are not always rapidly changing over time. However, it should be noted that factors shaping an individual's values could be also enhanced early in life and are learned linked to values.

Environmental values were operationalized in the context of the two major environmental values model (2-MEV) (Bogner & Wiseman, 2006). Bogner and Wiseman (2006) intended to develop a valid and reliable instrument based on a conceptual framework for the age group of adolescents since these individuals are significant due to their roles in environmental protection. These authors pointed out that environmental values of this population could not be represented in the items specifically designed for adults. In their model, two uncorrelated higher-order dimensions, namely, utilization and preservation, emerged. Preservation is regarded as an ecocentric reflection of environmental conservation and protection, whereas the second dimension, utilization, is a measure of anthropocentrism indicating the use and exploitation of the natural resources. Values structures depicted in the 2-MEV model have attracted attentions and have been placed in research in the field of both education and psychology of sustainable development. There is a strong consistency among research studies (e.g., Boeve-de Pauw & Van Petegem, 2011; Milfont et al., 2010) in that preservation values have an explanatory power in explaining pro-environmental behaviours. In other words, people with

more ecocentric inclinations for environmental conservation have higher tendency to engage in pro-environmental actions. Utilization values, contrary to our expectations, were reported to have no significant or only weak negative association with such behaviours. Thus, enhancement of preservation values while reducing utilization values seems as favourable outcomes of ESD and EE practices.

Since there is a need to describe and shape values in early ages of a human life, it is inevitable to portray the nature of basic belief patterns known as value orientations associated with values. As presented in the context of Value-Belief-Norm (VBN) Theory proposed by Stern et al. (1999), value orientations which are depicted under the term of environmental worldviews reflecting belief component are affected by values. Similar to the location of value orientations between values and norms in the VBN Theory, a previously proposed framework known as 'cognitive hierarchy model' by Fulton et al. (1996) forms the baseline of associations among values, value orientations (basic belief patterns), attitudes and norms, behavioural intentions, and behaviours. Accordingly, behaviours are numerous, quick to change, and specific to situations, whereas values and value orientations one step above are slow to change and few in number. The value orientations in EE and ESD literature have focused on the human-nature relationship specifically, and researchers in this field (Dunlap et al., 2000; Hills, 2002) associated the term of environmental worldviews with an individual's belief regarding human's relationship with nature. Thus, unlike values, value orientations refer to a domain of interest such as human-nature relationship (Li & Ernst, 2015) and human-forest relationships (Ritter & Dauksta, 2012).

Our understanding of value orientations has been evolved and shaped with respect to research conducted more than a half century. Among these research studies, investigation of urban youth's value orientations from a cross-cultural perspective (between American and

Chinese participants) by Li and Ernst (2015) deserves a scrutiny of the implications provided for EE and ESD practices. Urban youth's value orientations regarding human-nature relationship emerged under five categories, namely, humanistic, interdependence, stewardship, use, and dominion. *Humanistic* category refers to human's love of nature with deep affection. *Stewardship* category focuses on human's worldviews pertinent to responsibility for protecting and respecting nature. *Interdependence* category represents the belief in that we as humans should protect nature because we need the materials (oxygen, food, water, etc.) provided by nature. On the other hand, *use* category reflects human's use of nature to meet basic needs or to enjoy. *Dominion* category seems similar to *use* but additionally emphasizes that humans have the right to exert power on nature to obtain whatever they want or need. Li and Ernst (2015) reminded that starting from Tbilisi Declaration (UNESCO, 1978), environmental education (EE) aims to help individuals attain a set of values and feelings of concern for the environment. Thus, considering the significance of values and value orientations in EE and ESD, it appears that an individual with humanistic, interdependence, or stewardship value orientations in the domain of human-nature relationship is more preferable than the ones holding use or dominion value orientations. Supported by the report from the World Commission on Environment and Development, EE should seek to enhance children's sense of respect and feelings of responsibility for nature and other humans which is a requirement to make the necessary transition to sustainable development.

Considering their significance in the causation of pro-environmental behaviours, it seems meaningful to place knowledge, values, value orientations, and motivations as learning outcomes in EE and ESD initiatives as Green Schools and Eco-Schools project. These theoretical models also have an influence on research in the field of EE and ESD while determining the effects of these educational programs in various countries.

12.4 Becoming a Model Community Through Whole School Approach: Climate Action Schools

The theoretical models summarized above guided practitioners and researchers on possible learning outcomes in the context of EE and ESD to educate future generations engaging in some actions for a sustainable community. However, how these learning outcomes could be facilitated is a challenge. Thus, high-quality experiences in school environment might help children in creating their own sustainable community. After The Rio+20 Conference in 2012 (also known as United Nations Conference on Sustainable Development), it was documented that all educational sectors including formal and non-formal education are encouraged to develop high-quality practices on their campuses including their communities with the active engagement of students, teachers, local administrators, and partners (United Nations, 2012). A special emphasis remarked teaching sustainability issues as an integrated element across all disciplines. Thus, teaching and learning need a change in terms of roles of individuals in school community including students and teachers in that process. Furthermore, different fields in cross-curricular education such as nutrition education, environmental education, and education for democracy can carry us to the targets of ESD.

Sustainable development and the related components could be linked to the ESD through 'adjectival educations' (UNESCO, 2012). In the UNESCO, 2012 report, ESD was placed at the intersection of six adjectival educations which covered climate change education, human rights education, consumer education, development education, environmental education, and disaster risk reduction education. While having a closer look into ESD practices in primary schools, this chapter focused on climate change education that is highly associated with the other adjectival educations.

UNESCO (2015) published *Getting Climate-Ready: A Guide for Schools in Climate Action* indicating the central role of schools in helping children grasp the causes and consequences of climate change. This guideline intends to aid schools while educating these individuals in developing the appropriate values and skills to take necessary actions in the transition to sustainable development. It is declared that the number of schools adopting whole school approaches to climate action is increasing across the globe every day. This publication clarifies the whole school approach to climate change as taking necessary actions to reduce climate change in every aspect of school community. Accordingly, climate action requires inclusion of school governance, teaching and learning, campus facilities, and operations, establishing community partnerships and the broader communities. Active engagement of all people from different components of school community, namely, principals, teachers, students, families, and school staff, is the key for the whole school approaches to succeed in tackling with climate change.

The guidelines covered in UNESCO 2015 publication on climate action enlighten the path for school community members in Eco-Schools, Green Schools, and the others following whole school approach. Considering these guidelines, as an initial step under the subheading of 'school governance', sustainable schools invite everyone in the school community to put an effort towards climate action targets. It seems inevitable to share responsibilities with respect to roles in school life, but setting up a team is an effective way to monitor the process. This team is required to include members in- and outside the school such as student representatives from different grade levels (*Eco-Warriors in Eco-Schools*), representative parents and teachers from each subject, and individuals from local community with different skills and background knowledge to coordinate for climate actions. The role of the team members is to make a plan for climate actions, coordinate implementations, assess school progress in work, and make the necessary revisions in climate action plan.

Teaching climate change-related issues in all subject areas is one step further in climate action for a sustainable school. Climate change is among the most significant global challenge that humans have struggled with. Environmental, social, scientific, political, and economic factors interplay with each other in the fight against climate change. This situation implies a responsibility for teaching climate change-related issues by teachers in all subject areas. Table 12.1 presents examples of how sustainable schools teach students climate change in various subjects. Within a school community, different initiations to help students understand reasons and impacts of climate change and engage in responsible actions attract our attention since science or social sciences do not appear as only subjects placing goals pertinent to climate change. Consistent with the suggestions in UNESCO (2015) guidelines, the students of these sustainable schools create posters in visual arts, observe consequences of climate change in equator in geography, create tables regarding the habitats of animals in their local area in science, select appropriate designs for efficient use of energy in mathematics, etc.

As reported in the guidelines in UNESCO (2015) document, learning climate change-related issues in all subjects should be enriched through action-oriented processes. Primary teachers could empower their students in terms of knowledge of action strategies and critical and creative thinking skills so that these students will be donated with the necessary components to be successful in climate action. Furthermore, primary school students can use their knowledge and skills while designing and making their projects to achieve the transition to sustainable development. Different living spaces on school campus, for instance, buildings, gardens, and cafeterias, could be regarded as the places where students' own original climate action projects are designed for and implemented. To exemplify, planting more trees, flowers, fruits, and vegetables in the schoolyard is an effective way to create outdoor learning area for climate change mitigation. Table 12.2 summarizes us a climate action journey of a Green School under the energy theme. How these students are taught about actions for energy usage, how they learn through actions with a focus on energy conserva-

Table 12.1 Climate change-related issues in every subject: schools and their teaching activities

METU College Primary School, awarded with Eco-School Green Flag in 2019, is located in Ankara, Turkey, and presented its report on the topic of 'Light Pollution' with the Eco-Principle stated as 'Let's use the light right, illuminate the future' (ODTÜ Geliştirme Vakfı Özel İlköğretim Okulu, 2011)

Subjects	Grade level	Description of the activity
Mathematics	4th grade	Bring various light sources to the classroom to discuss the location, amount, and angle of the light with respect to the usage situation. Determine the conditions that light pollution is caused. Examine the examples of situations where the light illuminates at an angle of 45-90-180-360^0, and discuss about the unnecessary, excessive, and incorrect use of light
Social sciences	5th grade	Address light pollution as one of the problems emerging with the population growth after migration and unplanned urbanization. The students were asked to prepare texts or visual materials such as drawing pictures or taking photographs to compare the lighting in rural areas and urban lighting

Scoil Íde is a primary school for girls in Artane, Dublin. They were awarded their Green Flag for Energy in 2016 (Green-Schools, An Taisce Environmental Education Unit, 2021)

Mathematics	All classes	Calculating energy usage at school, comparing energy usage over time, making graphs to represent the comparison

Foxfield Primary School located in London, UK, received the Bronze Eco Schools Award (Foxfield Primary School, 2021)

Geography	2nd grade	Identify features of a locality, as well as naming the seven continents and the five oceans. Compare what the weather is like in different countries as well as observing climate change on the equator.
Science	4th grade	To deepen students' understanding of habitats, they are expected to research how and why certain animals live in particular areas and how they adapt to the climate and environment. Use classification tables while examining the local area to identify the types of animals widely found near this school and ask if there are common factors with the types of animals found

Alp Primary School located in Konya, Turkey, is titled with 'Nutrition-Friendly School' and published an action plan on climate change in 2019 (Alp Eğitim Kurumları, 2019). *It was also awarded with Eco-School Green Flag in 2019*

Visual arts	1st and 2nd grade	After watching a short video about global warming and climate change, the students are expected to draw pictures reflecting their feelings about this issue. Create some posters indicating causes and impacts of climate change to be presented in the Eco-School board
Visual arts	5th grade	Two-dimensional studies are made in cubic forms of animals whose habitats are destroyed as a result of global warming
Language	5th grade	Write stories about the impacts of global warming and climate change

tion in school buildings, and how they learn from these actions are hidden behind the lines of their journey.

Students spend most of their lives at school. It is not just a learning environment for them. School is like a home where students reflect their values, worldviews, and beliefs, especially in terms of climate change. Transformation of these attributes into climate-friendly actions could be reinforced by making school campus a laboratory where they can observe, design, make, and enjoy the learning opportunities. Table 12.3 presents two primary schools acting as a model of sustainable community. Latin America's first sustainable school in Uruguay seeks to achieve its targets

Table 12.2 A climate action journey of Scoil Íde primary school: energy theme (Green Schools, An Taisce Environmental Education Unit, 2021)

Artane, Dublin, Ireland   *Scoil Íde is a primary school for girls. With the practices for energy, they were awarded their Green Flag in 2016 Green Code: "We're Scoil Íde's Clean Green Queens!"*
They set up a team including the teachers, the principal, students from junior infants to sixth class, and the caretaker. The team members meet regularly to monitor the process, examine the problems arising if any, and take decision on their following actions. For their next step on Environmental Review, the members check the school's meter and determine the amount of gas used as their energy source for heating. Next, the necessary appliances and the ones which are no longer in use were determined through an appliance audit. Furthermore, the places that energy maybe wasted in are detected. The team members conducted a survey which indicated limited knowledge of energy in the school community. The students are empowered to take some actions while designing projects on renewable and non-renewable energy resources, creating visual displays of energy use in school, reminding teachers to print on both sides of the paper, etc. By using Sustainable Energy Authority of Ireland lesson plans for 5th and 6th classes, the teachers help students develop different scientific skills and provide opportunities to design and make sustainable energy projects such as wind turbines, solar owens, and energy audit of the school. The students acting as 'energy wardens' and 'pupil of the month' check the amount of energy used in classrooms. They take a role in monitoring that the appliances are shut down when not in use. Every school member is informed about the monthly energy usage by presenting graphs on school board.

Table 12.3 Self-sustaining primary schools as a model of sustainable community

Latin America's first sustainable school is located in Uruguay. There are about 60 children attending the school regularly (Uruguay Natural, 2016)
The school appears 'special' for its students as it does not depend on non-renewable energy sources. School building uses solar panels for electricity and air passages. The water comes from the rain. It was designed by US Architect Michael Reynolds, who is an expert on green buildings. Majority of this school building was built with recycled materials. They used car tires, bottles, and aluminium cans that replace the bricks. The temperature inside is regulated by a natural air conditioner. The building is oriented towards north so it is heated by the sun. The sun also heats the embankment and at night in winter it liberates the heat.   Environmental lessons are taught regularly, and the teachers link the courses to topics such as nature, climate change, and recycling. The students are learning in practice. They grow vegetables and fruits in the school garden and learn to care for their environment.
Self-sustaining eco-school: Killinchy Primary School with Green Flag Award in UK (MyNI Life, 2018)
They are different in terms of their waste reduction and gardening initiations. They have different types of bins, but green caddy bins are placed in the classroom to collect food waste. They also use big compost bins so that food waste breaks down and turns into compost. This compost is used in many different ways around the school. The students have their own wild garden which has some trees and flowers. They used the compost from the bins to plant fruits and vegetables such as tomatoes, pumpkins, and radishes. The students designed and made a system to collect rainwater and use it to water the crops and flowers in the wildlife area. Necessary skills and knowledge for these actions are a part of school curriculum, but these actions also become a part of their everyday lives. The school is self-sustaining with respect to their composting activities and the way that they grow their own vegetables.

by paying some special attention to its curriculum, school garden, and green buildings with sustainable energy and water usage. Killinchy Primary School in Newtownards, UK, has an emphasis on organic waste reduction with composting and gardening activities like growing own vegetables, fruits, and trees. Students take climate actions in school community on a regular basis, and it becomes a lifestyle for them. In other words, these actions are not carried out as a part of a one-off learning activity in primary education.

12.5 Implications on Learning Outcomes of Sustainable Schools: Where Are We on Our Way?

According to UNESCO document on ESD Implementation Framework beyond 2019 (UNESCO, 2019), ESD seeks to encourage individuals to engage in transformative actions for sustainable development. It also emphasizes that how this transformation occurs is so complex, but we are aware of the stones that we need to touch upon to empower learners in this transformation process. Acquisition of knowledge and information with experiential learning, attaching importance to certain values and worldviews, and developing critical thinking skills to reflect on individual values and attitudes are required for fundamental changes to occur for a sustainable community. Here, community refers to political, cultural, and social ties as well as physical integrity which cultivates young generations' individual and collective values and in turn facilitates transformative actions.

In order to portray how much we proceed on our way in creating schools as a model of sustainable community, scientific evidences collected about the impacts of ESD initiations on learning outcomes provide us some significant clues. Literature on impacts of ESD initiations covers comparisons of two groups, namely, schools as participants or non-participants of sustainable school programs in terms of their students' knowledge, values, behaviours, and other attributes. For instance, in Slovenia, Krnel and Naglic (2009) showed that the students engaging in eco-school activities know more about environmental issues than the students that do not. However, non-significant differences between these groups were reported in terms of their environmental attitudes and behaviours. As an experimental research study in Turkey, Özsoy et al. (2012) designed and implemented environmental activities with the cooperation of eco-school teachers, administrators, and the public university in their local area. The researchers and their partners followed the Eco-School program framework that covers seven steps guiding their journey through sustainability transformation. The effects of eco-school implementations were assessed by making comparisons with the regular public schools in terms of environmental knowledge, uses, concerns, and attitudes. Significant differences between experimental and control groups indicated the effects of eco-school implementations on the related variables. A nationwide study in Taiwan conducted by Olsson et al. (2019) determined the effects of Green School Partnership on students' sustainability consciousness. It was reported that Green Schools and regular schools have a similar effect on the students' sustainability consciousness. More specifically, Green School implementations do not facilitate this attribute. Research studies conducted in Flanders (Boeve-de Pauw & Van Petegem, 2011, 2013), in Iceland (Hallfredsdottir, 2011), and in Sweden (Olsson et al., 2016) revealed favourable impacts of certification programmes on knowledge of environmental issues as a cognitive construct but failed to indicate non-cognitive impacts.

More recent evidences were provided by Boeve-de Pauw and Van Petegem (2018), and these researchers elaborated previous evaluations of ESD initiations in terms of learning outcomes. These authors show that the eco-school project in Flanders has a significant power in decreasing students' utilization values but preservation values could not be cultivated. Furthermore, the eco-school project succeeds in increasing knowledge and results in a stronger controlled motivation. These attributes are unfortunately the outcomes with a small or no significant potential in fostering pro-environmental behaviours (Boeve-De Pauw & van Petegem, 2013). Such findings provide some indications of why creating primary schools as a model of sustainable communities still remains a challenge for us. We should leave our understandings of the linear causality of pro-environmental behaviours that 'increased knowledge leads to more favourable attitudes which in turn, results in a change of behaviours'.

Through Eco-School or Green School Programme, across the globe, there are thousands of primary schools which have declared that they accept whole school approach to reach the targets of ESD. However, we should have an in-depth insight into factors intervening as barriers and facilitators while translating primary schools into sustainable communities. For instance, the degree of participation in actions (involvement of every member in school community), interaction with nature (physical, emotional, cognitive), and local community involvement are among the issues that we need to evaluate if we intend to observe lifelong impacts of ESD initiations.

References

Alp Eğitim Kurumları [Alp Educational Institutions]. (2019). *Özel Konya Alp Ortaokulu 2019–2020 Eğitim Öğretim Yılı Eko-Okullar Programı Eylem Planı [Konya Alp College 2019–2020 Academic Year Eco-Schools Programme Action Plan]*. https://www.alpe-gitimkurumlari.k12.tr/assets/docs/Ortaokul20192020.pdf. Accessed 10 June 2021.

Anyolo, E. O., Kärkkäinen, S., & Keinonen, T. (2018). Implementing education for sustainable development in Namibia: School teachers' perceptions and teaching practices. *Journal of Teacher Education for Sustainability, 20*(1), 64–81. https://doi.org/10.2478/jtes-2018-0004

Boeve-de Pauw, J., & Van Petegem, P. (2011). A cross-cultural study of environmental values and their effect on the environmental behaviour of children. *Environment and Behavior, 45*(5), 551–583. https://doi.org/10.1177/0013916511429819

Boeve-De Pauw, J., & van Petegem, P. (2013). The effect of eco-schools on children's environmental values and behavior. *Journal of Biological Education, 47*(2), 96–103. https://doi.org/10.1080/00219266.2013.764342

Boeve-De Pauw, J., & van Petegem, P. (2018). Eco-school evaluation beyond labels: The impact of environmental policy, didactics and nature at school on student outcomes. *Environmental Education Research, 24*(9), 1250–1267. https://doi.org/10.1080/13504622.2017.1307327

Bogner, F. X., & Wiseman, M. (2006). Adolescents' attitudes towards nature and environment: Quantifying the 2-MEV model. *Environmentalist, 26*(4), 247–254. https://doi.org/10.1007/s10669-006-8660-9

Corcoran, P. B., Weakland, J., & Wals, A. E. J. (2017). *Envisioning futures for environmental and sustainability education*. Wageningen Academic Publishers.

Deci, E. L., & Ryan, R. M. (1985). *Intrinsic motivation and self-determination in human behaviour*. Plenum Press.

De Young, R. (1996). Some psychological aspects of reduced consumption behaviour: The role of intrinsic satisfaction and competence motivation.

Environment and Behaviour, 28, 358–409. https://doi.org/10.1177/0013916596283005

Dietz, T., Fitzgerald, A., & Shwom, R. (2005). Environmental values. *Annual Review of Environment and Resources, 30*, 335–372. https://doi.org/10.1146/annurev.energy.30.050504.144444

Dunlap, R. E., Van Liere, K. D., Mertig, A. G., & Jones, R. E. (2000). New trends in measuring environmental attitudes: Measuring endorsement of the new ecological paradigm: A revised NEP scale. *Journal of Social Issues, 56*(3), 425–442. https://doi.org/10.1111/0022-4537.00176

Dzerefos, S. (2020). Reviewing education for sustainable development practices in South African eco-schools. *Environmental Education Research, 26*(11), 1621–1635. https://doi.org/10.1080/13504622.2020.1809637

Foundation for Environmental Education (FEE). (2021). *Engaging the youth of today to protect the climate of tomorrow*. https://www.ecoschools.global/how-does-it-work. Accessed 05 May 2021.

Fulton, D. C., Manfredo, M. J., & Lipscomb, J. (1996). Wildlife value orientations: A conceptual and measurement approach. *Human Dimensions of Wildlife, 1*(2), 24–47. https://doi.org/10.1080/10871209609359060

Foxfield Primary School. (2021). *Bronze eco schools award*. https://www.foxfield.org.uk/123/latest-news/article/318/bronze-eco-schools-award. Accessed 10 June 2021.

Green School Alliance (GSA). (2021). *Green school alliance*. https://www.greenschoolsalliance.org/home. Accessed 04 May 2021.

Green-Schools, An Taisce Environmental Education Unit. (2021). *Green schools ireland*. https://greenschoolsireland.org/case-studies/scoil-ide-artane/?portfolioCats=46. Accessed 10 June 2021.

Green-Demers, I., Pelletier, L. G., & Menard, S. (1997). The impact of behavioural difficulty on saliency of the association between self-determined motivation and environmental behaviours. *Canadian Journal of Behavioural Science, 29*, 157–166. https://doi.org/10.1037/0008-400X.29.3.157

Hallfredsdottir, S. (2011). *Eco-schools: Are they really better? Comparison of environmental knowledge, attitudes and actions between students in environmentally certified schools and traditional schools in Iceland* (Master's thesis). Reykjavik University, Reykjavik, Iceland.

Hickman, G., & Riemer, M. (2016). A theory of engagement for fostering collective action in youth leading environmental change. *Ecopsychology, 8*(3), 167–173. https://doi.org/10.1089/eco.2016.0024

Hills, M. D. (2002). Kluckhohn and Strodtbeck's values orientation theory. *Online Readings in Psychology and Culture, 4*(4), 2307–0919. https://doi.org/10.9707/2307-0919.1040

Hofman, M. (2015). What is an education for sustainable development supposed to achieve – A question of what, how and why. *Journal of Education for*

Sustainable Development, 9(2), 213–228. https://doi.org/10.1177/0973408215588255

Kollmuss, A., & Agyeman, J. (2002). Mind the gap: Why do people act environmentally and what are the barriers to pro-environmental behaviour? *Environmental Education Research, 8*(3), 239–260. https://doi.org/10.1080/13504620220145401

Krnel, D., & Naglic, S. (2009). Environmental literacy comparison between eco-schools and ordinary schools in Slovenia. *Science Education International, 20*, 5–24.

Li, J., & Ernst, J. (2015). Exploring value orientations toward the human-nature relationship: A comparison of urban youth in Minnesota, USA and Guangdong, China. *Environmental Education Research, 21*(4), 556–585. https://doi.org/10.1080/13504622.2014.910499

McKeown, R., & Hopkins, C. (2007). International network of teacher education institutions: Past, present and future. *Journal of Education for Teaching, 33*(2), 149–155. https://doi.org/10.1080/02607470701259408

Milfont, T. L., Duckitt, J., & Wagner, C. (2010). The higher order structure of environmental attitudes: A cross-cultural examination. *Interamerican Journal of Psychology, 44*, 263–273.

MyNI Life. (2018, May 10). Self-sustaining eco-school. *YouTube*. https://www.youtube.com/watch?v=JfKD0_cRr7c. Accessed 14 June 2021.

ODTÜ Geliştirme Vakfı Özel İlköğretim Okulu [METU College Primary School]. (2011). *2010–2011 Eğitim Öğretim Yılı 1. Dönem Etkinlik Raporu [2020-2021 Academic Year 1st Semester Activity Report]*. https://docplayer.biz.tr/40550524-Odtu-gelistirme-vakfi-ozel-ilkogretim-okulu-egitim-ogretim-yili-1-donem-etkinlik-raporu-dur-saygilarimizla.html. Accessed 10 June 2021.

Olsson, D., Gericke, N., & Chang-Rundgren, S. (2016). The effect of implementation of education for sustainable development in Swedish compulsory schools– assessing pupils' sustainability consciousness. *Environmental Education Research, 22*(2), 176–202. https://doi.org/10.1080/13504622.2015.1005057

Olsson, D., Gericke, N., Boeve-de Pauw, J., Berglund, T., & Chang, T. (2019). Green schools in Taiwan – Effects on student sustainability consciousness. *Global Environmental Change, 54*(1), 184–194. https://doi.org/10.1016/j.gloenvcha.2018.11.011

Özsoy, S., Ertepınar, H., & Sağlam, N. (2012). Can eco-schools improve elementary school students' environmental literacy levels? *The Asia-Pacific Forum on Science Learning and Teaching, 13*(2), Article 3.

Robottom, I., & Hart, P. (1995). Behaviorist EE research: Environmentalism as individualism. *The Journal of Environmental Education, 26*(2), 5–9. https://doi.org/10.1080/00958964.1995.9941433

Ritter, E., & Dauksta, D. (2012). Human-forest relationships: Ancient values in modern perspectives. *Environment, Development and Sustainability, 15*(3), 645–662. https://doi.org/10.1007/s10668-012-9398-9

Schwartz, S. H. (1992). Universals in the content and structure of values: Theoretical advances and empirical tests in 20 countries. *Advances in Experimental Social Psychology, 25*, 1–65. https://doi.org/10.1016/S0065-2601(08)60281-6

Stern, P. C. (2000). New environmental theories: Toward a coherent theory of environmentally significant behavior. *Journal of Social Issues, 56*(3), 407–424. https://doi.org/10.1111/0022-4537.00175

Stern, P. C., Dietz, T., Abel, T., Guagnano, G. A., & Kalof, L. (1999). A value-belief-norm theory of support for social movements: The case of environmentalism. *Human Ecology Review, 6*, 81–97.

United Nations. (2012). *Report of the United Nations conference on sustainable development*. https://undocs.org/en/A/CONF.216/16. Accessed 25 May 2021.

UNESCO. (1978). *Intergovernmental conference on environmental education Tbilisi (USSR)*. https://www.gdrc.org/uem/ee/EE-Tbilisi_1977.pdf. Accessed 24 May 2021.

UNESCO. (2012). *Shaping the education of tomorrow*. https://desd.in/UNESCO%20report.pdf. Accessed 02 June 2021.

UNESCO. (2015). *Getting climate-ready: A guide for schools on climate action*. https://www.unesco.de/sites/default/files/2019-03/Getting_Climate-Ready-Guide_Schools.pdf. Accessed 01 June 2021.

UNESCO. (2019). *Framework for the implementation of education for sustainable development beyond 2019*. https://unesdoc.unesco.org/ark:/48223/pf0000370215. Accessed 07 June 2021.

Uruguay Natural. (2016). *Uruguay opens Latin America's first sustainable school*. https://marcapaisuruguay.gub.uy/en/uruguay-opens-latin-americas-first-sustainable-school/. Accessed 12 June 2021.

Elvan Sahin is Associate Professor of Science Education at Middle East Technical University, Turkey. She received her PhD (2008) and MS (2005) in science education, and BS (2002) in chemistry education from Middle East Technical University, Turkey. Her main research interests are education for sustainability, environmental literacy, and climate change education.

Part III

Assessment in ESD: Measuring ESD Learning Outcomes

Learning Our Way Forward and How We Might Assess That

13

Paul Vare

Abstract

This chapter explores the question of learning outcomes and their assessment in the context of education for sustainable development (ESD). Starting with a brief review of different dimensions of ESD, approaches to curriculum planning and purposes of assessment, the chapter turns to the current tendency to focus on competences, what has been described as the Competence Turn. After looking at the rationale for this focus, the chapter presents an example of a competence framework for educators of sustainable development developed by a European-funded project, *A Rounder Sense of Purpose*. The chapter goes on to explore the question of assessment in more detail making the distinction between assessment *of, for* and *as* ESD. The latter is exemplified by assessment of experiential learning linked to the arts, outdoor education and pupil-led community projects. The chapter concludes by highlighting how meaningful assessment of ESD is only likely to be achieved when rigid school structures shift in response to the need for education to broaden its purpose to embrace sustainability.

Keywords

Competences · Education for sustainable development · Assessment · Purpose of education · RSP

13.1 Introduction

Everywhere we turn we are beset with intractable social and environmental problems that threaten the very habitability of the Earth. Having got ourselves into this situation, we now face the urgent task of learning our way through, if not out of it. Education may assist us in preparing the next generation to recognise the fundamental flaws in our concepts of development; however, this has not been education's central role hitherto. As Schumacher puts it, in that oft-cited quotation, 'If still more education is to save us, it would have to be education of a different kind' (Sterling, 2001, p. 21). Education for sustainable development (ESD) is widely seen by its proponents as the different kind of education that Schumacher had in mind. Yet when it comes to assessment we seem to return to old habits, as if forgetting that education was supposed to be different to this. In this chapter, I wish to share something of my own understanding of ESD and explore one particular strand of its development, which we might call the Competence Turn. After a brief explanation of one example of an ESD competence frame-

P. Vare (✉)
School of Education and Humanities, University of Gloucestershire, Cheltenham, UK
e-mail: pvare@glos.ac.uk

G. Karaarslan-Semiz (ed.), *Education for Sustainable Development in Primary and Secondary Schools*, Sustainable Development Goals Series, https://doi.org/10.1007/978-3-031-09112-4_13

work, I turn to the question of assessment in ESD – an aspect that highlights the difference (or lack thereof) between ESD and mainstream education.

The very phrase education *for* sustainable development (ESD) is fraught with tensions. On the one hand, we want ESD to be distinct from 'mainstream' education, to make a difference in terms of its impact on our learners and society. Yet on the other hand, most teachers I've met would wish to expand young people's horizons and offer them choices in life, not sell them a set of prescribed, 'correct' behaviours.

One way of navigating this double bind is by seeing ESD as a combination of two approaches (Vare & Scott, 2007), each underpinned by its own philosophy:

- ESD 1: promoting/facilitating changes in what we do including promoting (informed, skilled) behaviours and ways of thinking, where the need for this is clearly identified and agreed – this is learning *for* sustainable development.
- ESD 2: building capacity to think critically about (and beyond) what experts say and to test sustainable development ideas including exploring the contradictions inherent in sustainable living – this is learning *as* sustainable development.

The first approach comes with a definable body of knowledge, albeit subject to rapid change, as such it can be aligned closely with what Kelly (2009) terms *curriculum as content*. The second approach is more complex; it requires facilitation rather than direct instruction and calls for a high degree of sensitivity in order to contribute to a transformation in how learners understand their position or possibilities within their own context. This reflects a co-created approach to curriculum that Kelly (2009) terms *curriculum as process*. By combining ESD 1 and ESD 2, it is hoped that we might produce agents of change who are both informed and critical and who simultaneously (a) are aware of – and skilled in – a range of possible actions and (b) can discern which of these, if any, would be most appropriate for them to take, given their current situation. The

phrase 'produce agents' indicates another item from Kelly's typology of curriculum approaches, that is, *curriculum as product*.

Curriculum as product focuses on the outcomes of education, e.g. what the leaner knows or can do, rather than specific curriculum content or even pedagogical concerns. In this regard, it lends itself to the identification of measurable outcomes; this is one reason for its popularity, particularly in the form of competence-based learning.

13.2 The Competence Turn

According to Hodge (2007), competency[1]-based education and training (CBET) became popular initially in the USA in the late 1950s as a means of holding teachers and teacher educators accountable. This in turn was a response to perceived deficits in education highlighted by the USA's falling behind the USSR in the space race. This approach views processes as predictable and manageable; it also fits well with behaviourism, which was the dominant theory of learning at the time (Shephard, 2022). The way in which this approach lends itself to measurable verification with apparent ease resonates with the all-pervasive managerialism that has come to characterise national education systems in the wake of neoliberal policy environments that have become a global phenomenon since the 1980s (Harvey, 2005). Indeed, outcomes are now 'used as criteria for the productivity of entire educational systems' (Klieme et al., 2008, p. 3).

Naturally this approach found its way into environmental education with the publication of a set of 'competencies' for teachers as part of the International Environmental Education Programme (IEEP) series of 'green books' (Wilke et al., 1987). This was significant because it was aimed at educators rather than the popula-

---

[1] I draw no distinction between these terms here, and I use whichever form is used by cited authors and publications. For a discussion on the difference between 'competence' and 'competency', see the Introduction to Vare et al. (2022).

tion as a whole. Broader sets of competences for learners include those defined by the Paris-based Organisation for Economic Cooperation Development (OECD) under its *Definition and Selection of Competencies* (DeSeCo) project, which aimed to identify the competencies necessary for individuals to confront the challenges of balancing economic growth with environmental sustainability and social equity (OECD, 2002). This work provides the foundation of the *OECD Learning Compass 2030*, a project that aims to identify 'transformative competencies' for young people (Rychen, 2019). A set of key competencies in sustainability, developed originally as learning outcomes of sustainability science students (Wiek et al., 2011), have become highly influential and were used as a basis for UNESCO's key competencies in education for sustainable development (Rieckmann, 2018) and a recent international Delphi Study on key competencies in sustainability (Brundiers et al., 2021).

Meanwhile in 2005, the United Nations Economic Commission for Europe (UNECE) published its Strategy for Education for Sustainable Development that called on Member States to 'develop the competence within the education sector to engage in ESD' (UNECE, 2009, p. 21). The international organisation ENSI (Environment and School Initiatives) answered this call with its CSCT model, i.e. *C*urriculum, *S*ustainable development, *C*ompetences, *T*eacher training (Sleurs, 2008). This was followed by a framework of 39 ESD competences for educators developed by a UNECE expert group (UNECE, 2012).

The Competence Turn in ESD has since provided us with some potentially useful, research-informed frameworks of educator competences for sustainability including the KOM-BiNE model (Rauch & Steiner, 2013), the work of Bertschy et al. (2013), A Rounder Sense of Purpose (Vare et al., 2019) and the work of Timm and Barth (2021).

them – A Rounder Sense of Purpose or RSP (Vare et al., 2019) – I will briefly explain the rationale behind it. The idea came initially from some of us who had been involved in the UNECE process as we were unhappy with the large number of competences in that framework, which teachers themselves were finding unwieldy. RSP sought to distil the UNECE competences into a tighter framework with carefully defined learning outcomes and materials to support the development of the competences among educators and student teachers. All of this is available on the RSP website at https://aroundersenseofpurpose.eu.

The result is a framework of 12 competences (Table 13.1), each with three learning outcomes and a number of underpinning components plus a bank of activities that are linked to the UN Sustainable Development Goals as well as each of the 12 RSP competences.

Although RSP uses the language of competences, the framework's name belies its broader intention, that is, to question the current narrow purpose of formal education as it is understood in policies that focus on its (albeit important) economic value.[2] The name – A Rounder Sense of Purpose – thus represents a deliberate challenge to the whole notion of reducing education to sets of competences; the reason the term 'competence' was used at all was to help the project to engage readily with current debates, to indicate the provenance of the framework and to find favour with potential donors who would recognise this term given its current ubiquity. Most of the teaching activities on the RSP website reflect a constructivist pedagogy that allows and encourages learners to build on what they already know. Many of the activities are exploratory, discursive and invite creativity of thought; in this way they reflect an ESD 2 approach. They also include steps that call upon learners to conduct their own research, thereby adding to their knowledge base in situations where they are overseen by the educator who can guide, pose questions and high-

13.3 A Rounder Sense of Purpose

At this point it may be helpful to reveal the contents of one of these competence frameworks, and given my close involvement with one of

[2]A striking example of this can be found in the UK where the Government's *Projected Completion and Employment from Entrant Data* (Proceed) records the nature of jobs (and income) secured by higher education alumni and uses this as a key measure of the 'quality' of education offered by each institution.

Table 13.1 The Rounder Sense of Purpose framework

Thinking holistically	Envisioning change	Achieving transformation
Systems The educator helps learners to develop an understanding of the world as an interconnected whole and to look for connections across our social and natural environment and consider the consequences of actions	*Futures* The educator helps learners to explore alternative possibilities for the future and to use these to consider how behaviours might need to change	*Participation* The educator helps learners to contribute to changes that will support sustainable development
Attentiveness The educator helps learners to understand fundamentally unsustainable aspects of our society and the way it is developing and increases their awareness of the urgent need for change	*Empathy* The educator helps learners to respond to their feelings and emotions and those of others as well as developing an emotional connection to the natural world	*Values* The educator develops an awareness among learners of how beliefs and values underpin actions and how values need to be negotiated and reconciled
Transdisciplinarity The educator helps learners to act collaboratively both within and outside of their own discipline, role, perspectives and values	*Creativity* The educator encourages creative thinking and flexibility within their learners	*Action* The educator helps the learners to take action in a proactive and considered manner
Criticality The educator helps learners to evaluate critically the relevance and reliability of assertions, sources, models and theories	*Responsibility* The educator helps learners to reflect on their own actions, act transparently and to accept personal responsibility for their work	*Decisiveness* The educator helps the learners to act in a cautious and timely manner even in situations of uncertainty

light misconceptions. Thus ESD 1 is also built in to the activities.

One serious hazard of taking a competence-based approach to ESD lies in the very reason that competences are so popular; that is, they define pre-determined learning outcomes. This presents a double bind: to put competences into practice, they need to be context bound and have specific outcomes, yet ESD is characterised by the need to prepare young people to engage in transforming our current unsustainable context. Educators for sustainability recognise that our best hope lies in educating citizens to be open to unforeseen conditions, to learn our way forward into an unknowable future. If our pedagogy is to be aligned with this purpose, how can it be focused on steering our learners towards carefully prescribed 'correct' solutions?

Taking this need for flexibility and creativity into account, the RSP competences are presented in the form of an artist's palette.[3] This invites the educator to combine the competences in ways that are unique to their context reflecting who they and their learners are in any given time and place. In this way the palette offers an emergent rather than a linear approach, suggesting that each learning episode will have its own unforeseen outcomes, which brings us to the question of assessment. We did not break down the 12 RSP competence areas into skills, values, knowledge and so forth because this would atomise learning into discrete components; this would undermine the notion of holistic thinking that characterises sustainable development. Such an approach would render the individual components meaningless, a point well made by Westera (2010), who critiques the tendency to break down components of competences, such as skills:

> Consequently, the entanglement of the skills-hierarchy and the competence-hierarchy produces a complex, confusing and inconsistent conceptual system that cannot be taken seriously. (*Ibid*, p. 85)

In response to this issue, some RSP partners have chosen to focus on quality criteria[4] and avoid the

---

[3]https://aroundersenseofpurpose.eu/framework/palette

[4]This approach reflects that taken in the past by ENSI; see Breiting et al. (2005).

issue of measurement altogether. The RSP framework is not accompanied by a single proposal for assessment, rather it recognises that assessment will need to be diverse and potentially quite complex if it is to be aligned with the subject matter of sustainable development itself. In the UK we have used a combination of approaches discussed below. What follows is my attempt to tease out some of the key issues for consideration when thinking about assessing ESD.

13.4 Purposes of Assessment

Before exploring how assessment might be conducted, we should first clarify why we wish to do it at all. Summative assessment *of* learning is carried out after a programme of study; it generally attracts most attention because it is used to judge students' final level of achievement and is of interest to stakeholders beyond the school. In the face of this dominant mode of assessment, seminal work by Black and Wiliam (1998) highlighted how formative assessment, i.e. a form of assessment *for* learning, has been shown to generate significant learning gains among students. A natural progression of this approach has been to involve students in their own assessment, building their skills, knowledge and predisposition for learning in the process – this is assessment *as* learning.

These three approaches can in turn be mapped against different *purposes* of assessment; Bloxham and Boyd (2007) identify four purposes:

1. *Certification* – a means of qualifying a learner by recognising their level of achievement, also used to discriminate between students. This is assessment *of* learning.
2. *Student learning* – 'promoting learning by motivating students, steering their approach to learning and giving the teacher useful information to inform changes in teaching strategies' (Ibid, p. 31). This is assessment *for* learning and can also be assessment *as* learning.
3. *Quality assurance* – often an external function or an internal moderation to check the

validity and consistency of assessment. Very much assessment *of* learning.
4. *Lifelong learning capacity* – using assessment to develop learners' broad capabilities (or competence) to continue with their own learning in the longer term. A form of assessment *as* learning.

From an ESD perspective, we may wish to emphasise assessment *as* learning because this represents a form of learning to learn, what we might call sustainable learning; it is thus aligned constructively with the aims of learning for sustainability. However, assessment *as* learning simply isn't the norm in mainstream schools and requires practice on the part of the learners as well as the educators (Kostons et al., 2012). Against this concern, we should recall that ESD is supposed to be 'education of a different kind' so the idea that assessment *as* learning is not widespread should encourage us to emphasise this approach. On the other hand, if ESD is to become embedded in the core business of schools, then it will need to cover all of the purposes outlined above, including their related use of assessment *of* learning, all of which highlights the fact that there is no 'silver bullet' by which we can solve the assessment question.

13.5 Means of Assessment in ESD

There is no clear consensus on how ESD relates to its forebearer, environmental education (Sterling, 2010), which can only add to the confusion about what ESD actually is. It is little wonder therefore that there is no clear framework for how to assess it. Given that sustainability itself is concerned with addressing 'wicked' problems for which there are no clear-cut solutions, education in this context would do well to reflect this complexity. This means that it cannot consist entirely of predetermined procedures with clearly measurable outcomes. Rather than focusing on expected results, as high-stakes assessment procedures tend to do, our assessment of ESD will need to embrace unexpected, emerging and unintended outcomes.

In such cases, *measurement* is not a particularly helpful concept. Rather than attempting to quantify these things, we will need to be concerned with quality, both of the outcomes and of the thinking that led to them. This requires us to define quality criteria against which to assess ESD (Breiting et al., 2005) using indicators derived from those criteria. This opens the way for dimensions such as participation, empathy or congruence with sustainability principles (e.g. equity, positive environmental impact) to be included in the assessment regime.

With ESD goals being so broad and far reaching, meaningful assessment has always been a challenge, and so attempts have tended to be piecemeal. Currently, specific information on which tools are best for yielding evidence for different aspects of ESD is dispersed across the literature, so more work is required to bring this together. One promising development has come in the form of a systematic literature review covering research on assessing competences in sustainability (Redman et al., 2021); this identifies eight distinct types of tools:

- Scaled self-assessment
- Reflective writing
- Scenario/case test
- Focus group/interview
- Performance observation
- Concept mapping
- Conventional test
- Regular course work

These in turn fall into one of three broad categories or meta-types of assessment procedure:
1. Self-perceiving-based
2. Observation-based
3. Test-based
This provides a useful framework for gathering together future research; it can also offer a menu from which practitioners can select the most appropriate tools for their purposes. Interestingly, the study by Redman et al. (2021) reveals that the most frequently used tool (occurring in over half of all studies reviewed) is scaled self-assessment. While this may be an effective form of assessment *for* learning, it has limited value as a means

of providing assessment *of* learning given that individuals cannot be expected to grade themselves for external certification purposes. The key therefore is to marshal a range of assessment approaches in order to provide a multi-dimensional picture of the attributes that we are attempting to assess.

13.6 Assessment *of, for* and *as* ESD

The search for clear, simple assessment procedures is understandable; that is how our education systems work. They teach facts and impart skills that students are expected to repeat, manipulate or demonstrate depending on the sophistication of the assessment tool. This, however, is unlikely to be the best preparation for the kind of adaptable, self-directed learner that will be required as environmental conditions become more challenging over the twenty-first century. Fortunately, as a result of the Competence Turn, there is already some consensus on the likely learning outcomes that a more sustainable future will demand (Rychen, 2019; Wiek et al., 2011; Rieckmann, 2018; Brundiers et al., 2021). The task before us is to link the range (and it will need to be a wide range) of available assessment tools to those outcomes. It is beyond the scope of this chapter, or indeed any one publication at present, to propose a 'best fit' set of procedures from the myriad possibilities. What I will leave you with is a few thoughts that take us beyond the most widespread approaches to assessment that are familiar in virtually every school. The terms of assessment *of, for* and *as* learning introduced earlier will help us here; only this time they are *of, for* and *as* learning for sustainability.

13.6.1 Assessment of and for ESD

In my own professional context, working in a higher education institution, students are required to demonstrate predetermined learning outcomes linked to a specific academic standard. Assessment in such a context is unlikely to reflect the complexity discussed above, but it can offer opportunities to assess whether learning *about* and *for* sustainability has taken place. Over the course of 4 years of offering non-accredited courses, we have developed an assessment approach that is now used on an accredited programme for student teachers.

By reviewing students' reflective journals, we identified nine aspects of learning that could be grouped under the three broad headings of Understanding, Action and Reflection. Within these, students report on how they have reflected on or applied their learning in their professional, social and/or private life, including where they have sought to develop sustainability learning or competences in others.

On this particular programme, the predetermined learning outcomes are linked to the 12 Rounder Sense of Purpose (RSP) competences. Seeking evidence of all 9 learning aspects for each of the 12 competences would sacrifice depth of engagement for breadth of coverage, so we have defined a meaningful indicator of the *extent* of a student's learning to be: where the student can provide evidence of at least four of the nine learning aspects under each competence, with at least one in each category (Understanding, Action and Reflection). We also seek evidence of each of the nine aspects in at least four competences.

To assess a student's *depth* of engagement, we analysed students' reflective journals and arrived at a series of exemplar statements that allowed us to construct a marking grid similar to those used on other university courses. The marker simply has to shade the descriptors that best describe the student's performance on a range of dimensions, such as presentation, criticality or use of literature, in order to build up a composite picture that helps them to define the appropriate grade to award the student. The actual work assessed can take the form of reflective journals, videos and formal essays. The critical feature of this approach is that it combines quantity with quality and can cover a wide range of evidence.

Despite this mixed methods approach, assessing learning outcomes in this way can lend itself to a linear model of education and can valorise weak forms of ESD that are unlikely to bring about much needed change. While we have witnessed our students undergoing transformative experiences, this has had little to do with their assessment. Yet transformative learning that might inspire learners to become agentic actors for change (Jickling & Sterling, 2017) is what is required if we are to respond to the complex and rapidly changing crises that face us all. I therefore turn to three examples where learning and assessment are integral to the learning process and fully aligned to familiar attributes of learning for sustainability.

13.6.2 Assessment as (Transformative) ESD

The grim reality of our situation in relation to environmental overshoot is often discussed in relation to scientific evidence (Bendell, 2018); this can be difficult to face both conceptually and emotionally. Engaging learners in feeling and sensing as well as understanding can play a crucial role in making concepts such as sustainability less distant or abstract (Jickling, 2017). Faced with complex, dynamic systems, ESD that focuses on experiential learning can provide learners with opportunities to combine different ways of knowing and valuing reality. Seeing the world anew in this way has a critical part to play in contributing to transformational learning (Dieleman & Huisingh, 2006). The Arts have a crucial role to play here. Their appeal to the senses; the way they can combine embodiment, cognition and intuition; their openness to the unexpected; and their playfulness, all provide fertile ground for encountering sustainability issues afresh. This suggests that the Arts have a transformative potential that can be tapped by educators willing to engage in the mystery and open-ended nature of aesthetic experience.

Following a 2-week workshop with artists and scientists at the Universitat Oberta de Catalunya (Barcelona, Spain), in which Maria Heras (Maria Heras 2021, Universitat Oberta de Catalunya, Barcelona, Spain, personal communication) worked with colleagues to create a science-based performance installation, the teaching team turned to the question of assessment. This is still under development, but Heras and colleagues are looking at the co-creation process involved in the artistic endeavour *and* at the impact on the audience. This is not the stuff of marking grids or prescribed learning outcomes, rather it suggests an immersive process in which learners and other stakeholders are changed. The significance of this example is not so much the specifics of the assessment but the way in which the artistic process is framed as a form of knowledge in itself rather than as a learning tool distinct from some independent message to be conveyed. Indeed, Østergaard (2019) warns us against seeing Art in an instrumental sense as it can quickly lose its potential for transformation if it is used in that way.

The tendency to reduce the potential power of a pedagogical approach by using it in an instrumental manner is often observed in the context of outdoor education. Without due attention being paid to the possibility of transformation, Lausselet and Zosso (forthcoming) suggest that outdoor education, while beneficial as a counterweight to classroom-based learning, can fail to achieve its potential in relation to ESD and remain a non-transformative example of, say, environmental studies. The relationship between transformation, outdoor education and student learning is under researched at present (Hill & Brown, 2014), but Lausselet and Zosso (Ibid) have shown that the potential is there if students can engage in real-world issues where the political dimension is foregrounded and where they can interact with stakeholders. In this way students do not simply learn about a place or an issue; they become a part of the story themselves.

Outdoor education provides opportunities to work with multiple forms of knowledge, and each student is likely to gain from the experience in ways that are different to their peers, so assessment will need to focus on the unforeseen and multi-dimensional outcomes that result from this engagement. This multiplicity of unforeseen outcomes could be embraced by constructing a meaningful framework of quality criteria in which to locate them.

Lastly, pupil-led community projects can also provide powerful and immersive experiences for pupils of all ages. A study of the impacts of such projects on secondary school pupils (Vare, 2021) identifies three broad categories of activity that contribute significantly to the projects' impacts, these are (i) making connections, (ii) taking action and (iii) engaging in planning. These three characteristics of the student learning emerged from interviews and workshops with the students and their teachers rather than being assessed explicitly. The assessment that did take place took the form of group presentations that allowed teachers to assess the quality of the students' logic in claiming that their project would have a positive impact within their communities. The presentations also, inter alia, revealed the strength (or otherwise) of community relationships, the quality of planning and the nature of the actions taken. Armed with the three-point framework (or quality criteria) above, a more structured assessment will be possible in the future.

All three of these examples, the Arts, outdoor education and pupil-led community projects, point to ways in which assessment of ESD per se might be achieved as well as assessment of a specific discipline. The examples illustrate the importance of working across disciplines including in trans-disciplinary ways, i.e. working with stakeholders from beyond academia. Meadows (2008) reminds us that reductionist thinking remains vital as we will always need discipline-based expertise, but sustainability also demands that we look beyond this tight academic framing; indeed for some, the whole notion of disciplines can be unhelpful:

> The notion of *un-disciplinarity* has been developed within the context of political ecology and environmental arts and humanities, which highlight how knowledge is created and reproduced within disciplines is inadequate to address the ecological challenges our societies are facing today. (Saratsi et al., 2019, p. 19, italics in original)

This implies that the current structures of schooling are likely to be inadequate for transformative ESD to take place in a widespread manner. Efforts to re-think schooling litter the history of education, but that may be because education has served its purpose very well in an era when that purpose was to inculcate the next generation into the logic (and roles) of the dominant society. Now that we can see just how unsustainable that society has become, surely it is time to re-think the structures that help to create it – and this time, we might give serious consideration to assessment from the outset.

13.7 Concluding Thoughts

If the goal of ESD is transformation, then it seems inevitable that experiential learning is going to have a large part to play – not forgetting that a lecture or reading a book can also be a transformative experience. Exploring the cognitive, physical *and* affective impacts of experience is therefore likely to play a significant role in the assessment of ESD. No single approach can hope to capture the multiple dimensions of this learning, which will include, for example, gains in knowledge, skills, connections, self-confidence and agency. If it is difficult to conceive of how this might sit within the formal education systems that we have today, then perhaps it really is time to adopt that 'education of a different kind' together with a different kind of assessment. As we can see, this will be complicated, but rather than looking for ready answers, we would do well to 'stay with the trouble' as Haraway (2015) puts it, remaining open to the complexity of our situation and the multiple possibilities that this affords.

References

Bendell, J. (2018). *Deep adaptation: A map for navigating climate tragedy*. IFLAS occasional paper 2.

Bertschy, F., Künzli, C., & Lehmann, M. (2013). Teachers' competencies for the implementation of educational offers in the field of education for sustainable development. *Sustainability, 5*(12), 5067–5080. https://doi.org/10.3390/su5125067

Black, P., & Wiliam, D. (1998). Assessment and classroom learning, assessment in education. *Principles, Policy & Practice, 5*(1), 774. https://doi.org/10.1080/0969595980050102

Bloxham, S., & Boyd, P. (2007). *Developing effective assessment in higher education: A practical guide*. McGraw-Hill/Open University Press.

Breiting, S., Mayer, M., & Morgensen, F. (2005). *Quality criteria for ESD-schools: Guidelines to enhance the quality of education for sustainable development*. ENSI/SEED & Austrian Federal Ministry of Education, Science & Culture. tinyurl.com/qlhe6aw

Brundiers, K., Barth, M., Cebrián, G., Cohen, M., Diaz, L., Doucette-Remington, S., Dripps, W., Habron, G., Harré, N., Jarchow, M., Losch, K., Michel, J., Mochizuki, Y., Rieckmann, M., Parnell, R., Walker, P. & Zint, M. (2021). Key competencies in sustainability in higher education—toward an agreed-upon reference framework. *Sustain Sci 16*, 13–29. https://doi.org/10.1007/s11625-020-00838-2

Dieleman, H., & Huisingh, D. (2006). Games by which to learn and teach about sustainable development: Exploring the relevance of games and experiential learning for sustainability. *Journal of Cleaner Production, 14*(9–11), 837–847. https://doi.org/10.1016/j.jclepro.2005.07.014

Haraway, D. J. (2015). Anthropocene, capitalocene, plantationocene, chthulucene: Making kin. *Environmental Humanities, 6*, 159–165.

Harvey, D. (2005). *A brief history of Neoliberalism*. Oxford: OUP

Hill, A., & Brown, M. (2014). Intersections between place, sustainability and transformative outdoor experiences. *Journal of Adventure Education & Outdoor Learning, 14*(3), 217–232. https://doi.org/10.1080/14729679.2014.918843

Hodge, G. (2007). The origins of competency-based training. *Australian Journal of Adult Learning, 47*(2), 179–209.

Jickling, B. (2017). Education revisited: Creating educational experiences that are held, felt, and disruptive. In B. Jickling & S. Sterling (Eds.), *Post-sustainability and environmental education* (Palgrave studies in education and the environment). Springer Nature.

Jickling, B., & Sterling, S. (Eds.). (2017). *Post-sustainability and environmental education* (Palgrave studies in education and the environment). Springer Nature.

Kelly, A. V. (2009). *The curriculum theory and practice* (6th ed.). Sage.

Klieme, E., Hartig, J., & Rauch, D. (2008). The concept of competence in educational contexts. In J. Hartig, E. Klieme, & D. Leutner (Eds.), *Assessment of competencies in educational settings. State of the art and future prospects* (pp. 3–22). Cambridge University Press.

Kostons, D., van Gog, T., & Paas, F. (2012). Training self-assessment and task-selection skills: A cognitive approach to improving self-regulated learning. *Learning and Instruction, 22*, 121–132. https://doi.org/10.1016/j.learninstruc.2011.08.004

Lausselet, N., & Zosso, I. (forthcoming). Bonding with the world: A pedagogical approach. In R. Jucker & J.

von Au (Eds.), *Outdoor-based learning - How can it contribute to high quality learning?* Springer.

Meadows, D. (2008). *Thinking in systems: A primer* (Edited by Wright, D.). Earthscan.

OECD. (2002). *Definition and selection of competencies (DeSeCo): Theoretical and conceptual foundations.* Directorate for Education, Employment, Labour and Social Affairs, Education Committee, DEELSA/ED/CERI/CD (2002) 9, 27.

Østergaard, E. (2019). Music and sustainability education – A contradiction? *Acta Didactica Norge, 13*(2), 2–20. https://doi.org/10.5617/adno.6452

Rauch, F., & Steiner, R. (2013). Competences for education for sustainable development in teacher education. *CEPS Journal, 3,* 9–24. https://doi.org/10.25656/01:7663

Redman, A., Wiek, A., & Barth, M. (2021). Current practice of assessing students' sustainability competencies: A review of tools. *Sustainability Science, 16*(1), 117–135. https://doi.org/10.1007/s11625-020-00855-1

Rieckmann, M. (2018). Learning to transform the world: Key competences in education for sustainable development. In A. Leicht, J. Heiss, & W. J. Byun (Eds.), *Issues and trends in education for sustainable development.* UNESCO.

Rychen, D. S. (2019). *Alignment with OECD definition and selection of competencies: Theoretical and conceptual foundations (DeSeCo) project.* OECD. https://www.oecd.org/education/2030-project/teaching-and-learning/learning/transformative-competencies/Thought_leader_written_statement_Rychen.pdf. Assessed 02 Oct 2021.

Saratsi, E., Acott, T., Allinson, E., Edwards, D., Fremantle, C., & Fish, R. (2019). *Valuing arts and arts research* (Valuing nature paper, 22). Valuing Nature [online]. https://valuing-nature.net/valuing-arts-and-arts-research. Assessed 01 Oct 2021.

Shephard, K. (2022). On the educational difference between being able and being willing. In P. Vare, N. Lausselet, & M. Rieckmann (Eds.), *Competences in education for sustainable development: Critical perspectives.* Springer.

Sleurs, W. (Ed.). (2008). *Competencies for ESD (Education for Sustainable Development) teachers, a framework to integrate ESD in the curriculum of teacher training Institutes.* Comenius 2.1 project 118277-c p-1-2004-b e-Comenius-c2.1. https://unece.org/fileadmin/DAM/env/esd/inf.meeting.docs/EGonInd/8mtg/CSCT%20Handbook_Extract.pdf. Assessed 01 Oct 2021.

Sterling, S. (2001). *Sustainable education, re-visioning learning and change.* Green Books.

Sterling, S. (2010). Living in the earth: Towards an education of our time. *Journal of Education for Sustainable Development, 4,* 213–218.

Timm, J.-M., & Barth, M. (2021). Making education for sustainable development happen in elementary schools: The role of teachers. *Environmental Education Research, 27*(1), 50–66. https://doi.org/10.1080/13504622.2020.1813256

UNECE. (2009). Learning from Each Other: The UNECE Strategy for Education for Sustainable Development, Geneva: UNECE. https://sustainabledevelopment.un.org/content/documents/798ece5.pdf

UNECE. (2012). *Learning for the future: Competences in education for sustainable development.* United Nations. https://unece.org/fileadmin/DAM/env/esd/ESD_Publications/Competences_Publication.pdf

Vare, P. (2021). Exploring the impacts of student-led sustainability projects with secondary school students and teachers. *Sustainability, 13,* 2790. https://doi.org/10.3390/su13052790

Vare, P., & Scott, W. A. H. (2007). Learning for a change: Exploring the relationship between education and sustainable development. *Journal of Education for Sustainable Development, 1*(2), 191–198.

Vare, P., Arro, G., de Hamer, A., Del Gobbo, G., de Vries, G., Farioli, F., Kadji-Beltran, C., Kangur, M., Mayer, M., Millican, R., Nijdam, C., Réti, M., & Zachariou, A. (2019). Devising a competence-based training program for educators of sustainable development: Lessons learned. *Sustainability, 11*(7), 1890. https://doi.org/10.3390/su11071890

Vare, P., Lausselet, N., & Rieckmann, M. (2022). *Competences in education for sustainable development: Critical perspectives.* Springer.

Westera, W. (2010). Competences in education: A confusion of tongues. *Journal of Curriculum Studies, 33*(1), 73–88.

Wiek, A., Withycombe, L., & Redman, C. L. (2011). Key competences in sustainability: A reference framework for academic program development. *Sustainability Science, 6,* 203–218.

Wilke, R. J., Peyton, R. B., & Hungerford, H. R. (1987). *Strategies for the training of teachers in environmental education* (International environmental education programme; environmental education series no. 25). UNESCO-UNEP.

Paul Vare leads postgraduate research in education and runs the Doctor of Education program at the University of Gloucestershire, UK, where he also coordinates international education projects. Currently, Paul is a member of the youth campaign Teach the Future's Adult Advisory Board, academic expert to the UN Economic Commission for Europe (UNECE) Steering Committee on ESD, and chair of the Continuing Professional Development Forum for the Universities' Council for Teacher Education (UCET), UK. In the past, he worked on community-based sustainability projects in sub-Saharan Africa before returning to Europe where he helped draft the UNECE Strategy on ESD.

Developing and Assessing Sustainability Competences in the Context of Education for Sustainable Development

14

Marco Rieckmann

Abstract

Education for Sustainable Development (ESD) is about enabling participation in the societal learning, communication and transformation processes that are required for sustainable development. This chapter provides an overview of the concept of ESD, its goals, pedagogical principles and methods and, above all, the assessment of ESD learning outcomes. ESD aims to develop sustainability competences and – as transformative learning – to promote critical reflection on individual and societal values. This requires an action-oriented, transformative pedagogy. In contrast to the intensive engagement of ESD research with the concept of key competences, research has so far been less focused on the development of methodological instruments for assessing competence development. The systematisation and differentiation of existing concepts of competence and the (further) development of competence models and suitable instruments for the objective measurement of competence are gaps in research that require further attention. Of particular importance is the comprehensive consideration of cognitive and non-cognitive competences.

M. Rieckmann (✉)
University of Vechta, Department of Education, Vechta, Germany
e-mail: Marco.Rieckmann@uni-vechta.de

Keywords

Education for sustainable development · Sustainability competences · Competence assessment

14.1 Introduction

Humanity has been exceeding the ecological limits of the planet for several decades now (Steffen et al., 2015); there is still no sign of a reversal of the trend. For example, anthropogenic greenhouse gas emissions are still rising worldwide; major species are continuing to become extinct, threatening the integrity of the planet and the Earth's ability to meet human needs (UN Environment, 2019). Major social challenges also persist. In particular, large parts of Africa have not yet succeeded in significantly reducing poverty (World Bank, 2020), and the coronavirus pandemic has led to an increase in poverty and inequality worldwide (Mahler et al., 2021).

In order to overcome these ecological and social challenges, a "Great Transformation" (WBGU, 2011) is needed, which must be achieved through sustainable development. The Sustainable Development Goals (SDGs) provide concrete guidelines for this necessary global transformation. These 17 goals form the core of the 2030 Agenda for Sustainable Development adopted by the UN General Assembly in 2015

(United Nations, 2015). Key features of the 2030 Agenda are its universality and indivisibility. All countries – from the Global South and the Global North – must align their own development efforts with the goal of promoting prosperity while protecting the planet. In this respect, all the countries of the world can be considered developing countries when it comes to the SDGs; all must take urgent action to promote sustainable development.

Since the late 1990s, there has been increasing reference to sustainable development in educational discourse as well as educational practice. It is in this context that the concept of Education for Sustainable Development (ESD) has been developed. ESD aims to enable people to participate in the societal learning, communication and transformation processes necessary for sustainable development, the implementation of the SDGs and thus the promotion of the "Great Transformation" (WBGU, 2011) (Rieckmann, 2018a). The relevance of ESD is explicitly recognised in the SDGs as part of Target 4.7:

> By 2030, ensure that all learners acquire the knowledge and skills needed to promote sustainable development, including, among others, through education for sustainable development and sustainable lifestyles, human rights, gender equality, promotion of a culture of peace and non-violence, global citizenship and appreciation of cultural diversity and of culture's contribution to sustainable development. (United Nations, 2015, p. 17)

At the same time, it is important to emphasise the central importance of ESD for all the other 16 SDGs. ESD enables all individuals to contribute to the achievement of the SDGs by equipping them with the knowledge and competences they need not only to understand what the SDGs are about but also to engage as informed citizens to drive forward the necessary transformation (UNESCO, 2017). The integration of ESD into all education systems was promoted worldwide by the UN Decade of Education for Sustainable Development (2005–2014). The activities of the Decade were continued under the auspices of UNESCO in the Global Action Programme (GAP) Education for Sustainable Development (2015–2019) and are now being continued with

the programme Education for Sustainable Development: Towards achieving the SDGs (2020–2030) – in short "ESD for 2030". The aim of this programme is to embed ESD structurally in all educational sectors and institutions (UNESCO, 2020). This chapter provides an overview of the concept of ESD, its goals, pedagogical principles and methods and, above all, the assessment of ESD learning outcomes.

14.2 Aims of ESD

14.2.1 Development of Sustainability Competences

ESD should enable all people to contribute to sustainable development. Therefore, it aims to develop key competences that are particularly relevant for sustainable development, but which most people still lack (Rieckmann, 2018a). It should enable individuals, "if they have appropriate goals, purposes or intentions" (translated from German), to act to promote sustainable development (de Haan et al., 2008, p. 117). It is thus about "opening up opportunities" (translated from German) (ibid., p. 123) and not about inculcating supposedly sustainability-compliant behaviour in learners. Learners should become "sustainability citizens" (Schank & Rieckmann, 2019) and should be able to consider questions relating to sustainable development and find their own answers. This emancipatory understanding of ESD (Vare & Scott, 2007; Wals, 2011) thus fulfils Klafki's (2007) requirement for general education.

In ESD discourse, the question of which key competences individuals should possess in order to be able to actively shape their own lives and the social environment in order to promote sustainable development has been under intensive discussion for a number of years (Rieckmann, 2018a). In Germany, reference is often made to the concept of *Gestaltungskompetenz* ("shaping competence"), which refers to the ability to "recognise problems caused by unsustainable development and effectively apply knowledge about sustainable development" (translated from German) (de Haan et al., 2008, p. 12). It comprises 12 sub-competences, including the competence to adopt perspectives, the competence to anticipate and the competence to cope with individual decisions (de Haan, 2010). The concept of *Gestaltungskompetenz* is thus characterised by competences "that enable the forward-looking and autonomous co-creation of sustainable development" (translated from German) (Michelsen, 2009, p. 84).

At the international level, there is also an explicit discussion of competence development through ESD (Brundiers et al., 2021; Rieckmann, 2018a; Wiek et al., 2011; Wiek et al., 2016). Wiek et al.'s Key Competencies in Sustainability (2016) bring together a range of concepts that have been the subject of international discussion and distinguish – with a special focus on sustainability science study programmes – five key competences: systems thinking, anticipation, normative competence, strategic competence and interpersonal competence. In addition, critical thinking and self-awareness competences are often cited as being particularly important for sustainable development; current international discourse on ESD thus prioritises the following eight sustainability competences (Rieckmann, 2018a; UNESCO, 2017; see also Brundiers et al., 2021):

- *Systems thinking competence:* the ability to recognise and understand relationships, to analyse complex systems, to understand how systems are embedded within different fields and on different scales and to deal with uncertainty
- *Anticipatory competence:* the ability to understand and evaluate multiple futures – possible, probable and desirable – to create one's own visions for the future, to apply the precautionary principle, to assess the consequences of actions and to deal with risks and changes
- *Normative competence:* the ability to understand and reflect on the norms and values that underlie one's actions and to negotiate sustainability values, principles, goals and targets – in the context of conflicts of interests and trade-offs, uncertain knowledge and contradictions

- *Strategic competence:* the ability to collectively develop and implement innovative actions that further sustainability at the local level and further afield
- *Collaboration competence:* the ability to learn from others; to understand and respect the needs, perspectives and actions of others (empathy); to understand, relate to and be sensitive to others (empathic leadership); to deal with conflicts in groups; and to facilitate collaborative and participatory problem-solving
- *Critical thinking competence:* the ability to question norms, practices and opinions; to reflect on own one's values, perceptions and actions; and to take a position in the sustainability discourse
- *Self-awareness competence:* the ability to reflect on one's own role in the local community and (global) society, to continually evaluate one's actions and prompt further action and to deal with one's feelings and desires
- *Integrated problem-solving competence:* the overarching ability to apply different problem-solving frameworks to complex sustainability problems and develop viable, inclusive and equitable solutions that promote sustainable development – integrating the above competences

However, ESD is not limited to the development of competences; as transformative education, it also aims at the "transformation of the individual 'self to world relationship' (Koller 2011, p. 16) in the context of a global perspective" (translated from German) (Scheunpflug, 2019, p. 66).

14.2.2 Addressing Values Relevant to Sustainability

ESD establishes a direct link between individual change and societal change and can thus be regarded as transformative education (Balsiger et al., 2017). A "conceptual change", i.e. the change of fundamental perspectives (attitudes, values, paradigms and worldviews) (Scheunpflug, 2019, p. 65), can and should be promoted through

ESD (Sterling, 2011). While transformative learning aims at empowering learners to question and change the way they see and think about the world, the related concept of transgressive learning (Lotz-Sisitka et al., 2015) goes one step further: it emphasises that learning in ESD should transcend the status quo and prepare learners for disruptive thinking and co-creation of new knowledge.

Even if, as emancipatory education, ESD is not about imparting certain pre-determined attitudes, values, paradigms and world views, it nevertheless focuses on the ideas of intra- and intergenerational justice (de Haan et al., 2008). Against this background, it always pursues the values-based goal of "sensitisation to the responsibility to survive" (translated from German) (Mokrosch, 2009, p. 38). At the same time, ESD aims to promote discussion for the purposes of "clarification of values" (translated from German) (ibid., p. 36) that are connected with the guiding principle of sustainable development, in particular with regard to the preservation of the natural basis of life, human dignity and justice. Understood in this way, ESD can contribute to a "change in values towards sustainability" (WBGU, 2011) without overwhelming learners. It also supports the development of reflexive competence. In other words, "Transformative learning must not be used to instrumentalise learners but to empower them for autonomous critical action" (Balsiger et al., 2017, p. 359). Therefore, ESD should be expected and required to contribute to critical discourse on values (Schank & Rieckmann, 2019; Rieckmann, 2020). It can and should provide stimuli for reflection on one's own values and to enable individuals to take a stand in the debate on values with regard to sustainable development (Balsiger et al., 2017; Schank & Rieckmann, 2019). There is thus also potential in broadening the horizon of learners' values. For example, Latin American discourses on *Buen Vivir* (Good Life) and on the rights of nature can be included in the debate on values (Rieckmann, 2017). In this way, Eurocentric ways of thinking can be expanded and critically reflected on.

However, transformation to promote sustainable development also requires structural and institutional change (WBGU, 2011). ESD not only should therefore focus on individual behaviour – and the acquisition of the necessary knowledge, competences and values – but should also address the question of structures, of the "Great Transformation" (ibid.). ESD should contribute to the (political) education of sustainability citizens who participate in transgressive learning and are empowered to question existing structures, think beyond them and thus contribute to structural and institutional transformation (Balsiger et al., 2017; Rieckmann, 2020; Schank & Rieckmann, 2019).

14.3 ESD Pedagogical Principles and Methods

Competences and values cannot be taught but must be developed by learners themselves (Weinert, 2001). ESD therefore requires an action-oriented, transformative pedagogy that is characterised by the following pedagogical principles (Künzli David, 2007; Lozano et al., 2017; Rieckmann, 2018a):

- Learner-centredness and accessibility
- A focus on action and reflection
- Transformative and transgressive learning
- A focus on participation
- Discovery learning
- Networked learning
- Prioritisation of vision
- The linking of social, self-referral and method-oriented learning with subject-related learning

In order to promote processes of transformative education in the context of ESD, learning environments are needed in which learners can become aware of and critically reflect on their own and others' assumptions (Mezirow, 1997) and in which they are encouraged to reflect on frames of reference in critical, deconstructive and transgressive ways in order to stimulate truly transformative learning processes that lead to "conceptual change" (Lotz-Sisitka et al., 2015; Rodríguez Aboytes & Barth, 2020; Sterling, 2011). The pedagogical principles represent general guidelines for the design of learning processes in ESD. Methods are needed that correspond to these pedagogical principles. Such methods are, for example (Rieckmann, 2018a; UNESCO, 2017):

- Collaborative real-world projects such as service-learning projects and campaigns on different sustainability issues
- Vision-building exercises such as future workshops, scenario analyses, utopian/dystopian story-telling, science-fiction thinking, forecasting and backcasting
- Analysing complex systems, for example, through community-based research projects, case studies, stakeholder analysis, actor analysis, modelling and systems games
- Critical and reflective thinking, for example, through fish-bowl discussions and reflective journals

These methods enable learners to become (co-)creators of their own learning process and thus to have a direct influence on the development of their own competences and values. They can be used to promote the development of a range of sustainability competences (Sprenger et al., 2016).

Partnerships and cooperation at local, national and international level also play an important role in the provision of diverse learning environments (UNESCO, 2017). In practical projects with local partners – for example, in the context of service learning – learners can learn about real-world challenges and benefit from partners' competences and experiences (Rieckmann, 2021).

Virtual cooperation and educational links between learners from different countries promote the exchange of knowledge and different views on a range of issues (Barth & Rieckmann, 2009). The methods mentioned can be used to deal with very different contents. However, the content should be

selected in such a way that it is suitable for achieving the relevant goals, for example, the development of a certain competence (Rieckmann, 2018b).

Working with the concept of ESD requires new pedagogical competences from teachers (Bertschy et al., 2013; Corres et al., 2020; Rieckmann and Barth, 2022). Ideally, teachers must have sustainability competences themselves and be able to support learners with the development of such competences. To do this, they need to have a critical understanding of sustainable development on the one hand and the pedagogical approach of ESD on the other. They need knowledge of innovative teaching and learning methods but also the competences to apply them. They also need the competences to support learners – for example, with projects, including critical reflection on one's own role as a teacher and seeing oneself more as a learning facilitator. The ESD competence framework developed in the EU project "A Rounder Sense of Purpose" provides detailed information on the requirements for educators in the context of ESD (Millican, 2022; Rieckmann and Barth, 2022).[1]

14.4 Assessing Sustainability Competences

In contrast to the intensive engagement of ESD research with the concept of key competences, research has so far been less focused on the development of methodological instruments for assessing competence development.

Against this background, the modelling of key competences relevant to sustainability is a central task. Before discussing current progress within ESD research on competence modelling and assessment, it is important to consider the basics of competence modelling and assessment in general. Particular attention will be paid to the assessment of key competences and the challenges involved.

---

[1] https://aroundersenseofpurpose.eu/framework/themodel/

14.4.1 Basics of Competence Modelling and Assessment

The empirical assessment of competences requires theoretically and empirically based competence models that enable the operationalisation of competences (Bräutigam, 2014; Koeppen et al., 2008, 2013). The models can be divided into three basic types: competence structure, competence level and competence development (Hartig & Klieme, 2006; Klieme & Leutner, 2006; Klieme et al., 2007). While competence structure models deal with the internal structure of a competence and describe the different elements – such as abilities, skills, knowledge, etc. – that are necessary to successfully master a situation or requirement, competence level models refer to the different levels of a competence. This means that they focus on the level at which a given requirement can be solved and how difficult this will be. Competence development models focus on the educational process. They ask which level in the learning process is achievable and which competences should be acquired at different stages of the educational process (Hammann, 2004; Klieme & Leutner, 2006). There remains a clear research gap with regard to the creation of competence models.

Klieme and Hartig (2007, p. 24) describe "the connection of pedagogical constructs, psychological competence models and measurement procedures" as "one of the most difficult issues in competence research" (translated from German). When assessing competences, a distinction is made between objective (standardised measurement) and subjective competence assessment (self-assessment) (Zlatkin-Troitschanskaia & Kuhn, 2010). Self-assessment is used more frequently than standardised measurement and is thus the most widespread method. Although these procedures involve less effort, there is criticism of their validity (Schaper et al., 2012) since they are indirect procedures "based on the self-perceptions and self-assessments of the persons surveyed" (Zlatkin-Troitschanskaia & Kuhn, 2010, p. 5). However, self-assessment can be appropriate, especially for "the assessment of the – equally important – non-cognitive competences [...] (e.g. social, volitional and motivational mindsets)" (ibid.: 5f) (translated from German).

The vast majority of research on objective competence measurement refers to subject-specific competences (Zlatkin-Troitschanskaia & Kuhn, 2010). However, trial instruments for measuring (subject-independent) key competences, such as critical thinking, have already been developed. One example is the Collegiate Learning Assessment (CLA) (Shavelson, 2013). Using the Collegiate Learning Assessment (CLA), the OECD Assessment of Higher Education Learning Outcomes (AHELO) programme (Tremblay et al., 2012; Tremblay, 2013) has also worked among other things on the further development and testing of an instrument for the assessment of key competences (such as critical and analytical thinking, problem-solving, written communication skills).

There is also scientific work on subjective competence assessment dealing with the assessment of key competences. In Germany, for example, the research work of the Higher Education Information System (HIS) includes studies on self-assessment of professional capacity (discipline-specific professional competences; social, presentational and methodological competences; organisational skills) (Schaeper & Briedis, 2004), and the National Education Panel (NEPS) surveys meta-competences and social competences (Schaeper, 2013). The overall focus on measuring and recording subject-specific (cognitive) competences – especially objective measurement of competences – impedes the acquisition of a more comprehensive understanding of learning outcomes. There is an obvious need to model and measure (subject-independent) key competences:

Transversal higher-order competencies such as critical-thinking, analytical reasoning, problem-solving or the generation of knowledge and the interaction between substantive and method-ological expertise are widely viewed as critical for the success of individuals in the information age. It is therefore important [...] to measure those skills, and not only cognitive knowledge, which are necessary for success in both academic and business contexts. (Tremblay, 2013, p. 117)

As with empirical research in general, there are also different – and complementary –methodological approaches in competence research: quantitative and qualitative competence research. Quantitative competence research emphasises the measurability and scalability of competences where competence assessment is based on an external perspective (Erpenbeck & von Rosenstiel, 2007) and uses methods such as experiments, tests and questionnaires (e.g. Klein et al., 2007; Shavelson, 2009; Tremblay et al., 2012). Qualitative competence research, by contrast, deals with the nature and quality of competence and uses methods such as unstructured observations, surveys and biographical and ethnographic approaches (e.g. Henze & van Driel, 2006). These are somewhat subjective techniques that take an interest in the internal perspective of learners (Erpenbeck & von Rosenstiel, 2007).

Two basic characteristics of competence constructs have to be considered in the context of a methodological approach to competence assessment: "First, they are [...] relatively complex compared with other cognitive constructs. Secondly, they develop in given contexts and are therefore situationally located" (translation from German) (Rost, 2008, p. 62). This means that individual competence structures "can only be recorded and measured in their complexity and their situational dependence" (ibid.). The assessment of key competences thus poses particular challenges for research: "particularly heavily abstracting and generalised key competences face the problem that key factors of these competencies are hardly measurable" (Barth, 2009, p. 85; cf. Harris, 2001). As key competences are cross-contextual, different methods have to be applied in different contexts in order to capture them (Barth, 2009).

14.4.2 ESD Research on Competence Modelling and Assessment

Research on measuring competences has so far played a rather minor role in Education for Sustainable Development. So far, only a few approaches have attempted to develop methodological instruments for recording competence development in the field of Education for Sustainable Development (Barth & Rieckmann, 2016; Redman et al., 2021).

In Germany, contributions to competence modelling and objective competence measurement have so far been made primarily by research projects on *Bewertungskompetenz* (decision-making competence) – within the framework among others of the DFG priority programme *Competence models for recording individual learning outcomes and for balancing educational processes* (2007–2013) (Bögeholz, 2007; Eggert & Bögeholz, 2010; Eggert et al., 2008; Gresch & Bögeholz, 2013; Lauströer, 2005; Rost et al., 2003). Furthermore, other DFG priority programme projects have looked at other key competences relating to sustainable development, such as (dynamic) problem-solving (Abele et al., 2012; Fleischer et al., 2010; Greiff, 2012; Greiff & Funke, 2010; Greiff et al., 2012; Leutner et al., 2012; Wüstenberg et al., 2012) and environmental literacy (Brügger et al., 2011; Kaiser et al., 2008; Roczen et al., 2010, 2014). There is also research on the modelling and measurement of systems thinking (Bräutigam, 2014). However, the projects in the DFG priority programme and the research on systemic thinking focus primarily on cognitive dispositions.

In ESD research also uses different methods for modelling and assessing competences (Redman et al., 2021): Some studies analyse competences quantitatively (e.g. Eggert et al., 2008; OECD, 2009; Rost et al., 2003) or qualitatively (e.g. Asbrand & Martens, 2013; Feierabend et al., 2013; Franz & Frieters, 2008; Gausmann et al., 2010; Wettstädt & Asbrand, 2014), while others use a mix of methods (e.g. Barth, 2007, 2008; Gardiner & Rieckmann, 2015; Hallitzky, 2008; Lauströer, 2008). Qualitative research can also contribute to the development of competence models, which can then serve as a basis for quantitative competence measurements (Asbrand & Martens, 2013). A particular challenge in the quantitative assessment of ESD competences is the development of suitable tasks:

[One] must assume [...] that the design of tasks for the measurement of ESD-specific competences [...] runs counter to the classical design, as it is also used to a large extent in the PISA tests: It will be difficult to implement clear or unambiguously correct solutions, homogeneous cognitive requirements and short processing times for tasks. Tasks must be formulated as "scenarios" in order, for example, to be able to depict the contextualized nature of skills within a situational framework. (translated from German) (Rode, 2013, p. 121–122; cf. Rost, 2008)

Suitable tasks are best developed using "'tailored testing' (difficulty levels of tasks tailored to the expected problem-solving competence of the subjects) and a multi-matrix design (where different groups of subjects are given different tasks of the same type)" (translated from German) (Rode, 2013, p. 122).

Due to a lack of well-founded competence modelling and suitable instruments for objective competence measurement, many studies of the development of competences in ESD have so far used students' self-assessment of their own competences (Redman et al., 2021). Studies on the development of *Gestaltungskompetenz* among school students (Rode, 2013) and sustainability-related competences by students (Barth et al., 2007; Bone & Agombar, 2011; Gardiner & Rieckmann, 2015; Oberle et al., 1997; Rieckmann, 2009; Shephard et al., 2011, 2013, 2014; Wiek et al., 2016) or university graduates (Brunner et al., 2010; Hansmann et al., 2010), for example, are of relevance here. Shephard et al. (2011, 2013, 2014), for instance, explore how changes in affective competence among students can best be assessed with various self-assessment instruments. Other instruments such as reflective writing, scenario/test case, focus group/interview, performance observation and concept mapping are also used, although much less than self-assessment (Redman et al., 2021).

The latest developments in competence measurement – especially psychometric models (Hartig et al., 2008) – have hardly featured in ESD competence research thus far. However, with research on *Bewertungskompetenz*, (dynamic) problem-solving, environmental com-petence and systemic thinking, some initial steps have been taken that need to be followed up.

However, the existing research on objective ESD competence measurement has so far focused primarily on cognitive aspects. This is true not only for ESD competence research but also for the DFG priority programme Competence Models (Klieme et al., 2010; Klieme & Leutner, 2006). In order to achieve a comprehensive understanding of competences as self-organisational facilities comprising both cognitive and non-cognitive elements, it is important to overcome this "cognitive bias within the concept of competence in empirical educational research" (translated from German) (Schaper et al., 2012, p. 15), taking both cognitive and non-cognitive aspects into account when modelling and measuring competences (Lind, 2009; Schaper et al., 2012; Schaeper, 2013). Volitional and motivational aspects are of great importance in terms of action to promote sustainable development and should therefore not be excluded (Shephard et al., 2011, 2013, 2014).

14.5 Conclusion

The development of key competences that enable individuals to actively shape the process of sustainable development is an essential goal of ESD. However, while there has been intensive debate on the concept of the key competences with regard to sustainability, thus far ESD research has made only limited attempts to model competence constructs and develop methodological instruments for assessing competence development. Research on the assessment of key sustainability-related competences is thus still in its infancy. The systematisation and differentiation of existing competence concepts and the (further) development of competence models and suitable instruments for objective competence measurement are research desiderata that will require attention in the coming years. Special attention should be paid to the comprehensive consideration of the cognitive and non-cognitive aspects of competence.

References

Abele, S., Greiff, S., Gschwendtner, T., Wüstenberg, S., Nickolaus, R., Nitschke, A., & Funke, J. (2012). Die Bedeutung übergreifender kognitiver Determinanten für die Bewältigung beruflicher Anforderungen. Untersuchung am Beispiel dynamischen und technischen Problemlösens. *Zeitschrift für Erziehungswissenschaft, 15*(2), 363–391.

Asbrand, B., & Martens, M. (2013). Qualitative Kompetenzforschung im Lernbereich Globale Entwicklung: Das Beispiel Perspektivenübernahme. In B. Overwien & H. Rode (Eds.), *Bildung für nachhaltige Entwicklung. Lebenslanges Lernen, Kompetenz und gesellschaftliche Teilhabe* (pp. 47–67). Verlag Barbara Budrich.

Balsiger, J., Förster, R., Mader, C., Nagel, U., Sironi, H., Wilhelm, S., & Zimmermann, A. B. (2017). Transformative learning and education for sustainable development. *GAIA – Ecological Perspectives for Science and Society, 26*(4), 357–359. https://doi.org/10.14512/gaia.26.4.15

Barth, M., & Rieckmann, M. (2009). Experiencing the global dimension of sustainability: Student dialogue in a European-Latin American virtual seminar. *International Journal of Development Education and Global Learning, 1*(3), 23–38.

Barth, M. (2007). *Gestaltungskompetenz durch Neue Medien? Die Rolle des Lernens mit Neuen Medien in der Bildung für eine nachhaltige Entwicklung*. Berliner Wissenschafts-Verlag.

Barth, M. (2008). Das Lernen mit Neuen Medien als Ansatz zur Vermittlung von Gestaltungskompetenz. In I. Bormann & G. de Haan (Eds.), *Kompetenzen der Bildung für nachhaltige Entwicklung* (Operationalisierung, Messung, Rahmenbedingungen, Befunde) (pp. 199–213). Springer.

Barth, M. (2009). Assessment of key competencies – A conceptual framework. In M. Adomßent, A. Beringer, & M. Barth (Eds.), *World in transition sustainability perspectives for higher education* (pp. 93–100). VAS.

Barth, M., Godemann, J., Rieckmann, M., & Stoltenberg, U. (2007). Developing key competencies for sustainable development in higher education. *International Journal of Sustainability in Higher Education, 8*(4), 416–430.

Barth, M., & Rieckmann, M. (2016). State of the art in research on higher education for sustainable development. In M. Barth, G. Michelsen, I. Thomas, & M. Rieckmann (Eds.), *Routledge handbook of higher education for sustainable development* (pp. 100–113). Routledge.

Bertschy, F., Künzli, C., & Lehmann, M. (2013). Teachers' competencies for the implementation of educational offers in the field of education for sustainable development. *Sustainability, 5*(12), 5067–5080. https://doi.org/10.3390/su5125067

Bögeholz, S. (2007). Bewertungskompetenz für systematisches Entscheiden in komplexen Gestaltungssituationen Nachhaltiger Entwicklung. In D. Krüger & H. Vogt (Eds.), *Theorien in der biologiedidaktischen Forschung: Ein Handbuch für Lehramtsstudenten und Doktoranden* (pp. 209–220). Springer.

Bone, E., & Agombar, J. (2011). *First-year attitudes towards, and skills in, sustainable development*. http://efsandquality.glos.ac.uk/toolkit/NUS_HEA_2011.pdf. Accessed 12 Sept 2021.

Bräutigam, J. I. (2014). *Systemisches Denken im Kontext einer Bildung für nachhaltige Entwicklung. Konstruktion und Validierung eines Messinstruments zur Evaluation einer Unterrichtseinheit*. phfr.bsz bw.de/files/412/DissertationBraeutigamJulia2014.pdf. Accessed 12 Sept 2021.

Brügger, A., Kaiser, F., & Roczen, N. (2011). One for all: Connectedness to nature, inclusion of nature, environmental identity, and implicit association with nature. *European Psychologist, 16*(4), 324–333. https://doi.org/10.1027/1016-9040/a000032

Brundiers, K., Barth, M., Cebrián, G., Cohen, M., Diaz, L., Doucette-Remington, S., Dripps, W., Habron, G., Harré, N., Jarchow, M., Losch, K., Michel, J., Mochizuki, Y., Rieckmann, M., Parnell, R., Walker, P., & Zint, M. (2021). Key competencies in sustainability in higher education—Toward an agreed-upon reference framework. *Sustainability Science, 16*(1), 13–29. https://doi.org/10.1007/s11625-020-00838-2

Brunner, S. H., Frischknecht, P., Hansmann, R. & Mieg, H. A. (2010). Environmental sciences education under the microscope – Do graduates promote a societal change towards sustainability? *Zürich*. https://ethz.ch/content/dam/ethz/special-interest/usys/department/documents/studium/umweltnaturwissenschaften/diverse/2010-Brunner-etal-Environmental-Sciences-education-under-the-microscope.pdf. Accessed 12 Sept 2021.

Corres, A., Rieckmann, M., Espasa, A., & Ruiz-Mallén, I. (2020). Educator competences in sustainability education: A systematic review of frameworks. *Sustainability, 12*(23), 9858. https://doi.org/10.3390/su12239858

de Haan, G. (2010). The development of ESD-related competencies in supportive institutional frameworks. *International Review of Education, 56*(2–3), 315–328. https://doi.org/10.1007/s11159-010-9157-9

de Haan, G., Kamp, G., Lerch, A., Martignon, L., Müller-Christ, G., & Nutzinger, H.-G. (Eds.). (2008). *Nachhaltigkeit und Gerechtigkeit. Grundlagen und schulpraktische Konsequenzen*. Springer.

Eggert, S., & Bögeholz, S. (2010). Students' use of decision-making strategies with regard to socioscientific issues. An application of the Rasch partial credit model. *Science Education, 94*(2), 230–258.

Eggert, S., Gausmann, E., Hasselhorn, M., Watermann, R., & Bögeholz, S. (2008). Entwicklung eines Messinstruments zur Analyse von Bewertungskompetenz bei Schüler(innen) sowie Studierenden. In Arbeitsgruppe für Empirische Pädagogische Forschung (Ed.), *Kompetenz: Modellierung – Diagnostik – Entwicklung – Förderung.* Tagungsband zur 71. Tagung der AEPF in Kiel vom 25.-27. August 2008 (p. 285). Kiel.

Erpenbeck, J., & von Rosenstiel, L. (2007). *Handbuch Kompetenzmessung. Erkennen, verstehen und bewerten von Kompetenzen in der betrieblichen, pädagogischen und psychologischen Praxis.* Schäffer-Poeschel.

Feierabend, T., Stuckey, M., & Eilks, I. (2013). Ansätze zur Analyse von Bewertungskompetenz in Gruppendiskussionen zum Klimawandel. In J. Menthe, D. Höttecke, I. Eilts, & C. Corinna (Eds.), *Handeln in Zeiten des Klimawandels. Bewerten lernen als Bildungsaufgabe* (pp. 171–181). Waxmann.

Fleischer, J., Wirth, J., Rumann, S., & Leutner, D. (2010). Strukturen fächerübergreifender und fachlicher Problemlösekompetenz – Analyse von Aufgabenprofilen. *Zeitschrift für Pädagogik, 56.* Beiheft, 239–248.

Franz, J., & Frieters, N. (2008). Kompetenzmodelle in Fortbildungen – pragmatische Wege. In I. Bormann & G. de Haan (Eds.), *Kompetenzen der Bildung für nachhaltige Entwicklung. Operationalisierung, Messung, Rahmenbedingungen, Befunde* (pp. 75–87). Springer.

Gardiner, S., & Rieckmann, M. (2015). Pedagogies of preparedness: Use of reflective journals in the operationalisation and development of anticipatory competence. *Sustainability, 7*(8), 10554–10575.

Gausmann, E., Eggert, S., Hasselhorn, M., Watermann, R., & Bögeholz, S. (2010). Projekt Bewertungskompetenz: Wie verarbeiten Schüler/-innen Sachinformationen in Problem- und Entscheidungssituationen Nachhaltiger Entwicklung – Ein Beitrag zur Bewertungskompetenz. *Zeitschrift für Pädagogik, 56.* Beiheft, 204–215.

Greiff, S. (2012). Assessment and theory in complex problem solving. A continuing contradiction? *Journal of Educational and Developmental Psychology, 2*(1), 45–56.

Greiff, S., & Funke, J. (2010). Systematische Erforschung komplexer Problemlösefähigkeit anhand minimal komplexer Systeme. *Zeitschrift für Pädagogik, 56.* Beiheft, 216–227.

Greiff, S., Wüstenberg, S., & Funke, J. (2012). Dynamic problem solving: A new measurement perspective. *Applied Psychological Measurement, 36*(1), 189–213.

Gresch, H., & Bögeholz, S. (2013). Identifying non-sustainable courses of action: A prerequisite for decision-making in education for sustainable development. *Research in Science Education, 43*(2), 733–754.

Hallitzky, M. (2008). Forschendes und selbstreflexives Lernen im Umgang mit Komplexität. In I. Bormann & G. de Haan (Eds.), *Kompetenzen der Bildung für nachhaltige Entwicklung. Operationalisierung,*

Messung, Rahmenbedingungen, Befunde (pp. 159–178). Springer.

Hammann, M. (2004). Kompetenzentwicklungsmodelle: Merkmale und ihre Bedeutung – dargestellt anhand von Kompetenzen beim Experimentieren. *Der mathematische und naturwissenschaftliche Unterricht, 57*(4), 196–203.

Hansmann, R., Mieg, H., & Frischknecht, P. M. (2010). Qualifications for contributing to sustainable development: A survey of environmental sciences graduates. *Gaia, 19*(4), 278–286.

Harris, B. (2001). Are all competencies measurable? An education perspective. In D. S. Rychen & L.-H. Salganik (Eds.), *Defining and selecting key competencies* (pp. 222–227). Hogrefe & Huber.

Hartig, J., & Klieme, E. (2006). Kompetenz und Kompetenzdiagnostik. In K. Schweizer (Ed.), *Leistung und Leistungsdiagnostik* (pp. 127–143). Springer.

Hartig, J., Klieme, E., & Leutner, D. (Eds.). (2008). *Assessment of competencies in educational contexts.* Hogrefe & Huber.

Henze, I., & van Driel, J. H. (2006). The development of experienced science teachers. Pedagogical content knowledge in the context of educational innovation. In I. Eilks, & B. Ralle (Eds.), *Towards research-based science teacher education.* Proceedings of the 18th Symposium on Chemical and Science Education held at the University of Bremen (pp. 99–112), 15–17 June 2006. Aachen.

Kaiser, F., Roczen, N., & Bogner, F. X. (2008). Competence formation in environmental education: Advancing ecology-specific rather than general abilities. *Umweltpsychologie, 12*(2), 56–70.

Klafki, W. (2007). *Neue Studien zur Bildungstheorie und Didaktik: Zeitgemäße Allgemeinbildung und kritisch-konstruktive Didaktik* (6th ed.). Beltz Verlag.

Klein, S., Benjamin, R., Shavelson, R., & Bolus, R. (2007). The collegiate learning assessment. Facts and fantasies. *Evaluation Review, 31*(5), 415–439.

Klieme, E., & Hartig, J. (2007). Kompetenzkonzepte in den Sozialwissenschaften und im erziehungswissenschaftlichen Diskurs. In M. Prenzel, I. Gogolin, & H. H. Krüger (Eds.), *Kompetenzdiagnostik. Zeitschrift für Erziehungswissenschaft,* Sonderheft 8/2007. Wiesbaden, 11–29.

Klieme, E., & Leutner, D. (2006). *Kompetenzmodelle zur Erfassung individueller Lernergebnisse und zur Bilanzierung von Bildungsprozessen.* Überarbeitete Fassung des Antrags an die DFG auf Einrichtung eines Schwerpunktprogramms. Universität Duisburg-Essen. http://kompetenzmodelle.dipf.de/pdf/rahmenantrag. Accessed 12 Sept 2021.

Klieme, E., Leutner, D., & Kenk, M. (2010). Kompetenzmodellierung. Eine aktuelle Zwischenbilanz des DFG-Schwerpunktprogramms. *Zeitschrift für Pädagogik.* Beiheft, *56,* 9–11.

Klieme, E., Maag-Merki, K., & Hartig, J. (2007). Kompetenzbegriff und Bedeutung von Kompetenzen im Bildungswesen. In J. Hartig & E. Klieme (Eds.), *Bildungsforschung Band 20: Möglichkeiten und Voraussetzungen technologiebasierter Kompetenzdiagnostik*. Bonn, Berlin.

Koeppen, K., Hartig, J., Klieme, E., & Leutner, D. (2008). Current issues in competence and modeling assessment. *Zeitschrift für Psychologie/Journal of Psychology, 216*(2), 61–73.

Koeppen, K., Hartig, J., Klieme, E., & Leutner, D. (2013). Competence models for assessing individual learning outcomes and evaluating educational processes – A priority program of the German Research Foundation (DFG). In S. Blömeke, O. Zlatkin-Troitschanskaia, C. Kuhn, & J. Fege (Eds.), *Modeling and measuring competencies in higher education* (Tasks and challenges) (pp. 171–192). SensePublishers.

Künzli David, C. (2007). *Zukunft mitgestalten: Bildung für eine nachhaltige Entwicklung – Didaktisches Konzept und Umsetzung in der Grundschule*. Haupt Verlag.

Lauströer, A. (2005). *Förderung von Bewertungskompetenz durch Bildung für eine nachhaltige Entwicklung*. Universitätsbibliothek.

Lauströer, A. (2008). *Bewertungskompetenz durch Bildung für eine nachhaltige Entwicklung: Evaluation einer Unterrichtseinheit zum Thema Massentourismus für die Sekundarstufe I*. VDM Verlag Dr. Müller.

Leutner, D., Fleischer, J., Wirth, J., Greiff, S., & Funke, J. (2012). Analytische und dynamische Problemlösekompetenz im Lichte internationaler Schulleistungsvergleichsstudien: Untersuchungen zur Dimensionalität. *Psychologische Rundschau, 63*(1), 34–42.

Lind, G. (2009). *Moral ist lehrbar. Handbuch zur Theorie und Praxis moralischer und demokratischer Bildung*. 2., überarbeitete und aktualisierte Auflage. Logos-Verlag.

Lotz-Sisitka, H., Wals, A. E. J., Kronlid, D., & McGarry, D. (2015). Transformative, transgressive social learning: Rethinking higher education pedagogy in times of systemic global dysfunction. *Current Opinion in Environmental Sustainability, 16*, 73–80. https://doi.org/10.1016/j.cosust.2015.07.018

Lozano, R., Merrill, M., Sammalisto, K., Ceulemans, K., & Lozano, F. (2017). Connecting competences and pedagogical approaches for sustainable development in higher education: A literature review and framework proposal. *Sustainability, 9*(11), 1889. https://doi.org/10.3390/su9101889

Mahler, D. G., Yonzan, N., Lakner, C., Castaneda Aguilar, R. A., & Wu, H. (2021). Updated estimates of the impact of COVID-19 on global poverty: Turning the corner on the pandemic in 2021? *World Bank*. https://blogs.worldbank.org/opendata/updated-estimates-impact-covid-19-global-poverty-turning-corner-pandemic-2021. Accessed 12 Sept 2021.

Mezirow, J. (1997). Transformative learning: Theory to practice. *New Directions for Adult and Continuing Education, 1997*(74), 5–12. https://doi.org/10.1002/ace.7401

Millican, R. (2022). A Rounder Sense of Purpose: Competences for Educators in Search of Transformation. In P. Vare, N. Lausselet & M. Rieckmann (Eds.), Competences in Education for Sustainable Development. Critical Perspectives (pp. 35–43). Springer International Publishing. https://doi.org/10.1007/978-3-030-91055-6_5.

Michelsen, G. (2009). Kompetenzen und Bildung für nachhaltige Entwicklung. In B. Overwien & H.-F. Rathenow (Eds.), *Globalisierung fordert politische Bildung: Politisches Lernen im globalen Kontext* (pp. 75–86). Verlag Barbara Budrich.

Mokrosch, R. (2009). Zum Verständnis von Werte-Erziehung: Aktuelle Modelle für die Schule. In R. Mokrosch & A. Regenbogen (Eds.), *Werte-Erziehung und Schule: Ein Handbuch für Unterrichtende* (pp. 32–40). Vandenhoeck & Ruprecht.

Oberle, B. M., Scholz, R. W., & Frischknecht, P. M. (1997). Ökologische Problemlösefähigkeit – Eine Herausforderung für die Ausbildung von UmweltnaturwissenschaflerInnen. *Gaia, 6*(1), 73–78.

OECD (Organisation for Economic Co-operation and Development). (2009). *Green at fifteen? How 15-year-olds perform in environmental science and geoscience in PISA 2006*. https://www.oecd.org/edu/innovation-education/centreforeffectivelearningenvironments-cele/42983010.pdf. Accessed 12 Sept 2021.

Redman, A., Wiek, A., & Barth, M. (2021). Current practice of assessing students' sustainability competencies: A review of tools. *Sustainability Science, 16*(1), 117–135. https://doi.org/10.1007/s11625-020-00855-1

Rieckmann, M. (2009). Developing Shaping Competence in Informal Setting at Universities. In M. Adomssent, A. Beringer & M. Barth (ed.), World in Transition Sustainability Perspectives for Higher Education (S. 78–84). VAS.

Rieckmann, M. (2017). Bildung für nachhaltige Entwicklung in der Großen Transformation – Neue Perspektiven aus den Buen Vivir- und Postwachstumsdiskursen. In O. Emde, U. Jakubczyk, B. Kappes, & B. Overwien (Eds.), *Schriftenreihe "Ökologie und Erziehungswissenschaft" der Kommission Bildung für eine nachhaltige Entwicklung der Deutschen Gesellschaft für Erziehungswissenschaft (DGfE). Mit Bildung die Welt verändern? Globales Lernen für eine nachhaltige Entwicklung* (pp. 147–159). Verlag Barbara Budrich.

Rieckmann, M. (2018a). Chapter 2 – Learning to transform the world: Key competencies in ESD. In A. Leicht, J. Heiss, & W. J. Byun (Eds.), *Education on the move. Issues and trends in education for sustainable development* (pp. 39–59). United Nations Educational, Scientific and Cultural Organization.

Rieckmann, M. (2018b). Chapter 3 – Key themes in education for sustainable development. In A. Leicht, J. Heiss, & W. J. Byun (Eds.), *Education on the move. Issues and trends in education for sustainable devel-*

opment (pp. 61–84). United Nations Educational, Scientific and Cultural Organization.

Rieckmann, M. (2020). Emancipatory and transformative global citizenship education in formal and informal settings – Empowering learners to change structures. *Tertium Comparationis: Journal für International und Interkulturell Vergleichende Erziehungswissenschaft, 26*(2), 174–186. https://www.waxmann.com/index.php?eID=download&id_artikel=ART104545&uid=frei

Rieckmann, M. (2021). Service Learning für nachhaltige Entwicklung. In A. Boos, M. van den Eeden, & T. Viere (Eds.), *CSR und Hochschullehre* (pp. 185–198). Springer. https://doi.org/10.1007/978-3-662-62679-5_9

Rieckmann, M. & Barth, M. (2022). Educators' Competence Frameworks in Education for Sustainable Development. In P. Vare, N. Lausselet & M. Rieckmann (Eds.), Competences in Education for Sustainable Development. Critical Perspectives (pp. 19–26). Springer International Publishing. https://doi.org/10.1007/978-3-030-91055-6_3.

Roczen, N., Kaiser, F. G., & Bogner, F. X. (2010). Umweltkompetenz – Modellierung, Entwicklung und Förderung. *Zeitschrift für Pädagogik, 56.* Beiheft, 126–134.

Roczen, N., Kaiser, F. G., Bogner, F. X., & Wilson, M. (2014). A competence model for environmental education. *Environment and Behavior, 46*(8), 972–992.

Rode, H. (2013). Kompetenzmessung in der Bildung für nachhaltige Entwicklung. Erste Ansätze. In M. Zschiesche (Ed.), *Klimaschutz im Kontext. Die Rolle von Bildung und Partizipation auf dem Weg in eine klimafreundliche Gesellschaft* (pp. 117–134). oekom.

Rodríguez Aboytes, J. G., & Barth, M. (2020). Transformative learning in the field of sustainability: A systematic literature review (1999–2019). *International Journal of Sustainability in Higher Education, 21*(5), 993–1013. https://doi.org/10.1108/IJSHE-05-2019-0168

Rost, J. (2008). Zur Messung von Kompetenzen einer Bildung für nachhaltige Entwicklung. In I. Bormann & G. de Haan (Eds.), *Kompetenzen der Bildung für nachhaltige Entwicklung. Operationalisierung, Messung, Rahmenbedingungen, Befunde* (pp. 61–73). Springer.

Rost, J., Lauströer, A., & Raack, N. (2003). Kompetenzmodelle einer Bildung für Nachhaltigkeit. In Praxis der Naturwissenschaften. *Chemie in der Schule, 52*(8), 10–15.

Schaeper, H. (2013). The German National Educational Panel Study (NEPS). Assessing competencies over the life course and in higher education. In S. Blömeke, O. Zlatkin-Troitschanskaia, C. Kuhn, & J. Fege (Eds.), *Modeling and measuring competencies in higher education. Tasks and challenges* (pp. 147–158). Springer.

Schaeper, H., & Briedis, K. (2004). Kompetenzen von Hochschulabsolventinnen und Hochschulabsolventen, berufliche Anforderungen und Forderungen für die Hochschulreform. *Informationen für die Beratungs- und Vermittlungsdienste, 2004*(23), 1–7.

Schank, C., & Rieckmann, M. (2019). Socio-economically substantiated education for sustainable development: Development of competencies and value orientations between individual responsibility and structural transformation. *Journal of Education for Sustainable Development, 13*(1), 67–91. https://doi.org/10.1177/0973408219844849

Schaper, N., Reis, O., Wildt, J., Horvath, E., & Bender, E. (2012). *Fachgutachten zur Kompetenzorientierung in Studium und Lehre. HRK – Hochschulrektorenkonferenz. Projekt nexus. Konzepte und gute Praxis für Studium und Lehre.* HRK.

Scheunpflug, A. (2019). Transformatives Globales Lernen – eine Grundlegung in didaktischer Absicht. In G. Lang-Wojtasik (Ed.), *Bildung für eine Welt in Transformation: Global Citizenship Education als Chance für die Weltgesellschaft* (pp. 63–74). Verlag Barbara Budrich.

Shavelson, R. J. (2009). *Measuring college learning responsibly. Accountability in a new era.* Stanford University Press.

Shavelson, R. J. (2013). An approach to testing & modelling competence. In S. Blömeke, O. Zlatkin-Troitschanskaia, C. Kuhn, & J. Fege (Eds.), *Modeling and measuring competencies in higher education. Tasks and challenges* (pp. 29–43). Springer.

Shephard, K., Harraway, J., Jowett, T., Lovelock, B., Skeaff, S., Slooten, L., Strack, M., & Furnari, M. (2014). Longitudinal analysis of the environmental attitudes of university students. *Environmental Education Research, 21*(6), 805–820.

Shephard, K., Harraway, J., Lovelock, B., Skeaff, S., Slooten, L., Strack, M., Furnari, M., & Jowett, T. (2013). Is the environmental literacy of university students measurable? *Environmental Education Research, 20*(4), 476–495.

Shephard, K., Smith, N., Deaker, L., Harraway, J., Broughton-Ansin, F., & Mann, S. (2011). Comparing different measures of affective attributes relating to sustainability. *Environmental Education Research, 17*(3), 329–340.

Sprenger, S., Menthe, J., & Höttecke, D. (2016). Methodenkonzeption und -einsatz. In M. K. W. Schweer (Ed.), *Bildung für nachhaltige Entwicklung in pädagogischen Handlungsfeldern: Grundlagen, Verankerung und Methodik in ausgewählten Lehr-Lern-Kontexten* (pp. 95–107). Peter Lang Pub Inc.

Steffen, W., Richardson, K., Rockström, J., Cornell, S. E., Fetzer, I., Bennett, E. M., Biggs, R., Carpenter, S. R., de Vries, W., de Wit, C. A., Folke, C., Gerten, D., Heinke, J., Mace, G. M., Persson, L. M., Ramanathan, V., Reyers, B., & Sörlin, S. (2015). Sustainability. Planetary boundaries: Guiding human development on a changing planet. *Science, 347*(6223), 1259855. https://doi.org/10.1126/science.1259855

Sterling, S. (2011). Transformative learning and sustainability: Sketching the conceptual ground. *Learning* and *Teaching in Higher Education, 5*(2010–11), 17–33.

Tremblay, K. (2013). OECD Assessment of Higher Education Learning Outcomes (AHELO). Rationale, challenges and initial insights from the feasibility study. In S. Blömeke, O. Zlatkin-Troitschanskaia, C. Kuhn, & J. Fege (Eds.), *Modeling and measuring competencies in higher education. Tasks and challenges* (pp. 113–126). Springer.

Tremblay, K., Lalancette, D., & Roseveare, D. (2012). *Assessment of higher education learning outcomes. Feasibility study report. Volume 1 – Design and implementation.* https://www.oecd.org/education/skills-beyond-school/AHELOFSReportVolume1.pdf. Accessed 12 Sept 2021.

UN Environment (Ed.). (2019). *Global environment outlook GEO-6: Healthy planet, healthy people.* Cambridge University Press. https://www.unenvironment.org/resources/global-environment-outlook-6. Accessed 12 Sept 2021

UNESCO. (2017). Education for sustainable development goals. *Learning Objectives.* UNESCO. http://unesdoc.unesco.org/images/0024/002474/247444e.pdf. Accessed 12 Sept 2021.

UNESCO. (2020). *Education for sustainable development: A roadmap. ESD for 2030.* UNESCO. https://unesdoc.unesco.org/ark:/48223/pf0000374802. Accessed 12 Sept 2021.

United Nations (UN). (2015). *Transforming our world: The 2030 Agenda for Sustainable Development.* Resolution adopted by the General Assembly on 25 Sept 2015. http://www.un.org/ga/search/view_doc.asp?symbol=A/RES/70/1&Lang=E. Accessed 12 Sept 2021.

Vare, P., & Scott, W. (2007). Learning for a change: Exploring the relationship between education and sustainable development. *Journal of Education for Sustainable Development, 1*(2), 191–198. https://doi.org/10.1177/097340820700100209

Wals, A. E. J. (2011). Learning our way to sustainability. *Journal of Education for Sustainable Development, 5*(2), 177–186. https://doi.org/10.1177/097340821100500208

WBGU – German Advisory Council on Global Change. (2011). *World in transition – A social contract for sustainability.* WBGU. https://www.wbgu.de/en/publications/publication/world-in-transition-a-social-contract-for-sustainability. Accessed 12 Sept 2021

Weinert, F. E. (2001). Concept of competence: A conceptual clarification. In D. S. Rychen & L. H. Salganik (Eds.), *Defining and selecting key c ompetencies* (pp. 45–65). Hogrefe & Huber.

Wettstädt, L., & Asbrand, B. (2014). Handeln in der Weltgesellschaft. Zum Umgang mit Handlungsaufforderungen im Unterricht zu Themen des Lernbereichs Globale Entwicklung. *ZEP – Zeitschrift für internationale Bildungsforschung und Entwicklungspädagogik, 37*(1), 4–12.

Wiek, A., Bernstein, M. J., Foley, R. W., Cohen, M., Forrest, N., Kuzdas, C., Kay, B., & Withycombe Keeler, L. (2016). Operationalising competencies in higher education for sustainable development. In M. Barth, G. Michelsen, I. Thomas, & M. Rieckmann (Eds.), *Routledge handbook of higher education for sustainable development* (pp. 241–260). Routledge.

Wiek, A., Withycombe, L., & Redman, C. L. (2011). Key competencies in sustainability: A reference framework for academic program development. *Sustainability Science, 6*(2), 203–218.

World Bank. (2020). *Reversals of fortune. Poverty and shared prosperity 2020.* World Bank. https://www.worldbank.org/en/publication/poverty-and-shared-prosperity. Accessed 12 Sept 2021.

Wüstenberg, S., Greiff, S., & Funke, J. (2012). Complex problem solving. More than reasoning? *Intelligence, 40*, 1–14.

Zlatkin-Troitschanskaia, O., & Kuhn, C. (2010). Messung akademisch vermittelter Fertigkeiten und Kenntnisse von Studierenden bzw. Hochschulabsolventen – Analyse zum Forschungsstand. Arbeitspapier WP 56. *Mainz.* http://www.wipaed.uni-mainz.de/ls/ArbeitspapiereWP/gr_Nr.56.pdf. Accessed 12 Sept 2021.

Marco Rieckmann is Professor of Higher Education Development in the Faculty I – Education and Social Sciences at the University of Vechta. He is the university's presidential advisor on sustainability and the representative of the German Educational Research Association (DGfE) in the Council of the European Educational Research Association (EERA). He is a member of the editorial board of the journal *Sustainability Science*. His main areas of research are higher education development, education for sustainable development, global education, and sustainable university development.

Application-Oriented Development of Outcome Indicators for Measuring Students' Sustainability Competencies: Turning from Input Focus to Outcome Orientation

15

Eva-Maria Waltner, Anne Overbeck, and Werner Rieß

Abstract

Since the Brundtland Report and the Agenda 21 conference in Rio, many Education for Sustainable Development (ESD) programs have been launched. However, until now, empirical data on the impact and outcome of ESD initiatives within educational settings is scarce. This chapter explores the assessment of sustainability competencies including cognitive, affective, and behavioral domains, by presenting different possibilities, results, and limitations of ESD assessment goals and frameworks. This contribution emerges from a collaboration of researchers on the operationalization and measurement of ESD outcomes at the University of Education Freiburg and the Otto-von-Guericke University Magdeburg. Data from a longitudinal measurement with students in secondary schools (grades 5–8, $n = 1324$, age 9–16) in the state of Baden-Württemberg was analyzed to gain a clearer picture of the development of students' sustainability competencies within 1 school year. This data shows that measuring the outcome of ESD teaching programs is possible. Using these empirical measures could thereby facilitate decision-making on ESD measures for many different levels.

Keywords

ESD · Measurement · Outcome · Sustainability competencies · Indicators

15.1 Introduction

Despite long and ongoing attempts to change toward a more sustainable society, environmental problems remain or have even worsened in many ways (e.g., Cai et al., 2016; Lenton et al., 2008; Schellnhuber, 2006). Based on these observations, a growing interest has emerged which calls for a shift from the Input focus of indicators (e.g., policies such as the international treaties on climate change or educational measures) to an Outcome orientation (e.g., actual changes in emissions or the development of sustainability competencies (SCs) on the students/stakehold-

E.-M. Waltner (✉) · W. Rieß
Department of Biology and Pedagogy of Biology,
University of Education Freiburg, Freiburg, Germany
e-mail: eva-maria.waltner@ph-freiburg.de

A. Overbeck
Personality and Social Psychology Division,
Otto-von-Guericke University Magdeburg,
Magdeburg, Germany

ers' level) (e.g., Burford et al., 2016; Waltner et al., 2018, p. 12). In this analysis, we focus on the school level of Education for Sustainable Development (ESD) measures. Though the students who are taught in our schools right now are not the decision-makers of today, they eventually will be the decision-makers of tomorrow. Some effects of the educational measures (e.g., teaching, whole institutional approach) might only be empirically verifiable in the long term or in general not clearly be attributable to a specific measure, due to the complexity of the interaction of many variables affecting, for example, sustainability awareness. However, from these circumstances it cannot be deduced or assumed that there are no characteristics at all which can be recorded empirically. These considerations show that when shifting the attention from the Input to the Output orientation of ESD measures, we might need more long-term assessments and additional method orientations to evaluate the impact. Until now, empirical data on the long- as well as the short-term impact of ESD initiatives within educational settings is scarce. This chapter explores the assessment possibilities of SCs, including cognitive, affective, and behavioral domains, by presenting different approaches and some exemplary results without the ambition of covering ESD assessment methods or the SCs in their entirety.[1]

Before being able to measure the desired outcomes, we need clarity about the goals which are targeted by ESD measures. In this article, the term ESD is used to describe the sum of all actions by which people seek to promote learners' (*or their own*) SCs, i.e., enabling them to facilitate a sustainable development (Waltner et al., 2019). ESD goal dimensions on the outcome level of learners can be divided into cognitive, affective-motivational, and behavioral facets as well as sub-competencies which cannot be clearly assigned to only one of these three areas. Input policies which want to achieve ESD goals on various levels of concreteness can be found in regional (e.g., educational plans, municipal pol-

icy) as well as on the inter- or supranational level (e.g., sustainable development goals (SDGs), Paris Agreement).

After defining the outcome which will be considered as relevant to assessing the success of ESD, the indicators which are developed on this basis have to be tested in extensive studies. This contribution emerges from a collaboration of researchers on the operationalization and measurement of ESD outcomes at the University of Education Freiburg and the Personality and Social Psychology Division of the Otto-von-Guericke University Magdeburg, Germany. The former mainly used the local educational plan as ESD goal orientation, the latter the global approach of the SDGs. The data from this first project, a longitudinal measurement with students in secondary schools (grades 5–8, $n = 1324$, age 9–16) in the state of Baden-Württemberg, will be presented. We will further discuss this approach in the light of generalizability of the measurement approaches and results.

The ultimate normative goal of ESD is the actual development of a more sustainable society through sustainable behavior or at least the facilitation of each individual to become a part of sustainable development – see, e.g., the definition of ESD on the local level: Education for Sustainable Development enables learners to make informed decisions and act responsibly for the protection of the environment and for a functioning economy and a just world society for current and future generations (Ministry of Education Baden-Württemberg, 2016, *translated by the authors*). This definition resembles many other internationally used definitions (e.g., Rieckmann, 2018; UNESCO, 2017). Consequently, good outcome measurement needs validity criteria to ensure that these gauges can assess the achievement of these goals – in our case, the competencies which enable the learners (among other things) to contribute to real-world change. To demonstrate how this can be achieved exemplarily, we will present two findings which emerged from the work on this data. First, we performed an ad hoc scale validation, using items from the newly developed ESD project, linking them to a well-established measurement instrument which has proven to be

---

[1] For a general discussion, please see, e.g., Redman et al. (2021).

predictive of real-world behavioral impact. Second, the Fridays for Future (F4F) participation of the surveyed students was used to draw further conclusions about environmental activism resulting from a high environmental attitude. This in turn can be interpreted as an exemplary behavioral manifestation of pro-sustainable behavior for a more sustainable society, which consequently serves to achieve the ESD goals.

15.2 Theory

15.2.1 Defining Fitting Indicators for ESD in Schools: From Input to Outcome Orientation

In their review on the development of indicators for educational reporting, Baethge et al. (2010, p. 15) define indicators as quantitative tools (or proxy variables) providing a simple and comprehensible status report on the quality or the state of art of a more complex, usually multi-dimensional system. This is in line with the general direction of indicator definitions. However, in some fields, they are a lot more policy oriented as, for example, in Bormann and Michelsens' "the Collaborative Production of Meaningful Measure(ment)s" (2010), where indicators "are considered to provide condensed information which can be transformed into knowledge relevant for decision making." Also depending on the literature, different types of indicators are mentioned. In the framework of this analysis, we would like to apply the classification of Tilbury and Janousek (2006) and the Expert Groups on ESD Indicators as they are used by and for the ESD field, but equally offer connectivity to other fields, as this partly follows the standard international context/input-process-output-outcome scheme (see e.g., DIPF, 2007, p. 5):

- Checklist Indicators: provide information on initial policy, legislation, regulatory, and governance measures taken by a government in order to implement the Strategy.
- Input Indicators: provide information on a broader spectrum of activities taking place in terms of the implementation of the Strategy (e.g., amount of public authority money invested in the ESD materials, proportion of public supported research on ESD).
- Output Indicators: provide information on the direct results of these activities (e.g., performance of trained teachers, number of businesses involved in ESD projects, number of educators who received training on ESD issues).
- Outcome Indicators: provide information on the possible impact of the implementation of the Strategy particularly on values, attitudes, and choices in favor of SD (e.g., learning outcomes resulting from ESD partnerships, community-based projects, and business involvement). (Tilbury & Janousek, 2006, p. 11)

As stated above, before empirical measurement is possible, the competency dimensions of interest must be defined with sufficient precision. Such a specification allows, in principle, the operationalization of the competency of interest in an appropriate measurement procedure (see Klieme & Hartig, 2007). In general, there is a quite common acceptance of the broad tripartite classification of ESD goals in terms of achieving competencies in the knowledge, attitudes, and behavioral dimensions (e.g., Rieckmann, 2018; Waltner et al., 2019). The regional educational plan (implemented in 2016 for Baden-Wuerttemberg) served as a basis for the subject-specific ESD goal orientation. Teaching plans and curricula as well as school/ESD experts (i.e., teachers, students, university staff) were consulted in the identification of the different ESD goals for this specific field, the secondary schools. The ESD research program at Otto-von-Guericke University Magdeburg took another approach, where the ESD goal dimensions where defined based on a more international framework (the SDGs). These approaches are only two possibilities for defining the content to be measured.

No matter what the goal framework might be (regional, international – education plan, policy document) in any case of (outcome) measurement, precise goal (content) specifications and operationalization are needed in order to capture the underlying construct.

15.2.2 Operationalization of ESD Outcomes, Theoretical Background, and Disciplinary Connectivity

Without operationalization, the need for ESD and more importantly the outcomes of ESD-related interventions (e.g., lessons, seminars, projects) cannot be determined empirically (Gräsel et al., 2012; Wiek et al., 2011). Until now, sustainability and ESD research still lack connectivity to well-grounded models and findings from related relevant disciplines; these include environmental psychology, competence research, or more specific fields such as the science of behavior or attitude research (see, e.g., Waltner et al., 2019). Connecting ESD research to well-established measurement procedures facilitates the integration of already operationalized facets of competencies (e.g., environmental attitude) into the larger construct of SCs (sustainability competencies). The findings in measuring environmental attitudes are pertinent for ESD research which is aiming at measuring the broader dimensions of SCs – including, for example, the fields of sustainability-related knowledge and attitudes toward sustainability or sustainability-related problem-solving. SCs thus contain supplementary dimensions to the ecological dimension, such as political, economic, or more socially oriented considerations.

For the cognitive dimension of a SC, numerous attempts of operationalization in the form of knowledge scales are already available. Most of them precisely capture only specific parts of sustainability knowledge (e.g., environmental knowledge) as a significant subset of sustainability knowledge (e.g., Frick et al., 2004; Maloney & Ward, 1973; McBeth et al., 2011; Roczen et al., 2014). Thus, the cognitive dimension is not only about knowledge but also about cognitive skills and abilities (Waltner et al., 2019). However, although environmental knowledge is found to be consistently and positively related to environmental attitudes, the relationship is not especially strong (e.g., Arcury, 1990).

In another study, the behavior prediction through different forms of knowledge was also quite low: According to Frick et al. (2004), "the low overall explained behavioral variance of 6% was comparable to other studies (e.g., Hines et al., 1986/87). Although apparently small, this link should not be underestimated, since influences of knowledge on behavior are thought to be indirect, which means that they are mediated by other variables (cf. Kaiser & Fuhrer, 2003)." Thus, environmental knowledge can be regarded as the basis for ecological behavior of individuals, while a person's appreciation for nature is a relevant motivational factor for promoting the actual behavior (e.g., Kaiser et al., 2011).

Accordingly, for the affective-motivational and behavioral dimension, a very close connection has been proven by various studies. Affective goal commitment or a positively valued sequence of actions is the core of every motivation. An action is not carried out if the costs are perceived as too high when compared to the affective goal commitment. This attitude-cost relationship is modeled in the Campbell paradigm (Kaiser et al., 2010), which implies a solid link between a person's attitudes and his or her behavior. Consequently, in the framework of the Campbell paradigm, behavioral self-reports are used as indicators for a person's attitudes. Operationalization approaches of this attitude and behavior dimension can be found in various measurement instruments which have been broadly applied, for example, in Michelsen et al. (2015) (Greenpeace Sustainability Barometer), Shepherd et al. (2009) (Sustainable Development Values scale), Bogner (2018) (2-MEV scale), and Kaiser (1998) (GEB). According to the criteria of specific objectivity (Kaiser et al., 2018), in the future it might be preferable to focus on the subjacent level of the latent construct of the scales instead of fighting a battle of individual items. The basis for this claim will be elaborated in the section "Validation of New Indicators - Establishing Impact-Relevance of ESD Outcome."

15.3 Methods

In the following, the operationalization, the measurement process, and the data from the longitudinal assessment in secondary schools will be presented. A total of 1318 students and 113 teachers in grades 5–8 were surveyed at 10 randomly selected schools in the state of Baden-Württemberg (B-W). The proportion of school forms corresponds to the students' ratio at secondary schools in B-W and therefore constitutes a stratified sample according to this distribution of the Baden-Wuerttemberg Statistical Office (2017). All schools included in the study did not have a specific ESD program. Data was collected at the beginning and at the end of the school year 2018–2019. This longitudinal data can be used for answering the research question of the development of SC (i.e., the development within this 1 school year) and thus allows us to draw conclusions about the outcome of the ESD policies in B-W. For survey development, well-established items and findings from empirical educational, ESD, and environmental research as well as environmental psychology and sustainability psychology were combined with ESD-curricular and subject-specific items drawn from a curricular analysis for B-W. The survey development process is described in more detail in Waltner et al. (2019). On-site surveys were conducted with the assistance of teachers, who received verbal and written briefings prior to the survey, and four individuals from the research project. Participants were assured full confidentiality and anonymity by assignment of a code to match the data sets. The test duration was 90 min (with a break after approximately 45 min). Altogether, data was collected in 79 school classes (through whole class assessments). For more details concerning the pilot phases, the structure of the questionnaire, number of items, reliability, and validation, please see Waltner et al. (2019). The survey contained several SC dimensions such as sustainability knowledge, attitudes, and behavior. The three before-mentioned dimensions of SC will be presented in the following.

15.3.1 Sustainability Knowledge

Sustainability knowledge was measured with 16 multiple-choice items with one correct and three distractor answers to reduce guessing probability (compared to a true/false format). The questionnaire assessed knowledge about environmental protection, pollution, and other sustainability-related relevant topics, such as recycling, plastic in the oceans, erosion, and fair trade. The students' knowledge was operationalized by a mixture of basic, more general (cross-curricular) and subject-specific items. The general items were questions about common knowledge related to sustainable development. The subject-specific sustainability knowledge items were derived from the specific educational plans (see Ministry of Education Baden-Württemberg, 2016) for the different grades (5–8). A sample item is given below:

> We use a lot of plastic in our country. Why is this plastic bad for the fish in the ocean?
> Please mark only one answer (the most applicable one).
> ☐ Because a lot of water is polluted during the production of plastic. The dirty water is led into the ocean and poisons the fish there.
> ☐ Because large quantities of fish get caught in plastic bags, sticking the gills together and suffocating the fish.
> ☐ The plastic often ends up in the habitat of the fish and rots there only very slowly. Sometimes the fish eat the plastic and can be injured.
> ☐ Because the plastic factories pollute the air. When it rains, the dirt gets from the air into the water and damages the fish there.

15.3.2 Attitudes Towards Sustainability and Self-Reported Sustainability-Related Behavior

The two dimensions, affective-motivational beliefs toward sustainability and self-reported sustainability-related behavior, were measured using a four-point Likert-type scale: (1) strongly disagree; (2) disagree; (3) agree; and (4) strongly agree. An uneven scale was avoided, as indecisive persons tend to choose a middle point (see, e.g., Chyung et al., 2017). These two scales consisted, respectively, of 16 evaluative verbal statements about the students' attitudes toward sustainability (e.g., *When I hear of cars that consume a lot of fuel, I get angry* or *the extinction of many animal and plant species makes me sad*) and of 13 self-reports about sustainability-related actions which the students could already perform or mostly decide themselves (e.g., *When I buy chocolate with my pocket money, I buy organic or fair-trade chocolate* or *On excursions, I take drinks in plastic bottles or disposable packaging, e.g., cans or PET bottles*).

15.4   Key Findings

In the framework of this chapter, we will not be able to describe in detail all findings which emerged from the project. We will, for example, not go into detail about the teacher-level data. However, the results on the development of SC at the student level are of course equally relevant for teachers, as they (can) influence the development of student competencies through appropriate instruction or educational actions. For the findings on the teacher level, see Waltner et al. (2020).

15.4.1 Sustainability Knowledge

Sustainability-related knowledge generally increases with age (Leeming et al., 1995; Waltner et al., 2019). We found this (positive) effect not only with increasing age across grade levels but also as an effect within the school year for each grade level separately. The sustainability knowledge increased statistically significantly across all grade levels within the school year $(F(1,1308) = 197.13, p < 0.001)$; partial $\eta^2 = 0.13$; see Fig. 15.1.

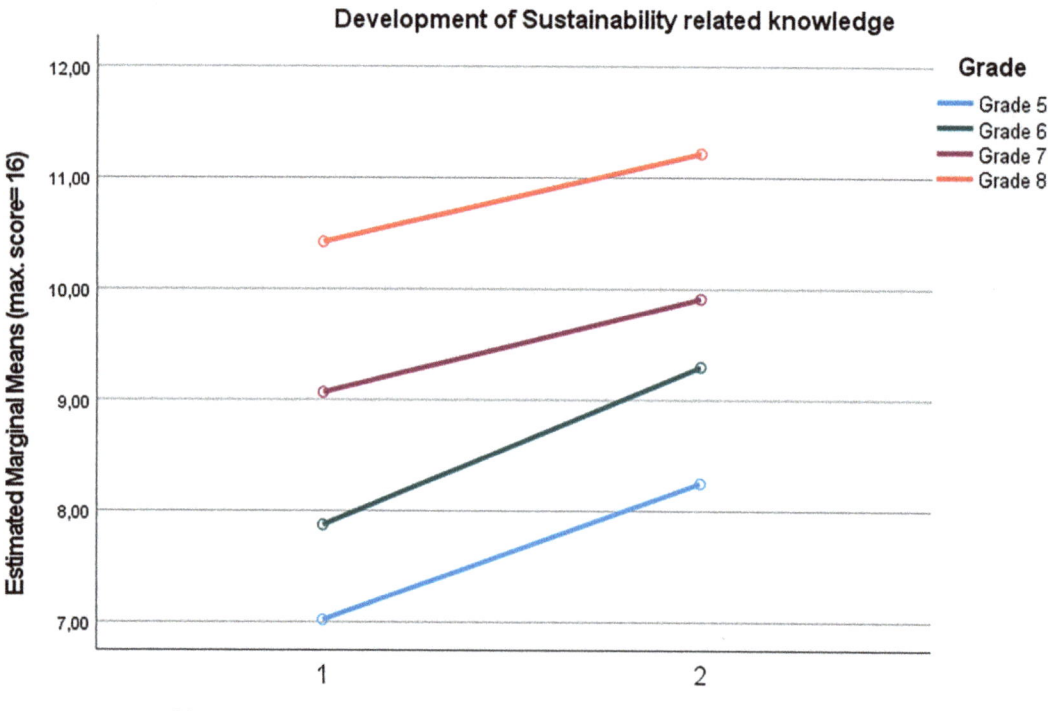

Fig. 15.1 Development of sustainability-related knowledge within 1 school year

15.4.2 Sustainability Attitudes

As mentioned above, there is a strong connection between sustainability attitudes and environmental attitudes which are a part of the broader concept of sustainability. Hence, the general tendencies should be comparable to findings of previous research in the field of environmental attitudes. According to previous research, younger children also tend to have a higher environmental attitude than older children (Krettenauer, 2017; Leeming et al., 1995; Liefländer et al., 2013). This holds true in our sample as well (see Fig. 15.2; grade 5 = lowest age group, grade 8 = highest age group). However, for the decreasing attitudes within 1 school year,

other research findings are still scarce. Sustainability attitudes in our sample decreased statistically significantly across all grade levels within the school year (F(1,1308) = 20,44, $p < 0.001$); partial $\eta^2 = 0.02$; see Fig. 15.2.

15.4.3 Sustainability Behavior

Self-reported behavior showed either a very slight but not significant decrease or increase depending on the different grades (see Fig. 15.3). Changes are not statistically significant across all grades (F(1,1318) = 0.21, $p = 0.649$); partial $\eta^2 = 0.00$.

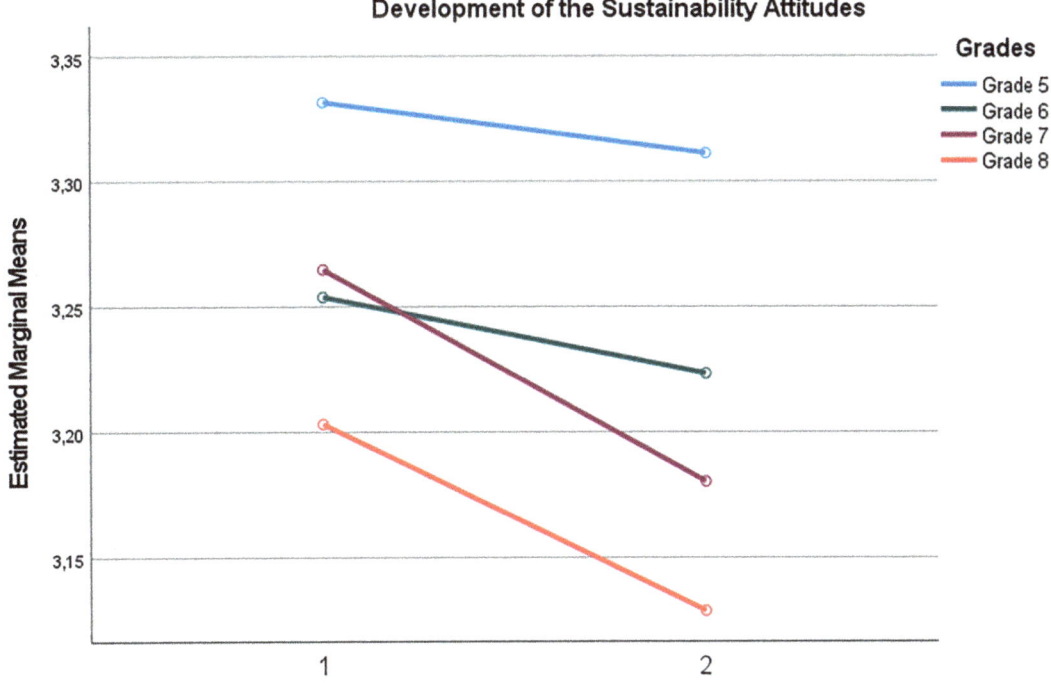

Fig. 15.2 Development of sustainability attitudes within 1 school year

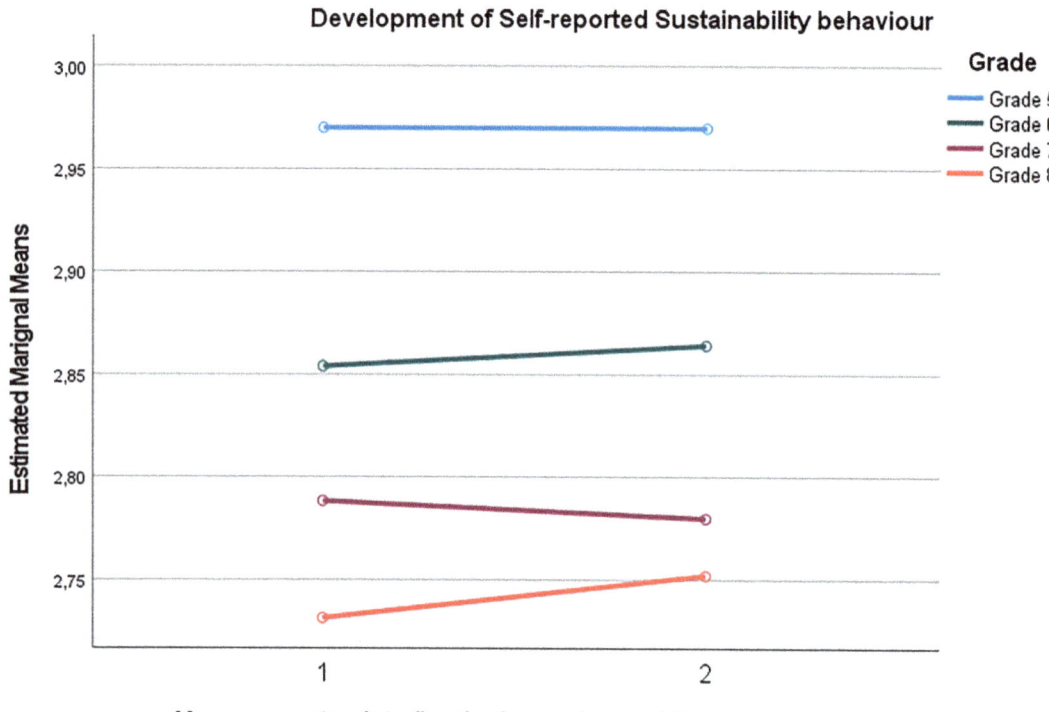

Fig. 15.3 Development of self-reported sustainability behavior within 1 school year

15.5 Validation of New Indicators-Establishing Impact-Relevance of ESD Outcome

If governments and other social agencies are to heed what psychologists have to say on [the issue of environmental behavior], then the validity of our statements becomes very important. (Lloyd, 1994, p. 132)

If we wanted to make failure-proof assumptions about the impact of any intervention (such as ESD), we would have to consult observational data on concrete behavior. For example, say we observe that a student who used to leave the light on when leaving his or her room for dinner changes this behavior after receiving information about the environmental harm of the waste of power; in this case, we can assume that the gain in knowledge had an impact on the student's actual behavior and thus his or her impact on the environment. In reality, self-reports are the more frequently used indicators of behavior. Working with self-reports has two relevant advantages

over observational data: first, they are easier to obtain, especially in large quantities; second (and based on the first advantage), they allow for a broader assessment of different behaviors (e.g., energy conservation behaviors, regional consumption behaviors, recycling behavior, etc.) at the same time and can be used to form aggregated measurements for behavioral classes (Kaiser et al., 2001). Research has shown, however, that there is a disparity between indicators based on self-reported behavior and objective (observable) behavior (e.g., Corral-Verdugo, 1997). The reasons for this have been widely discussed – a comprehensive review of this would go beyond the scope of this chapter. The main message which we want to draw from this insight is that indicators which are designed to assess the outcome of ESD have to be validated as to their congruence with real-life outcomes in the context of a shift to a more sustainable society or decisions; otherwise we might make wrong conclusions based on the data which we gather. This holds true for behavioral self-reports as well as for attitudinal indicators and knowledge-based scales. Two exemplary efforts at validating the indicators used in the presented research project are shown in the following section. The underlying debate about whether or not the actual change of behavior can and should be a goal of ESD interventions will be referred to in the discussion section of this chapter.

15.5.1 Validation Through the Connection to Established Indicators

In a first attempt at validation, we performed a conceptual examination, using items from the newly developed ESD measurement project presented here and linking them to a well-established measurement instrument which has proven to be predictive of real-world behavioral impact: the General Ecological Behavior Scale (GEB) (see, e.g., Kaiser & Wilson, 2004). To do this, we assembled an 18-item ad hoc scale from the large survey, which was conducted within the survey presented above (see the Methods section above). Eight of these items were evaluative verbal statements (classic attitude items), and ten items were behavioral self-reports. Following the Campbell paradigm (Kaiser et al., 2010), both item types can be used in order to assess an underlying (latent) attitude, which motivates verbal as well as behavioral responses toward the attitude object. Within the Campbell paradigm, a person's attitude becomes transparent in the amount of behavioral cost said person is willing to overcome in order to pursue their goal (Byrka et al., 2017). So far, research on people's attitudes toward the environment has demonstrated that the Campbell paradigm – and thus its conceptual account of individual behavior – holds true for approximately 95% of the people in a given society (Kaiser et al., 2014). For more information about the derivation of attitudes from various types of manifest indicators attitudes (e.g., evaluative verbal statements and behavioral self-reports), see Kaiser and Wilson (2019).

The GEB uses 50 behavioral self-report items to assess attitudes toward the environment

(Kaiser & Wilson, 2004). This operationalization has been used in countless studies (e.g., Evans et al., 2018; Overbeck & Kibbe, 2020; Otto & Kaiser, 2014), it has been translated into many languages, and it has proven to be predictive of real-life outcomes, such as the acceptance of nature preservation-related restrictions (Byrka et al., 2017) or a person's membership in an environmental organization (Arnold & Kaiser, 2018). We selected those (18) items for this validation attempt which we assumed would capture students' environmental attitudes as a substantial part of the attitudes toward sustainability concept as was fully assessed in the original questionnaire.

The goal of this validation was to show that the items which were newly developed within the project presented in this article prove to be valid predictors of environmental attitudes (as a subset of sustainability attitudes) – and in this way deduce that the development process of the project was successful. As a consequence, the competency differences which were assessed with this measurement instrument could point toward meaningful differences between the students which may have an actual impact on their future behavior. To investigate this, we conducted a validation study with $N = 154$ young adults (mean age = 30.21) in which we assessed the 18 items which we selected for our ad hoc measure alongside 47 GEB items. The correlation between the two measurements was $r = 0.75$ and when corrected for measurement error $r > 1$; this shows that the two scales measure the same latent construct, namely, environmental attitude. Additionally, we performed a joint Rasch calibration of all 65 items, which showed a person score reliability of rel. = 0.86 and a generally reasonable fit to the one-dimensional Rasch model; this indicates that all items assess the very same latent attribute, thus supporting our hypothesis.

On the other hand, our findings also provide support for the Campbell paradigm (see Kaiser et al., 2010) – in this paradigm, personal attitudes can be derived from verbal acts, such as expressions of appreciation for the environment and self-reports of past engagement in environmentally friendly behaviors (Kaiser et al., 2018). Our findings also show that it is not relevant with which specific items a latent attitude is assessed but that any number of reasonably well-phrased behavioral or verbal self-reports which are aimed at the attitude object in question can be used to infer the underlying disposition (i.e., attitude). This supports the call for a higher priority of specific objectivity within the validation criteria for measurements in general (for a detailed account, see Kaiser et al., 2018).

15.5.2 Validation Through Prediction of Impact-Relevant Behavior

In our second step of validation, the (voluntary and self-determined) Fridays for Future (F4F) participation of the surveyed students is used to draw conclusions about environmental activism resulting from a high level of environmental attitudes. The participation in this activist movement can be interpreted as a behavioral manifestation aimed at promoting a more sustainable society, which consequently serves to achieve the ESD goals. The participation in the F4F movement was recorded at the second assessment point of the project (in the summer of 2019) with a simple question regarding this activity within the last year: "Have you heard of the Fridays for Future Movement" with the answering options "no"($N = 283$), "yes" ($N = 868$), and "yes, and I have participated" ($N = 163$). In this case, the self-reported behavior was regarded as a valid proxy for actual behavior since self-reported behaviors seem to be especially reliable and valid when distinct facts are reported in a simple (dichotomous) way (e.g., Kaiser et al., 2003; Kormos & Gifford, 2014).

The data showed that students with higher sustainability attitudes in the first measurement had a significantly higher likelihood to participate in the F4F movement than students with a lower sustainability attitude. The same results could be shown for students' sustainability knowledge. Higher competency levels in both outcome variables could successfully

predict the participation in a real-life, impact-relevant behavior, i.e., F4F. This validation through a criterion outside of the measurement process is a crucial step in gaining an understanding of any attribute one wants to measure (Whitely, 1977). We can be relatively confident that our measurements meet their goals; this is because the newly developed scales from the survey presented here can predict actual behavior which aims at the targets which the assessed competencies are supposed to support.

15.6 Limitations and Conclusions

> Not everything that can be counted counts and not everything that counts can be counted.
> Attributed to Albert Einstein. (see e.g., McKee, 2004)

Even if one can agree with this quotation in principle, it should be noted that non-measurable target and implementation criteria are often subject to the risk of remaining on the level of general normative statements or vague or unrealized implementation stages and policies without achieving the wanted outcome. The exemplary outcome indicators shown in this study, with the longitudinal data at the level of the pupils, provide ESD stakeholders with a useful information base (cf. e.g., DIPF, 2007; Oekes, 1989). Methodologically quantitative research projects can make a very important contribution to the normative debate, through empirical insights. Moreover, this debate also involves the normative questions of whether all dimensions have the same importance or weighting or whether some sub-dimensions should be weighted more heavily in a latent construct, depending on how they are assessed for scholastic or societal relevance. Recalling the educational policy definition of ESD given at the beginning of this paper, education for sustainable development enables learners to make informed decisions and act responsibly for the protection of the environment and for a functioning economy and a just world society for current and future generations (Ministry of Education Baden-Württemberg, 2016). In this context, the acting component is clearly accentuated. So, one might ask the critical question: Do we need not only a shift from the input to the outcome orientation in the analytical/evaluative perspective but also a shift of attention from the purely cognitive to the behavioral components of SCs? Also, in terms of our research foci, we should not shy away from a call for more impact focused research. Indeed, the call for more impact focus within the sustainability sciences, such as environmental psychology, is getting more prominent (see, e.g., Nielsen et al., 2021).

15.6.1 Why Is Measuring Relevant for Teachers?

Studies on ESD implementation processes might help to further illuminate the interconnections between seemingly different realities of policy-makers and "the street ministers of education" (i.e., the teachers). In terms of Lipsky and other bottom-up theorists, they are also policy-makers "on the ground" as they are the very important stakeholders who implement the policies, i.e., in the classrooms (Lipsky, 2010; Waltner et al., 2020). In this way, the actors (in the sense of the bottom-up approach, also including teachers and students themselves or any ESD learners) can initiate desired developments in a more empirically based and targeted manner and do not only have to refer to plausible and normative considerations and assumptions. Decisions can therefore be taken based on empirical long-term data and do not only need to be based on intuitively based perceptions or observations which the teachers make in the classroom. Furthermore, this paper reported the development and validation of a measurement tool for an interdisciplinary SC. It is obvious that, in the further course of research, the specific contributions of the different subjects (biology, geography, economics, politics, social studies, etc.) to ESD should also be considered. Accordingly, instruments for facets of a subject-specific SC would then have to be developed and tested (Rieß et al., 2018; Waltner et al., 2019).

Our research adds to a growing understanding of SCs, their development, and the sustainability

and educational governance through policy-making. On this basis, appropriate evidence-based recommendations for the further development of ESD research and the implementation of ESD in school practice can be formulated. Through the possibilities of measurement presented and the data already generated, further insights into the successful implementation of ESD in schools and the associated conditions for success can be gained. Despite these advantages, some shortcomings should be mentioned. First, as implicated by the general definition of indicators, they only offer an approximation to the real world. Even though, as described in the section "Validation of New Indicators-Establishing Impact-Relevance of ESD outcome," we tried to validate our ESD outcome measures, they still remain a mere proxy for reality.

A second point which is important to mention in considerations about the limitations of indicators – and especially the wish to globally assess and compare the results of ESD or SDG outcomes – is the aspect of a reduced context sensitivity. Even for our study only focusing on one state in Germany, we needed to deal with the trade-off between regional specificity (or the aim of a detailed analysis of all the possible contexts and possible factors influencing SC) and the aim of getting a broad picture of the state of play and development. With the aim of a general assessment tool of ESD outcomes, e.g., in a global scope, this trade-off would become even more evident as the possibilities of observing all the details shrink. When aiming for global indicators which are easily replicable and comparable, detailed observations need to be sacrificed. Consequently, additional qualitative studies looking into further details of the SC development process would represent a useful complement.

We can currently still state, in agreement with Michelsen et al. (2012), that interlinkages hardly exist even in the fields of youth, learning, and education which are closer to ESD research. This holds true, not only within the different methodological fields but also in the different research domains, with regard to the networking and cooperation between different research and educational areas. Based on the experience of this collaboration, we would recommend a stronger interlinking of the already mentioned relevant scientific disciplines (such as (educational) psychology, educational science, environmental and sustainability sciences). Furthermore, a closer integration of experts from practice (pupils, teachers, students, etc.) with regard to the findings of ESD research which are important for school education – but also for society as a whole – seems to be an important goal.

In view of the impending global challenges, however, a global society which is serious about achieving the SDGs cannot afford to remain at the level of normative statements and vague target formulations. The corresponding possibilities of measurability will ideally stimulate benchmarking processes (concrete target formulations) (e.g., Sayed & Asmuss, 2013). An international monitoring system for the ESD implementation efforts of the various countries would be purposeful, in view of the future global challenges for humankind (Independent Group of Scientists appointed by the Secretary-General, 2019). Such an international orientation would facilitate the exchange of insights in the implementation processes of ESD and would also offer possibilities for international cooperation (see, e.g., Reid et al., 2006).

Acknowledgments The authors gratefully acknowledge all project members of the research groups and other supporters. In addition, we would like to thank the school principals, contact persons, and all other school stakeholders for their valuable contributions during the different project phases.

Funding This research was partly funded by the Ministry of Education, Youth, and Sports of Baden-Wuerttemberg, the Ministry of Environment of Baden-Wuerttemberg, and the Stiftung Naturschutzfonds [*Nature Conservation Foundation*] (2-8802.00-BNE/50 and 73-8831.21/54691-1743L).

References

Arcury, T. (1990). Environmental attitude and environmental knowledge. *Human Organization, 49*(4), 300–304. https://doi.org/10.17730/humo.49.4.y6135676n4 33r880

Arnold, O., & Kaiser, F. G. (2018). Understanding the foot-in-the-door effect as a pseudo-effect from the perspective of the Campbell paradigm. *International Journal of Psychology, 53*(2), 157–165. https://doi.org/10.1002/ijop.12289

Baethge, M., Brunke, J., Döbert, H., Fest, M., Freitag, H.-W., Fitzsch, B., Fuchs-Rechlin, K., Kerst, C., Kühne, S., Scharfe, S., Skripski, B., Wieck, M., & Wolter, A. (2010). Indikatorenentwicklung für den nationalen Bildungsbericht "Bildung in Deutschland": *Grundlagen, Ergebnisse, Perspektiven [Development of indicators for the national education report "Education in Germany": Foundations, results, perspectives]*. BMBF.

Bogner, F. X. (2018). Environmental values (2-MEV) and appreciation of nature. *Sustainability, 10*(2), 1–10. https://doi.org/10.3390/su10020350

Bormann, I., & Michelsen, G. (2010). The collaborative production of meaningful measure(ment)s: Preliminary insights into a work in progress. *European Educational Research Journal, 9*(4), 510–518.

Burford, G., Tamás, P., & Harder, M. (2016). Can we improve indicator design for complex sustainable development goals? A comparison of a values-based and conventional approach. *Sustainability, 8*(9), 1–38. https://doi.org/10.3390/su8090861

Byrka, K., Kaiser, F. G., & Olko, J. (2017). Understanding the acceptance of nature-preservation-related restrictions as the result of the compensatory effects of environmental attitude and behavioral costs. *Environment and Behavior, 49*(5), 487–508. https://doi.org/10.1177/0013916516653638

Cai, Y., Lenton, T. M., & Lontzek, T. S. (2016). Risk of multiple interacting tipping points should encourage rapid CO_2 emission reduction. *Nature Climate Change, 6*(5), 520–525. https://doi.org/10.1038/nclimate2964

Chyung, S. Y. Y., Roberts, K., Swanson, I., & Hankinson, A. (2017). Evidence-based survey design: The use of a midpoint on the likert scale. *Performance Improvement, 56*(10), 15–23. https://doi.org/10.1002/pfi.21727

Corral-Verdugo, V. (1997). Dual 'realities' of conservation behavior: Self-reports vs observations of re-use and recycling behavior. *Journal of Environmental Psychology, 17*, 135–145.

DIPF- Leibniz-Institut für Bildungsforschung und Bildungsinformation. (2007). Das weiterentwickelte Indikatorenkonzept der Bildungsberichterstattung [*The advanced indicator concept of educational reporting*]. http://www.bildungsbericht.de/de/forschungsdesign/pdf-grundlagen/indikatorenkonzept.pdf. Accessed 10 June 2021.

Evans, G. W., Otto, S., & Kaiser, F. G. (2018). Childhood origins of young adult environmental behavior. *Psychological Science, 29*(5), 679–687. https://doi.org/10.1177/0956797617741894

Frick, J., Kaiser, F. G., & Wilson, M. (2004). Environmental knowledge and conservation behavior: Exploring prevalence and structure in a representative sample.

Personality and Individual Differences, 37(8), 1597–1613. https://doi.org/10.1016/j.paid.2004.02.015

Gräsel, C., Bormann, I., Schütte, K., Trempler, K., Fischbach, R., & Asseburg, R. (2012). Perspektiven der Forschung im Bereich Bildung für nachhaltige Entwicklung [*Prospects for research in education for sustainable development*]. In Bundesministerium für Bildung und Forschung (Ed.), *Bildung: Vol. 39. Bildung für nachhaltige Entwicklung: Beiträge der Bildungsforschung* (pp. 7–25). Bundesministerium für Bildung und Forschung (BMBF) Referat Bildungsforschung.

Hines, J. M., Hungerford, H. R., & Tomera, A. N. (1986/87). Analysis and synthesis of research on responsible environmental behavior: A meta-analysis. *Journal of Environmental Education, 18*(2), 1–8.

Independent Group of Scientists appointed by the Secretary-General. (2019). The future is now. Science for achieving sustainable development. Global sustainable development report. https://sustainabledevelopment.un.org/content/documents/24797GSDR_report_2019.pdf. Accessed 14 June 2021.

Kaiser, F. G. (1998). A general measure of ecological behavior. *Journal of Applied Social Psychology, 28*(5), 395–422. https://doi.org/10.1111/j.1559-1816.1998.tb01712.x

Kaiser, F. G., & Fuhrer, U. (2003). Ecological behavior's dependency on different forms of knowledge. *Journal of Environmental Psychology, 52*(4), 598–613. https://doi.org/10.1111/1464-0597.00153

Kaiser, F. G., & Wilson, M. (2004). Goal-directed conservation behavior: The specific composition of a general performance. *Personality and Individual Differences, 36*(7), 1531–1544. https://doi.org/10.1016/j.paid.2003.06.003

Kaiser, F. G., & Wilson, M. (2019). The Campbell Paradigm as a behavior-predictive reinterpretation of the classical tripartite model of attitudes. *European Psychologist, 24*(4), 359–374. https://doi.org/10.1027/1016-9040/a000364

Kaiser, F. G., Frick, J., & Stoll-Kleemann, S. (2001). Zur Angemessenheit selbstberichteten Verhaltens: Eine Validitätsuntersuchung der Skala Allgemeinen Ökologischen Verhaltens [*On the appropriateness of self-reported behavior: A validity investigation of the General Ecological Behavior scale*]. *Diagnostica, 47*(2), 88–95. https://doi.org/10.1026//0012-1924.47.2.88

Kaiser, F. G., Doka, G., Hofstetter, P., & Ranney, M. A. (2003). Ecological behavior and its environmental consequences: A life cycle assessment of a self-report measure. *Journal of Environmental Psychology, 23*(1), 11–20. https://doi.org/10.1016/S0272-4944(02)00075-0

Kaiser, F. G., Byrka, K., & Hartig, T. (2010). Reviving Campbell's paradigm for attitude research. *Personality and Social Psychology Review, 14*, 351–367. https://doi.org/10.1177/1088868310366452

Kaiser, F. G., Hartig, T., Brügger, A., & Duvier, C. (2011). Environmental protection and nature as distinct attitudinal objects. *Environment and Behavior, 45*(3), 369–398. https://doi.org/10.1177/0013916511422444

Kaiser, F. G., Arnold, O., & Otto, S. (2014). Attitudes and defaults save lives and protect the environment jointly and compensatorily: Understanding the behavioral efficacy of nudges and other structural interventions. *Behavioral Sciences (Basel, Switzerland), 4*, 202–212. https://doi.org/10.3390/bs4030202

Kaiser, F. G., Merten, M., & Wetzel, E. (2018). How do we know we are measuring environmental attitude? Specific objectivity as the formal validation criterion for measures of latent attributes. *Journal of Environmental Psychology, 55*, 139–146. https://doi.org/10.1016/j.jenvp.2018.01.003

Klieme, E., & Hartig, J. (2007). Kompetenzkonzepte in den Sozialwissenschaften im erziehungswissenschaftlichen Diskurs [*Competencies' concepts in social sciences in educational discourse*]. In M. Prenzel, I. Gogolin, & H.-H. Krüger (Eds.), *Zeitschrift für Erziehungswissenschaft Sonderheft: Vol. 8. Kompetenzdiagnostik* (pp. 11–31). VS Verlag für Sozialwissenschaften / GWV Fachverlage GmbH Wiesbaden.

Kormos, C., & Gifford, R. (2014). The validity of self-report measures of proenvironmental behavior: A meta-analytic review. *Journal of Environmental Psychology, 40*, 359–371. https://doi.org/10.1016/j.jenvp.2014.09.003

Krettenauer, T. (2017). Pro-environmental behavior and adolescent moral development. *Journal of Research on Adolescence, 27*(3), 581–593. https://doi.org/10.1111/jora.12300

Leeming, F. C., Dwyer, W. O., & Bracken, B. A. (1995). Children's environmental attitude and knowledge scale: Construction and validation. *The Journal of Environmental Education, 26*(3), 22–31. https://doi.org/10.1080/00958964.1995.9941442

Lenton, T. M., Held, H., Kriegler, E., Hall, J. W., Lucht, W., Rahmstorf, S., & Schellnhuber, H. J. (2008). Tipping elements in the earth's climate system. *Proceedings of the National Academy of Sciences, 105*(6), 1786–1793. https://doi.org/10.1073/pnas.0705414105

Liefländer, A. K., Fröhlich, G., Bogner, F. X., & Schultz, P. W. (2013). Promoting connectedness with nature through environmental education. *Environmental Education Research, 19*(3), 370–384. https://doi.org/10.1080/13504622.2012.697545

Lipsky, M. (2010). *Street-level Bureaucracy: 30th Ann. Ed.: Dilemmas of the individual in public service* (30th Ann. Ed). Russell Sage Foundation.

Lloyd, K. E. (1994). Do as I say, not as I do. *The Behavior Analyst, 17*(1), 131–139.

Maloney, M. P., & Ward, M. P. (1973). Ecology: Let's hear from the people: An objective scale for the measurement of ecological attitudes and knowledge. *American Psychologist, 28*, 583–586.

McBeth, W., Hungerford, H. R., Marcinkowski, T., Volk, T. L., & Cifranick, K. (2011). *National environmental literacy assessment, phase two: Measuring the effectiveness of North American environmental education programs with respect to the parameters of environmental literacy: Final research report*. North American Association for Environmental Education. https://www.noaa.gov/sites/default/files/atoms/files/NELA_Phase_Two_Report_020711.pdf. Accessed 10 June 2021.

McKee, M. (2004). Not everything that counts can be counted; not everything that can be counted counts. *BMJ (Clinical Research Ed.), 328*(7432), 153. https://doi.org/10.1136/bmj.328.7432.153

Michelsen, G., Grunenberg, H., & Mader, C. (2012). Greenpeace Nachhaltigkeitsbarometer: Was bewegt die Jugend? [*Greenpeace sustainability barometer: What affects young people?*]. VAS-Verlag.

Michelsen, G., Grunenberg, H., Mader, C., & Barth, M. (2015). Greenpeace Nachhaltigkeitsbarometer 2015: Nachhaltigkeit bewegt die jüngere Generation [*Greenpeace sustainability barometer 2015: Sustainability affects the younger generation*]. VAS-Verlag.

Ministerium für Kultus, Jugend und Sport Baden-Württemberg. (2016). Bildungspläne Baden-Württemberg: Bildung für nachhaltige Entwicklung (BNE). [*Education plans for Baden-Württemberg: Education for Sustainable Development (ESD)*]. http://www.bildungsplaene-bw.de/Lde/Startseite/BP2016BW_ALLG/BP2016BW_ALLG_LP_BNE. Accessed 10 June 2021.

Nielsen, K. S., Cologna, V., Lange, F., Brick, C., & Stern, P. C. (2021). The case for impact-focused environmental psychology. *Journal of Environmental Psychology*. Advance online publication. https://doi.org/10.31234/osf.io/w39c5

Oekes, J. (1989). What educational indicators? The case for assessing the school context. *Educational Evaluation and Policy Analysis, 11*(2), 181–199.

Otto, S., & Kaiser, F. G. (2014). Ecological behavior across the lifespan: Why environmentalism increases as people grow older. *Journal of Environmental Psychology, 40*, 331–338. https://doi.org/10.1016/j.jenvp.2014.08.004

Overbeck, A. K., & Kibbe, A. (2020). Decoding activism: Examining the influence of environmental attitude and proactivity on environmental activism. *Umweltpsychologie, 24*(1), 183–190.

Redman, A., Wiek, A., & Barth, M. (2021). Current practice of assessing students' sustainability competencies: A review of tools. *Sustainability Science, 16*(1), 117–135. https://doi.org/10.1007/s11625-020-00855-1

Reid, A., Nikel, J., & Scott, W. (2006). *Indicators for Education for Sustainable Development: A report on perspectives, challenges and progress*. https://researchportal.bath.ac.uk/en/publications/indicators-for-education-for-sustainable-development-a-report-on. Accessed 13 June 2021.

Rieckmann, M. (2018). Chapter 2. Learning to transform the world: Key competencies in ESD. In UNESCO (Ed.), *UNESCO's education on the move series. Issues and trends in education for sustainable development* (pp. 39–59). UNESCO.

Rieß, W., Mischo, C., & Waltner, E.-M. (2018). Ziele einer Bildung für nachhaltige Entwicklung in Schule und Hochschule: Auf dem Weg zu empirisch überprüfbaren Kompetenzen [*The goals of education for sustainable development in schools and universities: Towards empirically verifiable competencies*]. *GAIA - Ecological Perspectives for Science and Society, 27*(3), 298–305. https://doi.org/10.14512/gaia.27.3.10.

Roczen, N., Kaiser, F. G., Bogner, F. X., & Wilson, M. (2014). A competence model for environmental education. *Environment and Behavior, 46*(8), 972–992. https://doi.org/10.1177/0013916513492416

Sayed, A., & Asmuss, M. (2013). Benchmarking tools for assessing and tracking sustainability in higher educational institutions. *International Journal of Sustainability in Higher Education, 14*(4), 449–465. https://doi.org/10.1108/IJSHE-08-2011-0052

Schellnhuber, H. J. (2006). Facing climate change: Tipping points and u-turns. In D. Buckland (Ed.), *Burning ice: Art & climate change; the Cape Farewell project was created by the artist* (pp. 112–113). Gaia Project.

Shepherd, D. A., Kuskova, V., & Patzelt, H. (2009). Measuring the values that underlie sustainable development: The development of a valid scale. *Journal of Economic Psychology, 30*(2), 246–256.

Statistisches Landesamt Baden-Württemberg. (2017). Allgemeinbildende Schulen in Baden-Württemberg [*General education schools in Baden-Württemberg*]. http://www.statistik.baden-wuerttemberg.de/Service/Veroeff/Statistik_AKTUELL/803417004.pdf Accessed 10 June 2021.

Tilbury, D., & Janousek, S. (2006). *Development of a national approach to monitoring, assessment and reporting on the decade of education for sustainable development: Stage 1: Identification of national indicators* [Summarising documented experiences on the development of ESD indicators and networking with expert groups on ESD indicators]. Australian Research Institute of Education for Sustainability and Australian Government Department of the Environment and Water Resources. http://aries.mq.edu.au/projects/esdIndicators/files/ESDIndicators_Feb07.pdf. Accessed 14 June 2021.

UNESCO. (2017). *Education for sustainable development goals: Learning objectives*. https://www.unesco.de/sites/default/files/2018-08/unesco_education_for_sustainable_development_goals.pdf. Accessed 14 June 2021.

Waltner, E.-M., Rieß, W., & Brock, A. (2018). Development of an ESD indicator for teacher training and the national monitoring for ESD implementation in Germany. *Sustainability, 10*(7), 2508. https://doi.org/10.3390/su10072508

Waltner, E.-M., Rieß, W., & Mischo, C. (2019). Development and validation of an instrument for measuring student sustainability competencies. *Sustainability, 11*(6), 1717. https://doi.org/10.3390/su11061717

Waltner, E.-M., Scharenberg, K., Hörsch, C., & Rieß, W. (2020). What teachers think and know about education for sustainable development and how they implement it in class. *Sustainability, 12*(4), 1690. https://doi.org/10.3390/su12041690

Whitely, S. E. (1977). Models, meanings and misunderstandings: Some issues in applying Rasch's theory. *Journal of Educational Measurement, 14*(3), 227–235.

Wiek, A., Withycombe, L., & Redman, C. L. (2011). Key competencies in sustainability: A reference framework for academic program development. *Sustainability Science, 6*(2), 203–218.

Eva-Maria Waltner studied at the University of Constance (Germany), Université du Québec à Montréal (Canada), and Université de Liège (Belgium), and graduated in political science, French, and pedagogy. She worked as an assistant for UNEP in Brussels and also in practical outdoor education. Eva-Maria completed her PhD in the project The implementation of Education for Sustainable Development (ESD) – a multilevel analysis of Sustainability Competencies (SC) in schools at the University of Education Freiburg (Germany) and is currently involved in the international The Monitoring and Evaluating Climate Communication and Education Project (MECCE).

Anne Overbeck is a member of the Division of Personality and Social Psychology at Otto-von-Guericke University Magdeburg. She studied psychology at Mannheim University (BSc) and later continued her studies at Magdeburg University (MSc, focus on environmental psychology). Her main scientific interests are sustainability, personality-psychology, and political participation. Anne has been working on research concerning the outcome-oriented assessment of ESD in students. In her PhD studies, she concentrates her research on the determinants of environmental activism.

Werner Rieß is Professor of Biological Education at the University of Education Freiburg (Germany). In his research projects, he particularly investigates the effects of ESD from kindergarten to university on systems thinking and other important goals (e.g., knowledge, attitudes, competencies, behavior of learners of different ages). In addition, his research examines the current state of ESD with the aim of creating indicators for ESD.

Assessing Learning Outcomes for Sustainability in Primary and Secondary Schools in the UK

16

Vasiliki Kioupi and Nikolaos Voulvoulis

Abstract

Education for Sustainable Development (ESD) is an integral component of Quality Education as stated in Sustainable Development Goal (SDG) 4 target 4.7. Its main aim is to develop in learners, knowledge, skills, attitudes, values and behaviours conducive to sustainable development. Sustainability was introduced as a curricular topic in primary and secondary schools in the UK after the Earth Summit in Rio in 2002; however, its effectiveness in achieving its learning outcomes (LOs) has not been systematically assessed. In this study, we present the application of a participatory framework for assessing LOs for sustainability. The framework was developed using systems thinking and applied in two case studies conducted in a primary and a secondary school in the UK that followed different approaches in integrating ESD into their curricula. The primary school introduced ESD as the thread that pervades and links all curricular subjects, whereas the secondary school introduced a new course on the SDGs. Both schools were found to be effective in developing the intended LOs in their students, while some weaknesses related to their approach were identified. The case studies demonstrated the tool's potential to measure and evaluate students' competences development and support the operationalisation of sustainability competences in primary and secondary education.

Keywords

Education for sustainability · Sustainability competences · Assessment of learning · Systems thinking · Primary and secondary schools

16.1 Introduction

Recent international commitments around sustainability, such as the Sustainable Development Goals and the Paris agreement targeting climate change, highlighted the central role of education in achieving their stated goals and targets (UNESCO, 2018, 2020). The SGDs specifically target education, state that ESD is an integral component of quality education and the learners should be empowered with knowledge, skills, attitudes, values and behaviours aligned with sustainability, citizenship, human rights, gender equality, cultural diversity and peace education (United Nations SDG4, 2021). This view of education through international agreements coincided with a shift in education policy that showed

V. Kioupi (✉) · N. Voulvoulis
Centre for Environmental Policy, Imperial College London, London, UK
e-mail: v.kioupi17@imperial.ac.uk

© The Author(s), under exclusive license to Springer Nature Switzerland AG 2022
G. Karaarslan-Semiz (ed.), *Education for Sustainable Development in Primary and Secondary Schools*, Sustainable Development Goals Series, https://doi.org/10.1007/978-3-031-09112-4_16

education systems moving towards evidence-based practices. This evidence-based orientation has been associated with the assessment of learning outcomes (LOs) or competences in learners as the means for improving the effectiveness of education offerings (Leutner et al., 2017).

In the UK, school assessments have already been used to collect evidence of students' alignment with key stage expectations, referred to as attainment targets. These targets specify the knowledge, understanding and skills related to specific subjects that learners are expected to have acquired by the end of an educational level and to be assessed against a predetermined set of criteria, to help improvement of the student, teacher and school and provide reliable information to the parent (Department for Education/ DfE, 2014). The first national strategy regarding sustainable development titled "Securing the Future" was rolled out in 2005 (Department for Environment, Food & Rural Affairs (DEFRA), 2005a), which coincided with the start of the UNESCO Decade for ESD in 2005 (UNESCO, 2014). In Chap. 2 of this strategy, education is included as a means to enable positive behavioural change that is critical for achieving the sustainable future envisioned by the UK government. Specifically, education can help learners form desirable habits for sustainability early on and these can be transformed to sustained behaviours throughout their lives (DEFRA, 2005b). This was aligned with an effort to make every school an environmentally sustainable school that teaches about sustainable development through the curriculum. One of the primary objectives of this plan was that "all learners will develop the skills, knowledge and value base to be active citizens in creating a more sustainable society". Another result of the strategy was the implementation of the National Framework for Sustainable Schools in 2006 to urge schools to consider sustainable development in teaching, learning, school management and community engagement (Reynolds & Scott, 2011).

An evaluation of the status of ESD in the UK as the Decade was approaching its end showed that although multiple ESD initiatives existed across the UK and they showed good practice in teaching, learning and teacher training, these were relatively small scale, mostly project based and within fixed timescales (UK National Commission for UNESCO, 2013). In terms of policy around ESD implementation, there was no uniform view or action on how ESD could be widely adopted in formal, informal and non-formal education, with significant variation among the nations of England, Northern Ireland, Whales and Scotland, and thus there was a need for a national strategic framework. A few years later, a second report (UK National Commission for UNESCO, 2017) assessed initiatives relating to the Global Action Programme (GAP) on ESD and to the SDGs as well. That report found that there were still many grassroots initiatives in schools, higher education, local community groups and businesses; however there was still no government framework within which those initiatives to be supported, flourish and their impact on achieving sustainable development to be assessed (UK National Commission for UNESCO, 2017).

In terms of evaluation of the effectiveness of implementing ESD in the UK, the national strategy developers were in favour of approaches that assessed learners' sustainability literacy to provide evidence. However, the resulting consultations of the UK government with its advisors led instead to the development of an ESD indicator that had to do with the institutional effectiveness of introducing ESD, based on a self-assessment instrument aimed at sustainable schools (Huckle, 2009). Considering the evidence-based orientation of education in general and ESD specifically, and the gap in reliably evaluating learner empowerment with sustainability competences required for them to become the future citizens of society, a framework for assessing LOs for sustainability was developed using systems thinking and described in detail in a previous publication (Kioupi & Voulvoulis, 2019). It outlines five steps as part of a participatory process of selecting and assessing LOs for sustainability and had been applied in higher education programmes (Kioupi & Voulvoulis, 2020).

In this chapter, we present the application of the framework in two case studies around the assessment of sustainability LOs in school education in the UK during the 2018–2019 school year: the first in a primary school academy, which has an eco-school status, and the second in an independent co-educational secondary school. Both schools have different approaches to curriculum implementation: the former having more freedom to develop its own curriculum being an Academy, satisfying the Education Act (2011) requirements at the same time, and the second being an independent school following the national curriculum but having the flexibility to introduce unique courses for innovative teaching and learning (UK Government, n.d.-a).

16.2 Case Studies: Framework Application

The case study approach was selected as the appropriate tool to demonstrate the application of the framework in school education, as it would enable the teachers and researchers to gain insights into the ESD practices employed in the schools (Lapan et al., 2012). It would further help the teachers and researchers collect various types of data, both qualitative and quantitative, so that when analysed in the context in which the curriculum, learning activities and assessment for sustainability LOs take place, they can provide the base for actionable decisions by the educational communities (Baxter & Jack, 2015). Lastly, it would make sure that the newly developed pedagogical assessment tools would meet the needs of the schools and capture the benefits and limitations of the framework. For both case studies, we received ethics approval from the Imperial College Research Ethics Committee (ICREC) with reference 18IC4498/14/05/2018 prior to the implementation of the research.

The steps of the applied framework for selecting and assessing sustainability competences can be found in Fig. 16.1. The framework uses a participatory approach, whereby the researchers

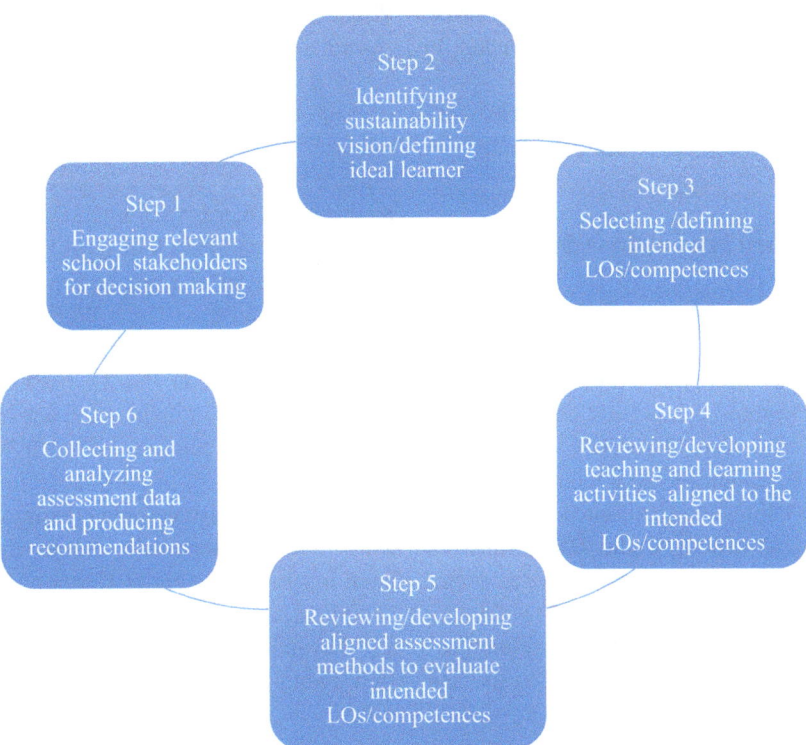

Fig. 16.1 The six steps of the applied framework for selecting and assessing sustainability competences in primary and secondary schools

work with the school stakeholders to implement every step of the framework; have open and meaningful discussions around their values, aims and objectives and research, teaching, learning and assessment methods; exchange feedback; use and analyse data transparently; and identify ways to improve practices (Bullock & Hitzhusen, 2015; Dlouhá & Pospíšilová, 2018). The first step of its application started with bringing together the relevant education stakeholders of the two schools and discussing on the sustainability vision they would like to achieve. This involved several meetings with the headmaster and teachers of the primary school and with the coordinator and teachers of the Global Goals course of the secondary school. In the next meetings, there were discussions around the competences needed for their learners to become the future citizens of this sustainable future, and on the pedagogies and assessments, they used to develop and evaluate the selected competences. In the final meetings, we reviewed those teaching and learning activities and assessment methods against their capacity to develop the selected competences in learners and worked with the teachers to revise them and to develop new ones where needed. In every step, we made sure that the decisions made reflected the realities, aims and needs of primary and secondary school teachers and students.

16.2.1 Primary School Case Study

The primary school is located in the home counties of England and is an academy school. Academy schools in England are publicly funded independent schools that are not obliged to follow the national curriculum (UK Government, n.d.-a). While academies have more freedom in terms of curriculum and term schedules, they are still required to adhere to the DfE's rules on admissions, special education needs and exclusions (UK Government, n.d.-a). The primary school is recognised as a leading eco-school, has held "Green Flag" status for over 4 years, and

more recently, in 2015, was chosen as one of nine schools in the UK to be a part of the eco-School's Ambassador programme (personal communication with head teacher). The school has identified seven sustainability themes to focus on as part of the eco-school programme, and these are *Energy, Healthy Living and Food, Recycling and Waste, Water, Biodiversity, Transport* and *Global Perspective* (personal communication with head master). These sustainability themes are interlinked with its curriculum, which follows the Harmony Framework. The Harmony Framework is a unique model of learning that guides how and when curriculum is taught at the school – including the core subjects of English, mathematics, science, computing and religious education (Dunne, 2020). This model of learning consists of four main concepts: values, principles of harmony, enquiries of learning and great works. The school's seven values are respect, kindness, honesty, responsibility, courage, forgiveness and joy (personal communication with head teacher). The values set by the school are then linked to the Principles of Harmony created by HRH the Prince of Wales and help incorporate themes of sustainability (Dunne, 2020).

Following the establishment of values and corresponding principles, the curriculum then incorporates enquiries of learning, which determine what core subjects are taught when the additional subjects such as ESD are included in the curriculum in six half-term periods (Dunne, 2020). Once an enquiry is complete, the students then engage in Great Works. The primary school uses Great Works as an opportunity for students to reflect on what they learned over the half term through a memorable activity or event (Dunne, 2020). Examples of Great Works completed at the school in the past include planting an orchard of fruit trees and creating a leaflet on solar energy (personal communication with head teacher). The entire Harmony Framework encompasses sustainability as a thread that links all the enquiries undertaken by the students. Discussing about the vision and mission of the school with the headmaster and the teachers, we realised it is deeply rooted in sustainability as it states:

> Sustainable living and learning is at the heart of the curriculum and everything we do at our school, with all our half termly year group learning enquiries directly linked to an element of sustainable living. We look to develop energy and environmentally conscious individuals who care about the world around them and understand what is required to sustain individual, team and global well-being. (Personal communication with head teacher)

This vision and mission definition of the school is aligned with our systemic framework's definition of a sustainable state citizen (Kioupi & Voulvoulis, 2019).

As per the adapted framework presented earlier, step 1 was taken to identify and engage the relevant stakeholders. Thus, we worked with the teachers and headmaster of the school to understand their ESD for the SDGs activities. The school had already developed a vision for sustainability and ideal learner profile, and thus we further discussed about it in our meetings (step 2). Next, we discussed around their intended LOs (step 3). The LOs envisaged pupils with affinity for sustainable living, having eco-conscience and showing care for the world around them. After that, we checked the constructive alignment among the LOs, teaching and assessment activities (steps 4 and 5) to evaluate if they enable the development of those LOs in pupils. Lastly, assessment data were collected, and results were analysed to draw insights and make recommendations. The process with specific adjustments made is described in more detail in this section.

In our initial discussions with the headmaster and the teachers of the school, they shared their aims for the Year 4, 5 and 6 curriculum for sustainability, which were directly linked with the concepts of food, water and energy. The interlinkages among food, energy and water are crucial for achieving sustainability and as concepts are challenging for students of young age to grapple (Barrutia et al., 2019; Opitz et al., 2017; Oztas & Oztas, 2017; Walshe, 2008). At the same time, balancing the water, food and energy nexus is a prerequisite for achieving the Sustainable Development Goals and has profound links with all of the goals (Simpson & Jewitt, 2019). A recent report found that because of urbanisation, population and consumption growth, the demands for energy, water and food would increase by 50%, 40% and 35%, respectively, by 2030 (Yillia, 2016).

Because of the importance of the water, food and energy nexus concept for environmental sustainability and after consultation with the teachers of the school, the focus of this study was placed on Years 4, 5 and 6. The assessment of Year 4, 5 and 6 pupils' knowledge, skills, attitudes and behaviours regarding food, water and energy systems would happen after they had participated in the relevant Enquiries of Learning during the school year of 2018–2019. The Year 4, 5 and 6 teachers selected specific LOs regarding sustainability they aimed their pupils to attain. For Year 4 students, they were knowledge of food production systems (conventional vs organic), where food comes from and what is food waste, what is produce seasonality and why it is important, skills in growing their own produce (vegetables) and appreciation of healthy food and behaviours that lead to consuming healthy food. For Year 5 the LOs included knowledge of water usage (direct and indirect), water footprint and how to decrease it, where does water come from and where does it go after we have used it, links between water use and vegetable production and consumption, attitudes towards responsible water use and reduction of wasteful behaviours. For the Year 6, the LOs included knowledge of energy as a physical quantity, its uses and measurement units, sources of energy, energy production, distribution and carbon footprint, skills in assessing the energy use of efficient and conventional electrical devices, monitoring and explaining the energy usage at school and adopting behaviours conducive to energy saving at school and at home.

The teachers participated in a discussion around the pedagogies and assessments used in Years 4, 5 and 6 to attain the intended LOs. The Year 4 pupils engaged in writing essays in topics such as organic food production, ethical farming methods and food miles. They also had to map countries and their products to understand how food travels and state their opinions in consuming local versus imported produce in terms of sustainability. Lastly, they had to describe 1 day in the life of a farmer and design and create the packaging and marketing material for a healthy snack as part of an arts project. During half term, the pupils were responsible for weighing and measuring food waste from the school kitchen and were asked to figure out ways to reduce waste before recycling it as compost. They also had an outdoor activity where the school gardener explained the importance of wildflower meadows for maintaining local biodiversity and the pupils identified important flower species. Another part of the outdoor lesson required pupils to split time between sieving compost, re-potting and watering seedlings. Other activities included in the outdoor lesson required students to weed a section of the garden, sow seeds and plant potatoes. The activities required active engagement in knowledge and skills development and were deemed appropriate for the students. However, there was no targeted assessment to evaluate the development of the LOs in the students.

The Year 5 students engaged in activities around river geography and ocean protection and were responsible for monitoring the school's water use through the Eco Driver tool on a weekly basis as well as measuring how much water is wasted at the end of each lunchtime to ensure maximum savings (personal communication with Year 5 teachers). As these learning activities were not entirely in alignment with the intended LOs, we worked with the teachers and the headmaster of the school to enrich the activities around water to meet the selected LOs. The resulting curriculum engaged the students in activities regarding direct and virtual water use in an average UK household per day. The students calculated their own water footprint and worked in groups to identify ways by which they could reduce their own direct and virtual/indirect footprint. They further examined the link between direct and virtual/indirect water use and food production and calculated the water needed to grow a vegetable locally versus growing it abroad and importing it. They discussed vegetable production in the greenhouses of Almeria, Spain, an area with arid climate and serious water stress to understand the practices involved in securing water for growing the vegetables. Lastly, they developed videos around water use in everyday life and in agriculture. In terms of the assessments, these were developed in consultation with the teachers and headmaster as the existing ones included pledges the students made around their personal water use.

As part of the energy sustainability theme in Year 6, the primary school aims to lower their energy consumption by relying on renewable energy sources from on-site solar panels as well as an on-site biomass boiler (personal communication with head master). In addition, the school strives to deploy energy-saving methods by creating weekly energy competition targets for each school building to motivate and educate the students on sustainable energy consumption. The school assigns the energy monitoring and tracking to the Year 6 students who use Eco Driver, an energy monitoring software system, on a weekly basis and share the results at school assemblies (Dunne, 2020). The Year 6 curriculum around energy was further enhanced through consultations with the teachers and the headmaster of the school, as it did not entirely capture the intended LOs. After the adaptations were made, it included inquiry-based activities around the use of energy at home, how energy is produced and distributed, what is 1 kWh and what kinds of activities you can perform with it, energy consumed in household activities by household appliances, debate over renewable and non-renewable sources of energy and personal carbon footprint calculation. Assessment tools were developed for this year's activities to capture the LOs attainment by the students.

The developed assessment tools were questionnaires assessing student cognitive (knowledge and skills), affective (attitudes) and behavioural (actions) dimensions of learning about food, water and energy in Years 4, 5 and 6. This classification of LOs was suggested by the school teachers and was found appropriate in the relevant literature for assessing LOs related to the SDGs in primary school education (UNESCO, 2017). The questions included a mix of open-ended, select the right choice, classification and Likert-scale questions as we wanted to capture the different types of knowledge, emotions and attitudes the students managed to develop (UNESCO, 2017). Year 4 students were also asked to draw storyboards to assess their attitudes around conventional and organic food. For these storyboards, the students were asked to draw pictures and explain with captions the life of a conventionally versus an organically grown tomato. Through providing visual explanations, students consolidate their learning, as they are required to do deeper processing of the information and produce more complete mental models (Bobek & Tversky, 2016). The analysis of the storyboards was based on contextualisation by use of the text descriptors on the drawings, segmentation that was implemented by design and qualitative coding of the themes presented in the segments of the storyboards and the emotions expressed (Loureiro et al., 2020).

Due to administrative complications, it was not possible to administer the questionnaire developed for Year 5 students, and thus the results are not reported as part of this study. The Year 5 and 6 students and teachers were also asked to complete a feedback form regarding the new activities that were introduced as part of the curriculum, and the results are reported. The questionnaires were administered as follows: a. 31 Year 6 students completed the energy questionnaire and student feedback form and 1 teacher completed the teacher feedback form, b. 26 Year

5 students completed the student feedback form and 1 teacher completed the teacher feedback form, and c. 59 Year 4 students completed the food questionnaire. The questionnaires comprised questions around the cognitive, affective and behavioural LOs targeted by the Year 4, 5 and 6 curricula. The storyboards were only distributed to Year 4 students. The analysis of the quantitative parts of the questionnaires was done using MS Excel software, and the analysis of storyboards and open-ended questions was done with NVIVO for qualitative analysis.

16.2.1.1 Year 6 Energy Questionnaire

The questionnaire for Year 6 can be found in the Appendix (Year 6). Here we report the main findings from the data collection and analysis. Students had good knowledge of everyday energy use, where and when they use it and are able to share examples of activities. Almost half of them were able to identify correctly an energy-saving light bulb as opposed to a conventional one based on energy rating data. Three quarters of the students managed to tag correctly at least eight energy sources as renewable or non-renewable, with biofuels often misunderstood as non-renewable.

Almost all students (except for two who did not know or gave irrelevant responses) supported that renewable energy sources are better for the environment and people. Most responses highlighted the positive aspects of renewable and the negatives of non-renewable energy. No negative aspects of renewables nor positive aspects of non-renewable energy were mentioned. The responses were framed as benefits for the environment if using renewables, such as reusability, no pollution, being natural and eco-friendly, and as drawbacks for people if using non-renewables, such as risk to health and leads to poverty, and the environment, such as climate change and generally harmful to the environment, without further explanation (Fig. 16.2).

Almost one third of the students could correctly identify that the electricity in their school

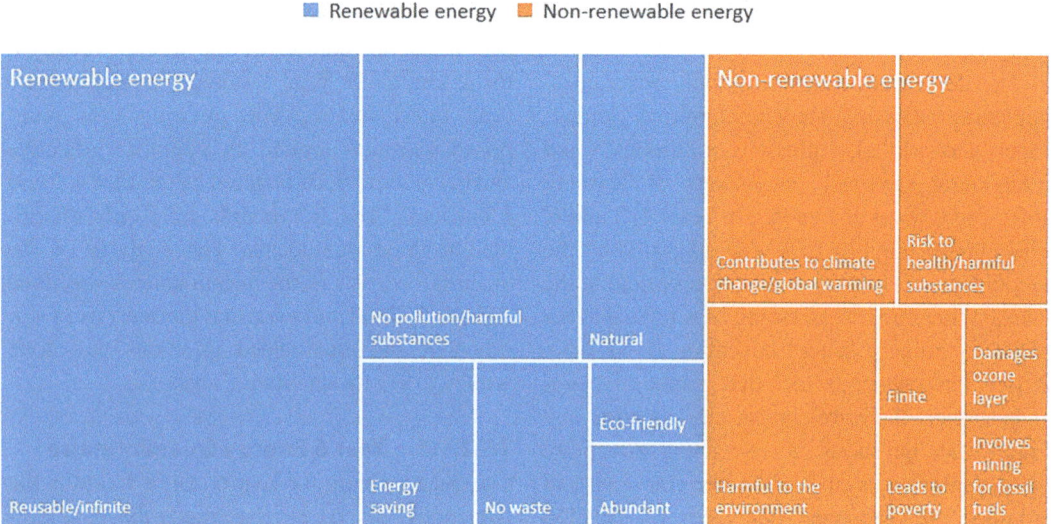

Fig. 16.2 Student responses to Q4: "Which energy sources are better for the environment and people: Renewable or Non-renewable? Explain why" by theme identified. The size of each box is proportionate to the number of students who included the theme in their response

came from the installed solar panels on the buildings roofs on sunny days, but gave no response for cloudy days. Only one third answered both parts of the question mentioning non-renewable, national grid, fossil fuels and main energy supply as the source of electricity on cloudy days. Out of those, only two were able to explain why. The majority of students thought that the school uses less electricity on weekends than on weekdays and said that this is because of less people in the buildings. Five explained further, giving reasons such as fewer lights are on, no heating, no smartboards and no computers are used, no lunches are cooked and so the kitchen is not in use.

Regarding engaging in energy-saving behaviours, most students mentioned: walking more, turning off lights, heating and other electric/electronic devices when not in use, use solar panels, *eat cold lunches, use natural light at home/school, use less your phone, TV and computer, do more outdoor activities, order meals online, switch to sustainable energy providers*, establish no-electricity/electricity-free days or hours every week, *use less water*, earth day participation, open windows instead of using fans, use more blankets instead of more heating, use energy

from wind turbines, *take shorter showers*, write instead of using PC and spend less time in front of screens. Some of the behaviours they suggested (words in italic) were not directly related with energy usage, and some were related with water usage; this shows that students made links between energy and water or energy and its sources. The results from the student and teacher feedback forms can be found in the Appendix (Year 6).

16.2.1.2 Year 5 Water Questionnaire Results

Unfortunately, the questionnaire was not administered to the students, as the responsible teacher was on maternity leave at the time and the other Year 5 teacher was very busy to do so. The results from the student and teacher feedback forms can be found in the Appendix (Year 5).

16.2.1.3 Year 4 Food Questionnaire

The top five concepts (83% of the responses) linked with organic food in the student open-ended responses were that it is free from chemicals and pesticides and thus better for the environment and human health, it is more expen-

Fig. 16.3 Student responses to Q2: "Why do you think organic food is good for you? Can you think of some problems or challenges with organic food?" by theme identified. The size of the boxes is proportionate to the number of students who included the theme in their response

sive than conventional food and animals are treated better in organic farming. Other less common responses mentioned that organic food is more natural, tastes better and has worse appearance and size than conventionally grown and travels less (food miles) to arrive to the consumer.

Most of the students' responses around the positives of organic food related with the absence of chemicals and pesticides that make it good for the environment and health as in Q2 (Fig. 16.3). Most of the negatives related with high prices that it is or can become infested and thus needs good washing before using or it expires quickly. Other less common responses included that organic food is tastier, natural, clean and fresh and that some problems with it are that it can be cross-contaminated by conventional food as it may be in close proximity to it, has smaller crop sizes, less variety and can produce more food waste as it spoils easily. One controversial aspect is that organic food was both associated with more and less food miles.

The majority of students managed to identify correctly the origin of at least eight out of ten fruits. Three quarters of the students could categorise correctly four to six of the fruit and vegetable according to harvest season; none correctly categorised all eight of them. Most students discussed the importance of recycling food waste in terms of reusability and minimisation of waste that ends up accumulating in the environment or home. The ecology topic resulted from student responses, which detailed different ecological processes including composting and soil composition and importance of cycling nutrients back to plants. The waste topic was a result of students who argued the importance of recycling in relation to the reduction of food waste and its uselessness if it is not reduced. The circular economy topic encapsulates responses, which detailed the need to reuse or repurpose things for the benefit of people and their activities, while the environment topic included any response, which mentioned positive or negative effects on the environment (Fig. 16.4).

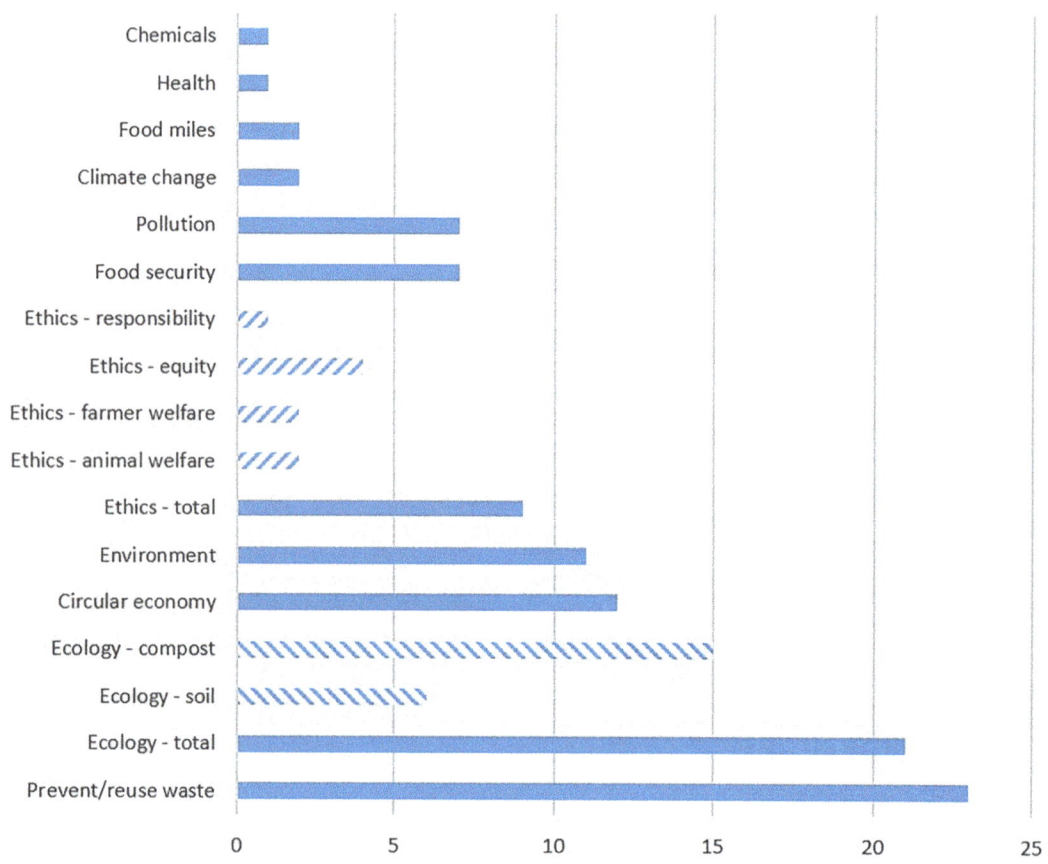

Fig. 16.4 Number of responses per topic introduced by students to respond to Q5: "Why is it important to recycle food waste?" The striped columns represent the sub-topics identified in student responses for the concepts ecology and ethics

16.2.1.4   Year 4 Storyboard Results

We collected 29 storyboards, as only 1 of the 2 Year 4 teachers was able to implement the activity with the students. After reading through the storyboards and analysing the drawings, we used open qualitative coding to record the themes that emerged in the storyboards. Then, we calculated the number of student responses that included each theme. The main themes that were identified in the Year 4 organic and conventional tomato storyboards are shown in Figs. 16.5 and 16.6.

For the organic food storyboard, the most frequently mentioned concept was ecology. This includes ecological processes such as the water and nutrient cycles, the soil community and root systems. In comparison, pesticides and bugs were the most frequently identified themes

regarding the conventional food. These two themes included the use of pesticides in any capacity to protect conventional food from harm as well as the presence and removal of bugs from conventionally farmed crops. Another theme related to that was the reference to pesticides in three cases as substances that boost the growth of crops, which constitutes a misconception of why and how pesticides are used in conventional farming (Fig. 16.6, pesticide confusion). Comparing the two storyboards, it is clear that all students referred to the organically farmed tomatoes as vegetables that grow because of important ecological processes in the soil, because of the sun and water they receive from farmers, whereas the conventionally grown tomatoes grow with pesticides

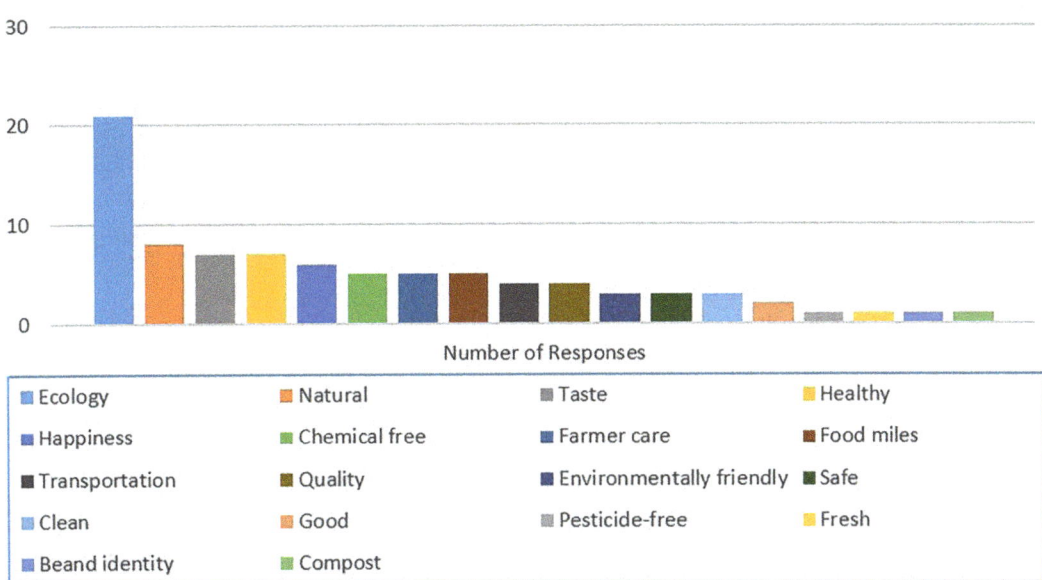

Fig. 16.5 Frequency of the main themes identified in student organic food storyboards

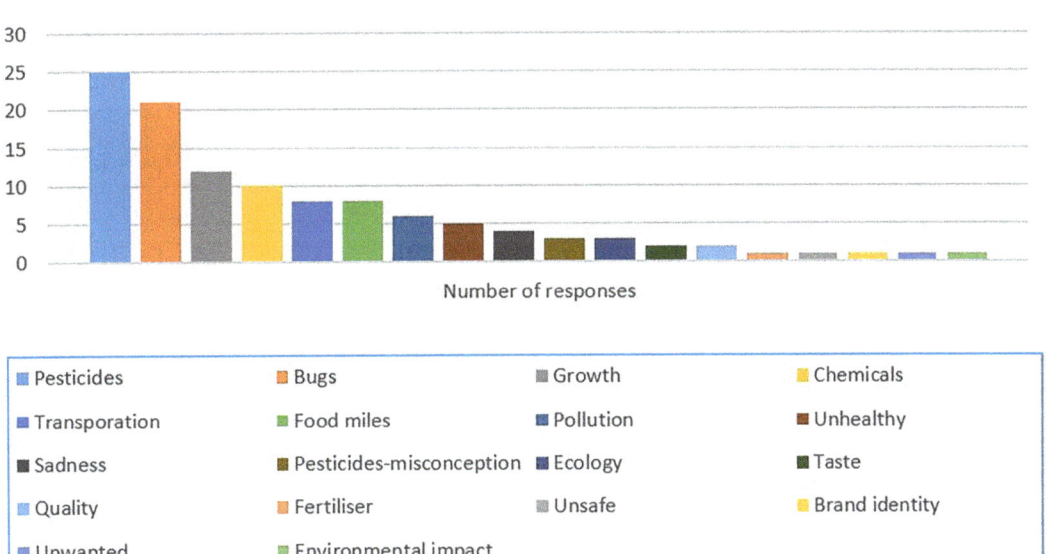

Fig. 16.6 Frequency of the main themes identified in student conventional food storyboards

that kill bugs and water that contains chemicals to help them grow faster. In almost all the storyboards, the final drawing shows a happy, healthy, safe and chemical-free organic tomato ready for consumption and, in contrast, an unhealthy, unsafe, unhappy and chemical-laden conventional tomato that is ready for consumption but is harmful for people (Fig. 16.7). In 8 out of the 29 storyboards, there was differentiation in the drawings for the organic and conventional ones. The differences most often referenced the origin of the seeds used for organic vs conventional farming, the latter being sourced from abroad, being "suspicious" or "secret". Often, the packaging in which the seeds arrive for organic is cardboard and for the conventional plastic, and the planting practices for organic seeds include good spacing and spread out planting to ensure the seeds grow in good conditions and for the conventional include packed planting and inadequate care. Other differences included transportation of conventional food most often by planes or lor-

ries, which cause pollution, and in one case, transportation by plane is referenced for organic tomatoes. Lastly, the students personified the tomatoes, showing them having faces that are smiling or are conveying positive feelings in the case of organic and crying, frowning or conveying negative feelings in the case of the conventional tomatoes.

16.3 Secondary School Case Study

The secondary school is a co-curricular independent school located in West London. Independent schools in the UK or also known as private schools charge fees to attend instead of being funded by the government (UK Government, n.d.-b). The secondary school has a strong ethos towards social inclusivity as it aims to offer opportunities to a well-rounded education to students from various backgrounds and operates a bursary award system (personal

Fig. 16.7 Example of a storyboard drawn by a Year 4 student depicting the journey from seed to table of an organic tomato and of a conventional one

communication with schoolteachers). As an independent school, it follows the UK curriculum and complements it with in-house developed courses to prepare the future citizens of the twenty-first century. The school has an educational vision around its reputation and the outstanding learning environment it aims to provide to students to help them become active citizens. The vision is quoted here:

> Our vision is to be the leading co-educational school in the United Kingdom, providing young people from all backgrounds with a life-changing education that equips and inspires them to make a positive impact on society and to excel in the wider world. (Personal communication with teacher coordinator)

Associated with its vision are the set of competences the school aims to develop in all of their students, which are the following:

- Independent learning and research
- Critical thinking, analysis and evaluation
- Written, oral and multi-media communication skills
- Collaborative working and problem-solving
- Intercultural knowledge and understanding
- Knowledge and understanding across a wide variety of disciplines
- Capacity for creativity and imagination
- A love of learning (personal communication with teacher coordinator)

These competences are complemented by the student learner profile that is used to measure the success of the school's education strategy. The ideal graduate of the school should be promoting international understanding, human rights and social justice, be sensitive to the importance of environment and sustainability, be responsible and respecting ethics, tolerant and open-minded, critical thinker and resourceful, hardworking and committed to act for a better world (personal communication with teachers of the school). The identified competences are in alignment with the competences for achieving the SDGs found in the international literature (Kioupi & Voulvoulis, 2019; UNESCO, 2017; Wiek et al., 2011).

The vision, educational aims, strategy and commitment to nurture sustainability-sensitive future citizens aspired by the school made it an appropriate candidate to apply our research methodology. We specifically worked with the coordinator and responsible teachers for the Global Goals course, which is offered to Year 9 students of the school. The Global Goals course runs throughout the school year and offers the opportunity to students to learn and act for the UN SDGs. During the autumn term, the students carry out an investigation on all the SDGs and identify a specific sustainability challenge with the aim being to find out the root causes of the problem. These challenges are related to the SDGs, but the students look at how problems are manifested across scales. The key is for students to find something they are passionate about and pursue it. Then, in the spring and summer terms, the students form teams and work on a specific challenge they feel passionate about and link it to one or more of the SDGs. They meet with the teachers and their groups fortnightly but also maintain collaboration and communication through online platforms and out of school meetings with their peers. In the meetings, they discuss progress and challenges and brainstorm solutions. The aim is by the end of the school year for each group to have engaged in at least one action that will benefit the community and help progress towards the targeted SDGs.

As per our framework, we engaged the coordinating teacher team of the Global Goals course (step 1) and worked with them to understand their school's vision towards sustainability and discuss about the competences that will contribute to that vision (step 2). The school was already advanced in having defined school-wide competences and LOs for the Global Goals course. The teachers identified the following competences (step 3), which are central to the realisation of the aims of the course and to prepare the students to be sustainability citizens:

- Systems thinking to allow them to understand the root causes of problems.
- Reflective thinking that allows them to be independent learners.

- Critical thinking that allows them to conduct valid research around the SDGs
- Self-regulation that allows them to cope with failure
- Collaboration to help them become team players
- Problem-solving and action to enable them develop creative and practical solutions and take action on them

The students engaged in a programme of learning activities throughout the school year to develop their knowledge, plan and implement their projects and present their outcomes in a school fair at the end of the school year. These learning activities included for the autumn term analysing problem scenarios, doing literature research around the SDGs and root causes of sustainability problems, learning and using PESTLE (political, economic, social, technological, legal and environmental) analysis to understand problems, engaging in presentations around the outcomes of problem scenarios and discussions to provide feedback to their peer's work. For the spring and summer term, the students mostly engaged in project-based learning, identifying challenges linked to the SDGs and planning, implementing and presenting their projects. This project-based learning was supported by activities for setting SMART (specific, measurable, achievable, realistic and timely) project goals, using the Double Diamond design process for defining problems and developing courses of action and the six hats technique for analysing a problem through multiple perspectives and thinking creatively about its solutions. All the activities were aligned with the selected competences' development in the students (step 4), but there was lack of appropriate assessment methods (step 5).

Regarding the need for assessment of the competences the students developed by participating in the Global Goals course and to reflect the experimental nature of this unique offering to the students of the school, the teachers suggested more "informal" types of assessing the LOs. After organising two meetings with the teachers and the coordinator of the course, the teachers highlighted the importance of empowering the

students to assess their own performance during the course, as they worked both independently and in groups and the course did not follow traditional teaching methods, but it was student-led. The teachers also highlighted that the students should have the opportunity to receive feedback on their final project presentation and that there was a need to evaluate their final product. After understanding the needs of teachers and students in terms of how to conduct the assessment and searching in the literature, we developed the assessment tools. The tools comprised a self-assessment questionnaire based on an adaptation of the self-efficacy scale suggested by Bandura (Bandura, 1994, 2006), to enable students evaluate their degree of agency; a team assessment questionnaire to allow students to evaluate their group work; and a peer assessment questionnaire to assess the final project product/presentation the students developed.

The self-assessment questionnaire included 27 questions that asked the students to rate their self-efficacy to perform specific tasks on a scale from 1 to 5, where 1 denotes strongly disagree with the statement and 5 denotes strongly agree, and 1 open-ended question about the role of the student in delivering the project work. The research shows that students of this age are able to use Likert-type scales to assess their own performance (Chambers & Johnston, 2002). The questions were carefully selected to represent the areas of competence the course aims to develop in the students, but there was no indication of which statements represented which competences. The team assessment questionnaire (Year 9, Table A.1, Appendix) comprised 21 statements around how they worked as a group, how they regulated group work and how they coped with difficulties. The students had to read each statement and select Yes, Partially or No to describe to what extent the statement described their group work. At the end, they had to complete an open-ended question about how they worked together as a group. The final assessment questionnaire was used by students to assess their peer's final project product/presentation and was based on six groups of criteria. The criteria examined were the research and development that went to the project; how realistic and relevant, innovative and creative, sustainable and scalable it was; if it was the outcome of collaboration; and how well it was communicated in the school fair. The students assigned 1 (for poor) to 5 stars (for top performance in a criterion).

We disseminated the self and peer assessment questionnaires to one class of the Global Goals course of 23 students. We also disseminated the final questionnaire to the entire Global Goals cohort to evaluate the project presentation during the fair and collected 123 completed forms. We analysed the results of the questionnaires quantitatively using MS Office Excel and IBM SPSS software. The results of the analyses are reported in the next section and are discussed with respect to the research framework.

16.3.1 Self-Assessment Questionnaire Results

The reliability of the self-assessment questionnaire was assessed using reliability analysis in SPSS, and the results show high reliability of the measure with Cronbach's a equal to 0.917, which is in the accepted value range of >0.7. The statements used in the self-assessment questionnaire were grouped according to the intended competences they described and can be found in Table 16.1. The analysis of the self-assessment questionnaire produced results that are shown in Table 16.1 as well.

The statements for which the highest score was assigned were "I can explain how the problem can affect my community, my country and the world". These statements are part of the systems thinking construct. The lowest score was assigned to the statement: "I can cope with complex problems". For all the statements, the students self-assessed between 3.7 and 4.5, which shows that they perceive they are advanced in those competences. We report cases where the lowest assigned value was 1, which means strongly disagree, as those highlight where the educators should place more emphasis. These statements' values are highlighted in orange in Table 16.1. These include statements about stu-

dents' ability to work in teams and collaborate with others, to cope with complex problems and failure and lastly to identify and combine information to understand the problem and how it links to the SDGs. The results for the entire classroom in terms of the competences assessed are shown in Fig. 16.8.

The students self-assessed higher in systems thinking and collaboration, while they perceive their weakest competence to be self-regulation. The intermediate competences were critical thinking, metacognitive/reflective thinking and problem-solving. There were six groups of students working on six projects around the SDGs in the class and used the team assessment questionnaire to assess their group work. We grouped the statements used in the team assessment questionnaire according to three competences, teamwork, difficulty coping as team and team regulation (Appendix, Year 9). The results of the analysis of the team assessment questionnaire per group of students are shown in Figs. 16.9, 16.10, and 16.11.

The combined results from the three figures show that teams 3, 4 and 6 had the highest teamwork competence, faced the least difficulties in coping with working as a group and had the highest ability to regulate their teamwork. However, teams 1, 2 and 5 had the biggest problems with group work, regulation and coping with difficulties, and the efforts of educators should focus on these groups to enable them to achieve better results.

In terms of the open-ended questions for the self and team assessment questionnaires, we present below the main themes that were introduced in the students' responses. In the self-assessment questionnaire, the students mentioned the specific roles they had in the group such as coming up with the initial idea, researching the topic, identifying existing and new solutions, communicating with external organisations to implement action, encouraging and motivating group members to continue with project work, mediating when problems in collaboration arose and developing the final prototype. In the team assessment questionnaire, the students of the groups that had problems with collaboration

Table 16.1 Descriptive statistics of the results of the self-assessment questionnaire and categorisation of each statement according to competence

	N	Range	Min	Max	Mean	Std. error	Std. deviation	Variance
I can identify problems related to the SDGs (CT)	23	4.00	1.00	5.00	4.2174	0.19838	0.95139	0.905
I can link the problem we identified to one or more SDGs (CT)	23	4.00	1.00	5.00	4.3043	0.20309	0.97397	0.949
I can explain why we selected the specific problem to work on (MC)	23	2.00	3.00	5.00	4.3913	0.15061	0.72232	0.522
I can identify sources of information related to the problem we identified (CT)	23	4.00	1.00	5.00	3.9130	0.20769	0.99604	0.992
I can select the most appropriate information to include in my work (CT)	23	3.00	2.00	5.00	4.1304	0.19177	0.91970	0.846
I can explain both the root causes and the effects of our chosen problem (ST)	23	3.00	2.00	5.00	3.8261	0.14947	0.71682	0.514
I can combine information from various sources to understand the problem (CT)	23	4.00	1.00	5.00	4.0000	0.18861	0.90453	0.818
I can explain how the problem affects my school (ST)	23	4.00	1.00	5.00	3.8261	0.22363	1.07247	1.150
I can explain how the problem affects my community (ST)	23	2.00	3.00	5.00	4.4783	0.12367	0.59311	0.352
I can explain how the problem affects my country (ST)	23	2.00	3.00	5.00	4.4348	0.13811	0.66237	0.439
I can explain how the problem affects the world (ST)	23	3.00	2.00	5.00	4.4348	0.16426	0.78775	0.621
I can cope with failure during doing my work for the Global Goals Course (SR)	23	4.00	1.00	5.00	3.8261	0.26414	1.26678	1.605
I can manage my own learning during the Global Goals Course (SR)	23	3.00	2.00	5.00	4.0435	0.19355	0.92826	0.862
I can mention existing solutions to the problem (mean) (PSA)	23	2.00	3.00	5.00	4.0652	0.15175	0.72777	0.530
I can propose new solutions to the problem (PSA)	23	3.00	2.00	5.00	3.8261	0.17391	0.83406	0.696
I can explain why the solution selected is appropriate for the problem (MC)	23	2.00	3.00	5.00	4.2174	0.15344	0.73587	0.542
I can identify the limitations of the solution we suggested (MC)	23	2.00	3.00	5.00	4.3913	0.15061	0.72232	0.522
I can collaborate with my team members (CO)	23	4.00	1.00	5.00	4.2609	0.22857	1.09617	1.202
I can understand my team members' needs (CO)	23	3.00	2.00	5.00	4.3913	0.16321	0.78272	0.613
I can cope with complex problems (SR)	23	4.00	1.00	5.00	3.6522	0.18446	0.88465	0.783
I can communicate our solution to other people effectively (PSA)	23	2.00	3.00	5.00	4.1304	0.18117	0.86887	0.755
I can work as part of a team (CO)	23	4.00	1.00	5.00	4.1739	0.20519	0.98406	0.968
I can develop a plan to implement the solution we suggested (PSA)	23	3.00	2.00	5.00	3.8261	0.16215	0.77765	0.605
I can reflect on my work and make changes if needed (SR)	23	3.00	2.00	5.00	4.1304	0.18117	0.86887	0.755
I can evaluate the effectiveness of our solution (mean) (MC)	23	2.00	3.00	5.00	4.0435	0.15372	0.73721	0.543
I can give constructive and helpful feedback to my team members about their work (CO)	23	3.00	2.00	5.00	3.9130	0.16530	0.79275	0.628
I am open to receive feedback from team members about my work (CO)	23	2.00	3.00	5.00	4.2609	0.16890	0.81002	0.656

ST systems thinking, *MC* metacognitive, *CT* critical thinking, *SR* self-regulation, *CO* collaboration, *PSA* problem-solving and action

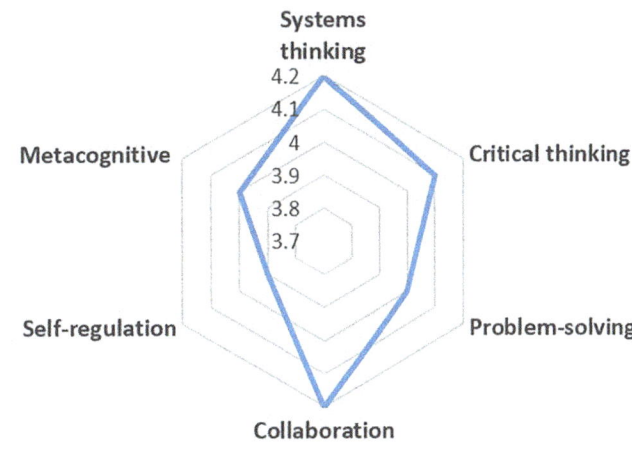

Fig. 16.8 Performance results for the six competences assessed in the secondary school case study

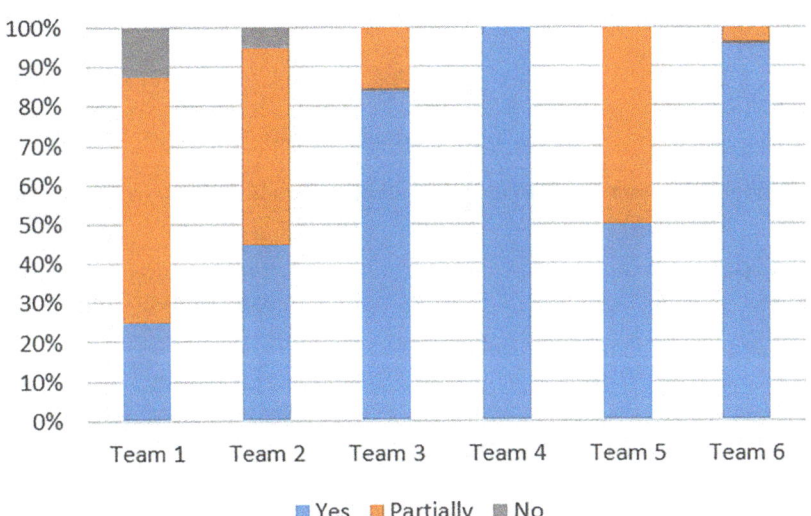

Fig. 16.9 Teamwork assessment results for the six groups of secondary school students (stacked columns)

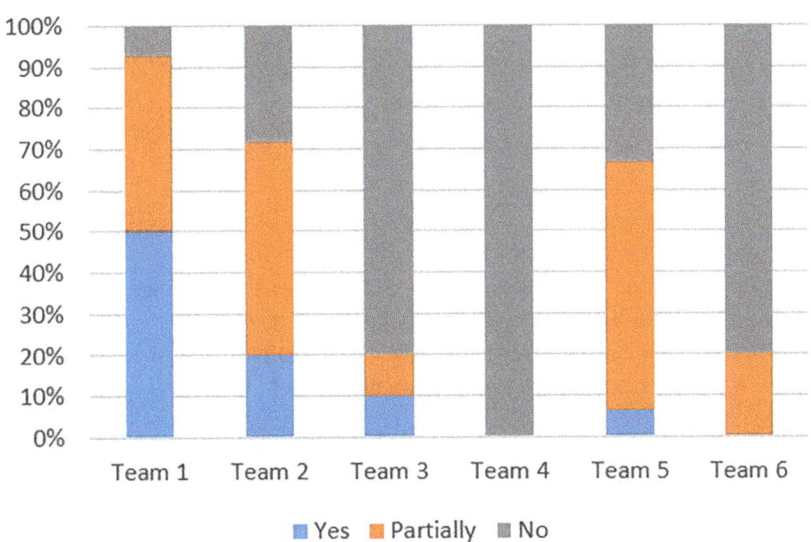

Fig. 16.10 Difficulty coping with the project as team assessment results for the six groups of secondary students (stacked column)

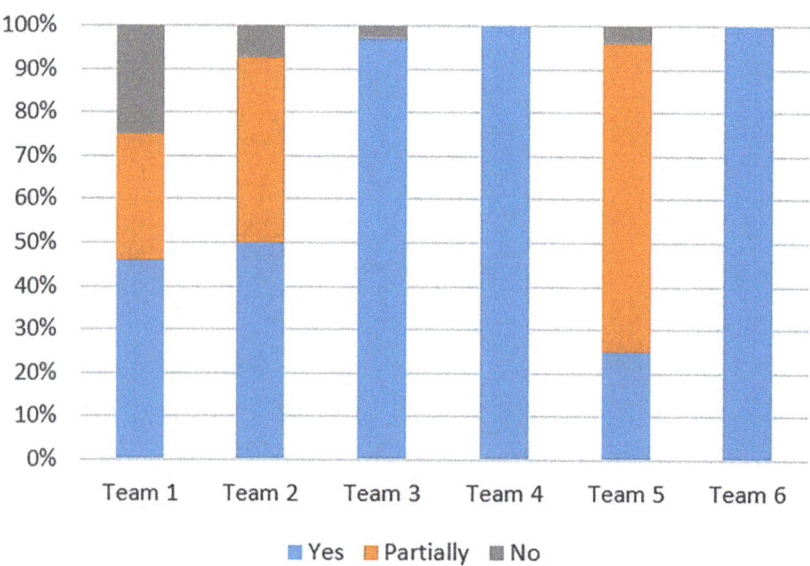

Fig. 16.11 Team regulation assessment results for the six groups of secondary students (stacked columns)

Fig. 16.12 Peer-review assessment results for the six projects of the student groups based on six criteria

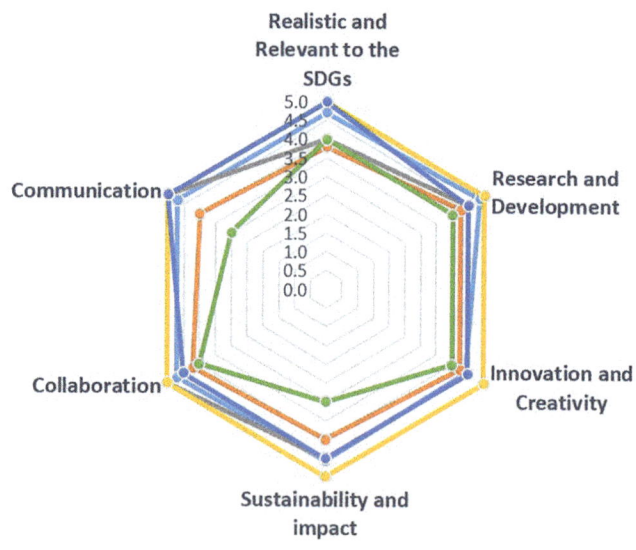

mentioned they argued a lot, they had difficulty supporting their opinions with arguments, they were slow as a group, arguing was sometimes fun and if they decided to solve their differences, they could be productive. The groups that had good collaborations mentioned that they discussed about their problems and challenges and listened to each other to help them overcome difficulties, they had chemistry as a group and they were pro-

ducing a lot of ideas which they were ready to compromise if the group in its entirety did not agree with them.

The results of the final assessment questionnaire, which was about peer-reviewing class's projects, are presented in Fig. 16.12.

The highest score of 5 stars was given to the project safety app, and closely behind were the projects plastic buffet and leftover cookbook; the lowest scores were given to the website for teen issues and reusable cup projects, which received low scores in sustainability and impact and communication and realistic and relevant to the SDGs, respectively. The open-ended positive feedback the peer reviews gave was around the creativity of the idea that groups went the extra length to develop outstanding prototypes and that the ideas were clear and clever, could be used by everyone and could produce big impact in solving the problems identified.

16.4   Discussion and Suggestions

The primary school results support that, in terms of science literacy, which is the main cognitive LO assessed through the questionnaires, Year 6 and Year 4 students actually seemed to have attained cognitive objectives of factual knowledge of energy and its uses at home and school, energy-saving products, renewable and non-renewable sources of energy and organic and conventional food systems and origin and seasonality of food and food waste. However, when asked to provide explanations and support their opinions with arguments, only a few students were able to do that. Conceptual and procedural knowledge around scientific concepts in energy and food systems need to be mobilised in students in order for them to gain deeper conceptual understanding and achieve scientific reasoning (Koerber et al., 2017). The Year 6 students that show the ability to explain concepts did so in the topics of "where energy is found" and "how it can be used" but were not able to link different forms of energy or discuss energy transformations, interpret energy data and explain "where energy comes from". They also showed misconceptions around ozone layer depletion being a consequence of using non-renewable energy sources. Year 4 students show some conceptual understanding of ecology concepts linked with soil processes such as decomposition of food waste, plant water absorption and the role of soil communities in plant growth but have misconceptions around the use of pesticides and other agrochemicals in conventional farming.

Some authors provide validated science literacy models that show three levels in the abilities of primary school students in understanding and reasoning scientifically, and these are the naïve, intermediate and scientific (Pollmeier et al., 2017). Our results show that for some concepts students are intermediate in their explanations, e.g. ecology concepts in food systems as they recognise the processes that are involved but do not provide accurate explanations or energy provision at school and energy performance of products as they are able to explain only part of the question that is related to their direct experiences and cannot provide explanations for parts they cannot control for. Some students have naïve conceptions that are not aligned with any scientific explanation, but these were only a few. A small proportion of Year 6 students used entirely scientific explanations in the open-ended questions. Those students were able to identify energy not only in devices we use but in our bodies and in the environment and were able to use the energy units to make informed decisions about energy saving and explain how solar energy is converted to other types of energy in solar panels and that the electricity grid provides energy from various sources.

Another aspect of teaching that became apparent through the assessment questionnaires was that of framing. Both Year 6 and Year 4 students showed biased responses in questions regarding comparisons of renewable and non-renewable and organic and conventional food products. Students were strongly in favour of the perceived "sustainable" option be it renewable energy or organic food and expressed strong emotions around it. In the case of energy sources, they only mentioned the positives of renewable and the negatives of non-renewable energy, and in the second case, they equated organic food with positive aspects, emotions and feelings and conventional with negative ones. When asked to think about problems with organic food, they managed to identify concepts around price, infestation by bugs, need for more washing than conventional, food miles it has to travel and its potential to generate more food waste as it expires more easily due to not being treated with chemicals. This shows their ability to think critically around organic food, but when it comes to their attitudes and emotions, they stay extremely positive stating that people should only consume organic food unless they are poor and thus not able to afford it. This is because of its perceived health and environmental benefits, which are not supported in the scientific literature, and the better farmer and animal welfare. Year 6 students show some ability of critical thinking in terms of the number of concepts they introduced in their open-ended responses, which is 2–3, but still overall think that non-renewables only cause harm to the environment and the health of people.

In terms of behavioural outcomes around energy, water and food systems, the Year 6 students identified many actions that can be taken to reduce the energy footprint at home and school, and some of them included direct links with reducing water usage, which shows an ability of linking different concepts and thinking holistically. This is also apparent in Year 5 responses around actions to reduce their water footprint, which include consuming locally produced and seasonal food. Most Year 4 students in terms of their behaviours stated they already eat organic food at least at school and that their parents will support them by changing their behaviours so they can eat organic food at home as well. A few students stated they already eat only organic food, while one mentioned they have never tried organic food. Two students also added that it is important to check the certification of the products you buy as in some case food that is claimed to be organic is not in reality.

Linking the results of the assessment to the vision the school is trying to achieve, we recommend that the school should place focus in developing conceptual understanding and scientific reasoning skills to students as well as their ability to think critically. This is important to achieve the energy and environmentally literate citizen that can make sustainable choices that are envisioned

by the school. Developing activities that enable students to engage in scientific thinking, inquiry and reasoning skills related to physical, chemical and ecological processes in energy, water and food systems would enable the students to achieve higher level of competence in science literacy (Zimmerman, 2007) and help them make informed decisions as future citizens (Bögeholz et al., 2017). Presenting the topics of the curriculum in a balanced way, allowing students to form their own opinions, enabling all voices to be heard and all perspectives to be explored would benefit their critical thinking skills (Cotton, 2006; Pauw et al., 2015). The assessment results also show that students can form strong attitudes, emotions and dispositions for sustainable behaviours that have roots in ethical beliefs of "doing the right thing" for people and for nature from a very young age. Nevertheless, this behavioural predisposition needs to be coupled with a strong foundation of science literacy and critical thinking so that students align their actions with both their beliefs and values and are able to consider multiple aspects of an argument and decide on what needs to be done.

The secondary school aimed to develop six core sustainability competences in students through the Global Goals course. The results of the self-assessment showed that all students perceived they significantly developed all of these competences (mean > 3.7), assigning higher scores to systems thinking, collaboration and critical thinking and lower to reflective thinking, self-regulation and problem-solving. After having discussions with the students, we realised that the open-ended format of the course although being of benefit to them posed come challenges as well. Mostly discussed were the difficulty coming up with realistic projects and completing them within the available time. Other aspects of concern were communicating with external stakeholders and getting them interested to help, redesigning their projects in cases of failure and being responsible for the entirety of their projects. Working as a team and receiving questions and feedback on their projects from teachers and their peers posed challenges for them as well.

In terms of working together, which was assessed through the team assessment questionnaire, the students showed high degree of teamwork, team regulation and coping with difficulties; however, some teams assessed their work lower but were able to identify what the problems were as well as coping strategies to solve them. All of the groups managed to complete their projects on time, and the final peer assessment showed that their work was of intermediate to high quality (scored 3–5 stars). The highest variability in marks was for the communication and sustainability and impact of the projects, which shows that these need to be paid attention to.

Regarding the school's vision to develop environmentally and sustainability minded learners who will show understanding of international affairs and will be responsible, hard-working, critical thinkers and will be committed to creating a positive impact on the world, our results show that these aims were actually obtained. The students mentioned during our discussions that the Global Goals course helped them open their minds to the sustainability challenges faced globally. They were able to make links between global challenges and local, national or community effects. However, in the self-assessment questionnaire, they scored low in linking how these challenges are related with their school life. Students also commented that the way to succeeding in completing their projects was to be responsible of them, working hard and only asking for teacher or parental support when they were faced with challenges they could not solve on their own. However, students had challenges working as a team, coping with failure and making alternative plans. All of the projects were able to highlight sustainability and real-world impact according to the peer-assessment results. Students felt highly creative throughout conceiving, planning and implementing their projects; however they felt that most of their original ideas were not realistic enough and had to rethink them.

As this was the first time the school implemented some form of assessment of the LOs of the Global Goals course, we think it will be beneficial for the school to keep and enrich the implemented assessments. This will assist both the teachers and students in terms of keeping track of their progress and identifying and addressing challenges throughout project implementation. Although the school is doing a very good job in using a variety of active learning methods to encourage the students to develop the intended competences and implement sustainable, realistic and impactful projects, they do not ensure continuity of those projects the following years, and thus the students become disengaged. It is crucial to find ways to scale those projects so that students can derive meaning from them, which is important for sustaining their engagement with ESD (Mickelsson et al., 2019).

After discussing some of these challenges with the teachers, we came up with the following:

- The students could study the work of scientists/entrepreneurs who tackle sustainability challenges and gain a better understanding of how they work to develop the solutions that exist.
- The students could overcome the identified difficulties by discussing with the previous cohort to identify which problems they tackled, how they coped and which solutions they provided and build on those.
- The teachers could encourage the school's administration and the local council to take up some of those projects so that they can be implemented on a larger scale the coming years and thus enhance student engagement.
- Lastly, the students could benefit from some classes on giving and receiving feedback because this will improve their interactions and reduce the stress they feel when others assess their work. They can have a class on strategies regarding coping with failure, as these will help them develop important life skills (Sarason & Sarason, 1981).

The application of the framework in the two schools confirmed the potential of ESD programmes for transforming visions, intended leaning outcomes, pedagogies and assessments towards sustainability. It also confirmed that the constructive alignment of all elements of the curriculum contributes to the development of student's sustainability competences. Thus, primary and secondary schools in general would definitely benefit from applying the framework and adapting it to meet their needs and priorities. Both schools had the advantage of flexibility in implementing curricula for sustainability and were quite advanced in terms of the ESD practices. This highlights that offering flexibility to schools to design their curricula would be an important step in advancing educational policy around ESD.

Acknowledgement The authors wish to thank the headmasters, teachers and students of the primary and secondary schools that participated in this study for their active engagement in this participatory research, their useful guidance and feedback and their collaboration. The authors also wish to thank the students of the master's programme in Environmental Technology Erika Starke and Xiao Zhu for helping with the data collection and analysis.

Appendix

Primary School

Year 6

Student Questionnaire

Question 1: Where and when do you use energy? List at least five things.

Question 2: Which light bulb saves more energy, A or B? Why?

Question 3: Which of the following energy sources are renewable and which are non-renewable? Put the following words in the correct column: Waves, oil, gas, wind, coal, sun, water, nuclear, biofuels

Question 4: Which energy sources are better for the environment and people: renewable or non-renewable? Explain why.

Question 5: Where does the electricity you use at your school come from: A. On sunny days? B. On cloudy days? Explain why.

Question 6: When do you use less electricity at school, on weekdays or weekends? Explain why.

Question 7: How can you save energy at school and at home? Give at least three tips for each.

Student Feedback

The students mostly felt great and confident about the activities as they had fun and learnt a lot about energy sources, how energy is used at school and how they can save energy and help the environment (empowered to take action and make a difference). Some students felt confused and less confident about their ability to use the Eco Driver tool. They would like their future learning to include comparing different schools' energy use, exploring the energy challenges in more depth, finding out how much energy and water the world uses, helping their city use less energy, calculating fossil fuel use/year, trying energy saving at home, exploring more types of energy, the financial gain/cost from using solar panels, how much energy solar panels produce, comparing changes of energy use in longer periods, having more time to work with Eco Driver and use it to make their own graphs. They identified as learning worth sharing with others the amount of energy the school uses every day, how to use the Eco Driver tool and that food, water and energy are linked.

Teacher Feedback

The teachers enjoyed the entire lesson and felt happy, excited and inspired to be part of it. They felt the most interesting part of it was discussing with the students the pros and cons of different energy sources, which made them consider about the wider argument of energy. They would have liked the activities to be more student-led and spread across more lessons, but felt the activities were very relevant to their teaching. They would like this conversation about energy to occur throughout the school. Lastly, they thought the activity was fair and balanced, the content well organised, the teaching methods appropriate and the students became engaged and enjoyed it; overall, it was successful to a great extent.

Year 5

Student Feedback

Students felt more aware about the water they use every day and learnt about the importance of water, some were sad to know how much water is wasted, but overall they felt excited about doing the activities as they learnt a lot and were surprised about the facts regarding direct and virtual human water consumption. They would like their future learing to include learning about rives and doing outdoor lessons, setting a water challenge for the whole school, doing robotics related to solving water challenges, helping other people who do not have access to water gain access by reducing consumption in areas where it is high, developing a water re-using building, reducing the amount of water people use to make things, helping save aquatic creatures and informing people about their water use. They identified as learning worth sharing with others: direct water use per day, water used to make the products we use every day such as vegetables and clothes, *we should eat seasonal food*, water, food and energy are linked in many ways, *the tomatoes we buy in the winter come from abroad and grow on more water, we use greenhouses to control vegetable growth, tomatoes in Spain are grown all year round and that food production uses a lot of water*. These last student learnings show their ability to link water use with food production and are worth exploring further.

Teacher Feedback

The teachers enjoyed the entire lesson and felt happy, interested and surprised. They believe students enjoyed the lesson very much as they liked learning the facts about water use but would have liked more hands-on activity. The content was relevant to their teaching and interesting as it was linked to real life. Their suggestion would be to split it up into more sessions so that the information provided is more manageable for the students. They think students developed their thinking, collaborated, explored new topics, came up with new ideas, estimated and linked learning to real life. Overall, the lesson was successful.

Year 4

Student Questionnaire

Question 1: What is organic food?

Question 2: Why do you think organic food is good for you? Can you think of some problems or challenges with organic food?

Question 3: Can you guess the origin of each fruit (UK or overseas)? Tick under the column you think is the best fit.

Question 4: Which of the following foods are harvested in the summer and which in autumn? Strawberries, tomatoes, lettuce, pumpkins, carrots, cabbage, apples, pears. Write in the appropriate space below.

Question 5: Why is it important to recycle our food waste?

Question 6: Circle how much you like eating the following fruits and vegetables.

Secondary School

Year 9

Table A.1 Competences assessed and statements used in the team-assessment questionnaire

Competence	Statement
Teamwork	All members of our team have clear roles
	Everyone contributes to the project
	Everyone does their own part of the work
	We have developed a plan about the project
	We plan our project steps together
	We all work to achieve our common goal
	We are able to present and communicate our work effectively
	We work well together
Difficulty coping as a team	We need plenty of support to work as a team
	Some team members are following their own ideas
	Some team members are not committed to the work
	We struggle with project complexity
	We have difficulty agreeing what to do
Team work regulation	We reflect on our project and make improvements
	We divide the work between us fairly
	We show high responsibility doing our work
	We are flexible to consider new directions for our work
	We listen to each other and include all opinions
	We overcome project difficulties by open discussion
	We encourage each other to do the work
	We overcome conflict in a peaceful way

References

Bandura, A. (1994). Self-efficacy. In V. S. Ramachaudran (Ed.), *Encyclopedia of human behavior* (Vol. 4, pp. 71–81). Academic Press.

Bandura, A. (2006). Guide for constructing self-efficacy scales. In F. Pajares & T. S. Urdan (Eds.), *Self-efficacy beliefs of adolescents* (pp. 307–337). Age Information Publishing.

Barrutia, O., Ruíz-González, A., Villarroel, J. D., & Díez, J. R. (2019). Primary and secondary students' understanding of the rainfall phenomenon and related water systems: A comparative study of two methodological approaches. *Research in Science Education.* https://doi.org/10.1007/s11165-019-9831-2

Baxter, P., & Jack, S. (2015). Qualitative case study methodology: Study design and implementation for novice researchers. *The Qualitative Report, 13*(4), 544–559. https://doi.org/10.46743/2160-3715/2008.1573

Bobek, E., & Tversky, B. (2016). Creating visual explanations improves learning. *Cognitive Research: Principles and Implications, 1*(1), 1–14. https://doi.org/10.1186/s41235-016-0031-6

Bögeholz, S., Eggert, S., Ziese, C., & Hasselhorn, M. (2017). Modeling and fostering decision-making competencies regarding challenging issues of sustainable development. In D. Leutner, J. Fleischer, J. Grünkorn, & E. Klieme (Eds.), *Methodology of educational measurement and assessment. Competence assessment in education: Research, models and instruments* (pp. 263–284). Springer International Publishing. https://doi.org/10.1007/978-3-319-50030-0_16

Bullock, C., & Hitzhusen, G. (2015). Participatory development of key sustainability concepts for dialogue and curricula at the Ohio State University. *Sustainability (Switzerland), 7*(10), 14063–14091. https://doi.org/10.3390/su71014063

Chambers, C. T., & Johnston, C. (2002). Developmental differences in children's use of rating scales. *Journal of Pediatric Psychology, 27*(1), 27–36. https://doi.org/10.1093/jpepsy/27.1.27

Cotton, D. (2006). Teaching controversial environmental issues: Neutrality and balance in the reality of the classroom. *Educational Research, 48*(2), 223–241. https://doi.org/10.1080/00131880600732306

de Pauw, J. B., Gericke, N., Olsson, D., & Berglund, T. (2015). The effectiveness of education for sustainable development. *Sustainability (Switzerland)*, *7*(11), 15693–15717. https://doi.org/10.3390/su71115693

DEFRA. (2005a). Securing the future – UK Government sustainable development strategy. http://www.defra.gov.uk/sustainable/government/publications/uk-strategy/index.htm. Accessed 16 July 2021.

DEFRA. (2005b). One future – Different paths – The UK's shared framework for sustainable development. http://www.defra.gov.uk/sustainable/government/publications/uk-strategy/framework-for%sd.htm. Accessed 16 July 2021.

Department for Education. (2014). Assessment principles. https://assets.publishing.service.gov.uk/government/uploads/system/uploads/attachment_data/file/304602/Assessment_Principles.pdf. Accessed 16 July 2021.

Dlouhá, J., & Pospíšilová, M. (2018). Education for sustainable development goals in public debate: The importance of participatory research in reflecting and supporting the consultation process in developing a vision for Czech education. *Journal of Cleaner Production*, *172*, 4314–4327. https://doi.org/10.1016/j.jclepro.2017.06.145

Dunne, R. (2020). Harmony: A new way of looking at and learning about our world (A teachers' guide). https://www.theharmonyproject.org.uk/product/harmony-a-new-way-of-looking-at-and-learning-about-our-world-a-teachers-guide/. Accessed 16 July 2021.

Huckle, J. (2009). Consulting the UK ESD community on an ESD indicator to recommend to government: An insight into the micro-politics of ESD. *Environmental Education Research*, *15*(1), 1–15. https://doi.org/10.1080/13504620802578509

Kioupi, V., & Voulvoulis, N. (2019). Education for sustainable development: A systemic framework for connecting the SDGs to educational outcomes. *Sustainability (Switzerland)*, *11*(21), 6104. https://doi.org/10.3390/su11216104

Kioupi, V., & Voulvoulis, N. (2020). Sustainable development goals (SDGs): Assessing the contribution of higher education programmes. *Sustainability*, *12*(17), 6701. https://doi.org/10.3390/su12176701

Koerber, S., Sodian, B., Osterhaus, C., Mayer, D., Kropf, N., & Schwippert, K. (2017). Science-P II: Modeling scientific reasoning in primary school. In D. Leutner, J. Fleischer, J. Grünkorn, & E. Klieme (Eds.), *Methodology of educational measurement and assessment. Competence assessment in education: Research, models and instruments* (pp. 9–17). Springer International Publishing. https://doi.org/10.1007/978-3-319-50030-0_3

Lapan, S. D., Quartaroli, M. T., & Riemer, F. J. (2012). *Qualitative research: An introduction to methods and designs*. Jossey-Bass.

Leutner, D., Fleischer, J., Grünkorn, J., & Klieme, E. (2017). Competence assessment in education: An introduction. In D. Leutner, J. Fleischer, J. Grünkorn, & E. Klieme (Eds.), *Methodology of educational measurement and assessment. Competence assessment in education: Research, models and instruments*

(pp. 1–6). Springer International Publishing. https://doi.org/10.1007/978-3-319-50030-0_1

Loureiro, K. S., Grecu, A., de Moll, F., & Hadjar, A. (2020). Analyzing drawings to explore children's concepts of an ideal school: Implications for the improvement of children's well-being at school. *Child Indicators Research*, *13*(4). https://doi.org/10.1007/s12187-019-09705-8

Mickelsson, M., Kronlid, D. O., & Lotz-Sisitka, H. (2019). Consider the unexpected: Scaling ESD as a matter of learning. *Environmental Education Research*, *25*(1). https://doi.org/10.1080/13504622.2018.1429572

Opitz, S. T., Blankenstein, A., & Harms, U. (2017). Student conceptions about energy in biological contexts. *Journal of Biological Education*, *51*(4). https://doi.org/10.1080/00219266.2016.1257504

Oztas, F., & Oztas, H. (2017). Pupils' understanding of food concept: The assessment of children's preconceptions ideas about food. *Journal of Education and Practice*, *8*(7). https://eric.ed.gov/?id=EJ1137571

Pollmeier, J., Tröbst, S., Hardy, I., Möller, K., Kleickmann, T., Jurecka, A., & Schwippert, K. (2017). Science-P I: Modeling conceptual understanding in primary school. In D. Leutner, J. Fleischer, J. Grünkorn, & E. Klieme (Eds.), *Methodology of educational measurement and assessment. Competence assessment in education: Research, models and instruments* (pp. 9–17). Springer International Publishing. https://doi.org/10.1007/978-3-319-50030-0_2

Reynolds, J., & Scott W. A. H. (2011). Sustainable schools in England: Background and lessons learned. *National Association of Field Studies Officers Annual Journal and Review*. ISBN: 978-1-901642-18-6. http://se-ed.co.uk/edu/wp-content/uploads/2012/10/Sustainable-Schools-in-England-background-and-lessons-learned.pdf. Accessed 16 July 2021.

Sarason, I. G., & Sarason, B. R. (1981). Teaching cognitive and social skills to high school students. *Journal of Consulting and Clinical Psychology*, *49*(6). https://doi.org/10.1037/0022-006X.49.6.908

Simpson, G. B., & Jewitt, G. P. W. (2019). The development of the water-energy-food nexus as a framework for achieving resource security: A review. *Frontiers in Environmental Science*, *7*. https://doi.org/10.3389/fenvs.2019.00008

UK Government. (n.d.-a). Types of school, academies. https://www.gov.uk/types-of-school/academies. Accessed 12 Apr 2021.

UK Government. (n.d.-b). Types of school, private schools.https://www.gov.uk/types-of-school/private-schools. Accessed 12 Apr 2021.

UK National Commission for UNESCO. (2013). Policy Brief 9: ESD in the UK - Current status, best practice and opportunities for the future. www.unesco.org.uk. Accessed 16 July 2021.

UK National Commission for UNESCO. (2017). Good practice in education for sustainable development (ESD) in the UK: Case Studies. www.unesco.org.uk. Accessed 16 July 2021.

UNESCO. (2014). *Shaping the future we want. UN decade of education for sustainable development 2005-2014 final report*. United Nations Educational, Scientific and Cultural Organization.

UNESCO. (2017). *Education for sustainable development goals. The global education 2030 agenda, learning objectives*. United Nations Educational, Scientific and Cultural Organization.

UNESCO. (2018). Issues and trends in education for sustainable development. https://unesdoc.unesco.org/ark:/48223/pf0000261445. Accessed 16 July 2021.

UNESCO. (2020). *Education for sustainable development: A roadmap*. United Nations Educational, Scientific and Cultural Organization. https://doi.org/10.1111/j.2048-416x.2009.tb00140.x

United Nations SDG4. (2021). Ensure inclusive and equitable quality education and promote lifelong learning opportunities for all. https://sdgs.un.org/goals/goal4. Accessed 16 July 2021.

Walshe, N. (2008). Understanding students' conceptions of sustainability. *Environmental Education Research, 14*(5), 537–558. https://doi.org/10.1080/13504620802345958

Wiek, A., Withycombe, L., & Redman, C. L. (2011). Key competencies in sustainability: A reference framework for academic program development. *Sustainability Science, 6*(2), 203–218. https://doi.org/10.1007/s11625-011-0132-6

Yillia, P. (2016). Water-energy-food nexus: Framing the opportunities, challenges and synergies for implementing the SDGs. *Österreichische Wasser- Und Abfallwirtschaft, 68*, 86–98.

Zimmerman, C. (2007). The development of scientific thinking skills in elementary and middle school. *Developmental Review, 27*(2), 172–223. https://doi.org/10.1016/j.dr.2006.12.001

Vasiliki Kioupi has completed her doctoral research in the Centre for Environmental Policy of Imperial College London (ICL) in education for sustainable development (ESD) with special emphasis on the development and assessment of learner sustainability competences from a systems thinking perspective. She received a prestigious President's PhD Scholarship from ICL to fund her research. She is an experienced science teacher, environmental educator, and teacher trainer and has worked for the Greek Ministry of Education as ESD coordinator. She is a fellow of the Schumacher Institute, an independent think tank for environmental, social, and economic issues. She has published her research in scientific and educational journals.

Nikolaos Voulvoulis, BSc (Hons), MSc, DIC, PhD, DMS, is Professor of Environmental Technology at Imperial College's Centre for Environmental Policy and deputy HoD. He is an international expert in environmental management, especially where science and engineering interface with public policy. His research targets the interactions and interdependencies between human and natural systems and focuses on processes and practices across many areas and diverse sectors. His work aims to facilitate policy development, support education initiatives, create an environment for innovation, and engage businesses, industry, and the public in the transformational shift required for our society to reach sustainability.

The Body in Mind: Ideas for Assessing Sustainability Literacy

17

Jennie Farber Lane and Armağan Ateşkan

Abstract

Achieving sustainable development involves whole body learning. The head (knowledge, critical thinking, systems thinking), the heart (empathy, dispositions, communication), the gut (agency, locus of control, advocacy), and the arms and legs (problem solving, cooperation, participation) all play a role in taking purposeful and meaningful steps for sustaining our future. To ensure the whole body works together as a wholistic system, it is important to examine each of the parts. Since the parts have unique roles, it follows that they be assessed in relevant and meaningful ways. This chapter provides an overview of different types of assessment, including formative and summative strategies. It discusses the importance of having both quantitative and qualitative means to gain insights into students' heads, hearts, and actions related to their understanding of, dispositions toward, and participation in achieving sustainable development.

Keywords

Assessment · Literacy · Portfolio · Rubric

J. F. Lane (✉) · A. Ateşkan
Bilkent University, Department of Educational
Sciences, Ankara, Türkiye
e-mail: jennie.lane@bilkent.edu.tr

17.1 Introduction

Imagine a senior graduating class who are certified as sustainability literate. What is it about these students that enabled them to earn this certification? What do they know? How do they behave? What actions have they taken? Nolet (2009) defines sustainability literacy as "ability and disposition to engage in thinking, problem solving, decision making, and actions associated with achieving sustainability" (p. 421). More than just content knowledge, he emphasizes that sustainability literate individuals are aware of and willing to constructively interact with natural and human systems. So, what can we look for if we wanted to certify that students are sustainability literate? As with many certifications, perhaps these students have passed certain benchmarks or have completed certain tasks that helped them become sustainability literate. It would be ideal if these students maintained a file in which they could record, monitor, and report their achievements. This chapter will be designed with such a file in mind. Here it will be called a portfolio, but other titles could be used, such as passbook or a digital library. One suggestion for organizing this portfolio is to divide it into sections based on the human body. Other researchers have noted the importance of considering the body in relation to literacy. Leander and Boldt (2013) explain that:

> [the] body is both material and incorporeal. Materially, we move within time and space as bod-

ies. As bodies, we perceive and register, consciously or unconsciously, some of the infinite patterns and variations in our environment. It is in the body that we locate the affective sensations of those registrations that are available to our consciousness, often making meaning of them by giving them form and significance as emotion, physical sensation, response, or energy. (p. 29)

There is the whole body, and there are also its parts. These parts include the head, the heart, the gut, and the arms and legs. Achieving sustainable development involves whole body learning. The head (knowledge, critical thinking, systems thinking), the heart (empathy, dispositions, communication), the gut (agency, locus of control, advocacy), and the arms and legs (problem-solving, cooperation, participation) all play a role in taking purposeful and meaningful steps for sustaining our future. Figure 17.1 shows a graphic representation of this whole body approach to sustainability literacy assessment. This figure can be used as a reference when reading the chapter as it lists selected assessment strategies presented and discussed in the sections below.

It should be noted that while assessment ideas are placed in each section, it is likely that each

idea can address multiple aspects of sustainability (i.e., other body parts). Just as the body is an integrated entity and all its parts need to work together in harmony, sustainability also involves an interaction of our thoughts, dispositions, and actions.

This chapter will provide assessment ideas for the whole body and the aforementioned body parts. The ideas are for primary and secondary school teachers to use to assess aspects of students' sustainability literacy. Teachers may focus on one or more ideas depending on their teaching situation; however, it is hoped they will consider a variety of approaches that will contribute to whole body learning. There will be assessment ideas for a variety of learning domains (cognitive, affective, behavior). These will include traditional strategies such as questionnaires, which are more quantitative as well as qualitative strategies including interviews and observations. Some of the assessments can be used during the learning process (formative), and others will be more cumulative (summative). Indeed, some of the assessment strategies take place while students are undertaking an activity rather than afterward;

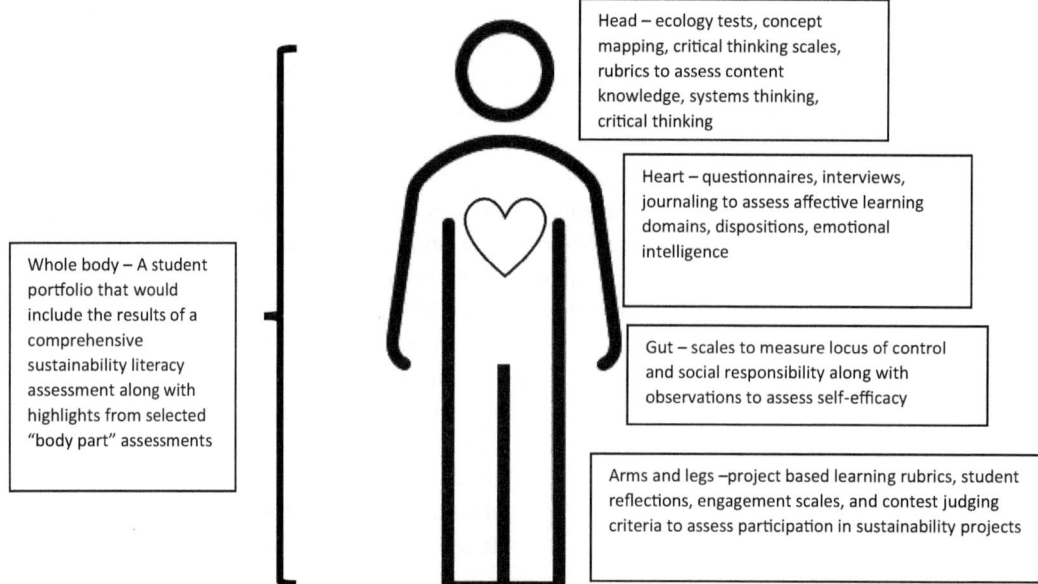

Fig. 17.1 Examples of different types of assessment strategies that can be used during a students' learning career and maintained in a sustainability literacy portfolio

in other words, more than assessment of learning, the suggested methods and strategies can be assessments *for* learning (Black et al., 2004). The ideas listed here are just a few examples; practitioners are encouraged to think of other ideas beyond the ones provided here.

17.2   Sum of the Parts: Comprehensive Assessment Ideas

The "whole body" section would provide a comprehensive assessment that certifies students as sustainability literate. Ideally, if students have kept a portfolio of their actions and activities, this would be a record of their accomplishments and achievements. The portfolio would showcase how students have developed their knowledge and skills and participated in actions to contribute to a sustainable world. Schools may be interested in having students complete a comprehensive assessment to show they have retained and can express their cumulative understanding of sustainability. For example, Waltner et al. (2019) developed and assessed an instrument to measure a variety of competencies related to sustainability. They considered knowledge, behaviors, affective-motivational beliefs, and other parameters. In this book, there are also additional chapters describing assessment strategies of sustainability competencies in primary and secondary education.

Students' scores on the sustainability literacy assessment could be reported in the "whole body" section of their portfolio. It should be noted that this comprehensive assessment may not be a single instrument. Researchers have noted that multiple strategies are important for gaining valid insights into aspects of responsible environmental behavior (Chao & Lam, 2011; Lange & Dewitte, 2019). While a school system may administer a "final examination" of sustainability literacy, it is important to consider the entire portfolio of assessment development of students' pro-environmental behavior.

Another idea is for students to complete a senior project that successfully addresses one or more of the UN goals for sustainable development. Completing a major project is common to many upper high school programs around the world. Many school programs have a culminating or senior project that students are to complete before they graduate. For example, the curriculum of the Turkish Ministry of Education includes a term project designed to help students advance their research and communication skills. To perform their improved skills, students are also expected to enhance their educational experience by completing a project which is created using their own knowledge and skills. The criteria to evaluate the project are prepared by each school's own teachers (Ministry of National Education (MoNE), 2014). The International Baccalaureate (IB) is an international education program that strives to foster critical thinking and global awareness among its student population. It offers education programs that institutions around the world implement as part of their school curriculum. The IB Diploma Program (DP) involves students in a variety of projects to help them develop and apply their knowledge and skills. For example, the Creativity, Activity, Service (CAS) is an experiential learning program. Students can express themselves artistically and through physical exercise, and they are also encouraged to become involved in community service work (International Baccalaureate Organization, n.d.-a). The IB DP's Extended Essay (EE) is an independent investigation on a topic chosen by the student (International Baccalaureate Organization, n.d.-b). For both CAS and EE, students often chose to explore environmental issues in their community.

A culminating project idea to assess sustainability literacy is an environmental issue investigation and action project. These types of projects were promoted in the early days of environmental education and could be revised and applied in current times. Environmental issue analysis involves students identifying and analyzing an environmental problem in their community (Ramsey et al., 1989). The issue analysis process involves doing background research, conducting surveys, interpreting data, and developing an action plan (Ramsey et al., 1989, 1992). An outcome of this process is for students to become actively involved in issue resolution. Hungerford and Peyton (1980) introduced five categories in which people can take action: Persuasion, Consumerism, Political Action, Legal Action, and Eco-management, the latter of which entails physical action or manual labor to improve the environment. Given that these activities are based on student performance, it follows that they be assessed by authentic assessment strategies. The students could be given their investigation in stages and be required to submit progress reports on each stage. Students can keep a reflective journal that would facilitate integrating action research into the experience. This project may be conducted individually or in a team. With teamwork, the project can also help develop and review collaboration and cooperation. A rubric can be used to guide student work and to assess their accomplishments. The rubric could include criteria such as the following:

- Strategies used to learn background information about the issue, such as collecting first- and secondhand data and organizing the information
- Description about the background of the issue, including identification of varying viewpoints and stakeholders
- Development of an effective and practical research question to guide the investigation
- Review of potential ideas to take action to address the issue
- Monitoring of the progress and reporting results

The comprehensive assessment of students can be a culminating activity in their schooling. Throughout their primary and secondary education, they will have opportunities to develop the knowledge and skills they need to undertake this project or to complete this assessment. In other words, if the comprehensive assessment activity involves the whole body, then educational experiences can build the capacity of various "body parts."

17.2.1 The Head

This section of the student sustainability literacy is perhaps the most traditional form of school learning and assessment. It features students' acquisition of facts and concepts related to sustainability. A strong knowledge of ecological concepts and principles is important for this section. Therefore, assessments would include instruments where students record their content knowledge, such as multiple-choice exams, case study analysis, and essays. Teachers could also conduct pre- and post-exams to monitor how student knowledge has changed as a result of instruction and learning experiences. Science teachers already have tests they use to assess students' ecological and scientific knowledge. They can enhance their strategies with test items from environmental literacy assessments such as the one developed by McBeth and Volk (2009). Maurer et al. (2020) developed an instrument that includes measurements of environmental knowledge. They may also want to consider the environmental paradigm scale developed by Dunlap (2008). The instrument focuses on environmental concern and gives insights into learners' ecological knowledge. In the sustainability literature, Schneiderhan-Opel and Bogner (2019) provide a study that involves a measurement of students' understanding of biodiversity. In their study, Hoppe et al. (2020) detail how teachers measure students' understanding of the nature and cycling of matter.

More than remembering facts, however, sustainability literacy encompasses advanced thinking skills, including systems thinking and critical thinking. Therefore, it is important to consider assessment strategies that examine how students construct knowledge, analyze information, and connect concepts. One idea to assess these thinking skills is through concept mapping. This strategy involves students in identifying and organizing concepts related to a topic; especially important is how students explain the relationships among the concepts. Students can be given a template for designing maps or they create their own, although greater insights into their thinking might be revealed by the latter (Eggert et al., 2017).

Concept maps and other aspects of systems thinking can be in turn assessed with a rubric, such as the one developed by Karaarslan and Teksöz (2019, 2020). In their studies, they conducted a comprehensive review of the literature to define systems thinking and then identified themes and associated skills. Each skill area has criteria to help assess the degree of mastery. Ateşkan and Lane (2018) adapted this rubric to assess learner concept maps in their study. The rubric was used to determine if the maps effectively identified components of a system and also connections among them and revealed understanding of more subtle connections and patterns (i.e., hidden dimensions [Assaraf & Orion, 2010]).

There are myriad ways to help students develop critical thinking skills. Formative assessments can be undertaken during instruction to monitor how students are refining and enhancing their ability to critique and question. Researchers have used a variety of tools to assess students' critical thinking as part of sustainability education. Ernst and Monroe (2004) used commercial assessments such as those available through Cornell University to examine if students' critical thinking changed after being involved in educational experiences that promoted sustainability. Heinrich et al. (2015) adapted critical thinking rubrics to explore how experiential learning strategies promoted students' ability to plan, implement, and evaluate their action projects.

One strategy for developing and assessing critical thinking skills is through debates (Healey, 2012; Narmaditya & Omar, 2019). This strategy also helps students build their argumentation skills and practice making and analyzing persuasive speeches (Iman, 2017). Proulx (2004) describes how students can develop scientific reasoning through debates. His main point is that as part of making and analyzing claims, students need to use critical thinking. He recommends that debates can motivate students to research and present their arguments. Kennedy (2007) provides an overview of different types of debates

and how they can be conducted in classrooms. Through her review of the literature, she also provides recommendations for assessing debates. She notes that teachers can look for how clearly students state their argument, the extent to which they cite recent and relevant resources, and how they address questions from counter arguments. The students should avoid overgeneralizations and oversimplification. She says teachers can also look for debating styles, such as eye contact and voice modulation. Teamwork and collaboration may also be part of the assessment criteria. She notes that students can also self-evaluate and assess their peers. As with most any assessment, it is important for students to know the criteria in advance to guide them through their planning and presentations.

17.2.2 The Heart

Knowing and thinking about sustainability is important. However, studies have shown that passion and caring about nature, society, and the future are essential drivers for willingness to be a constructive actor for a sustainable world (Chawla, 1999; Mobley et al., 2010; Osbaldiston & Sheldon, 2003; Vaske & Kobrin, 2001). There are many aspects of this affective domain of sustainability literacy. Often, they are referred to as attitudes or, more recently, dispositions. Researchers also discuss the importance of empathy and caring for sustainability literacy. Communication may be included in this domain as it is involved in person-to-person understanding and connections. Another reason for including communication is that it provides researchers with tangible and measurable means to assess dispositions and empathy. Given that one's dispositions and attitudes are personal and internal, it may be challenging to ascertain how another person thinks or feels until they say or do something. Therefore, a common way to assess aspects of the affective domain is through self-reporting instruments, such as questionnaires and opinionnaires. Many of the comprehensive sustainability literacy assessments discussed above include items where respondents share their opinions about topics on a scale. Recently, researchers have applied personality assessment tools that measure emotional intelligence to sustainability literacy (Di Fabio & Saklofske, 2019; Callea et al., 2019). Estrada et al. (2021) designed an instrument to explore relationships among emotional intelligence, compassion, engagement, and academic performance. While they did not find a direct relationship between emotional intelligence and academic achievement, their study confirmed the importance of compassion and engagement as mediators between emotional intelligence and learning. Therefore, educators can use assessment tools such as these available through the literature to explore how students report managing their emotions, which may be a contributor to their social well-being and academic performance. A shortcoming of these self-reporting instruments is that they rely on the honesty and intention of the respondent. Therefore, researchers and educators often apply other strategies to gain insights into this affective domain.

Similar to responding to a survey, people can participate in interviews to share their beliefs and attitudes about sustainability. Interview questions may be like written questions, but the oral format allows the participants to discuss the responses more extensively. Researchers can also explore different interviewing formats, such as focus groups and group interviews. One challenge with interviewing is to avoid biasing or influencing the responses. The questioner needs to make sure the questions are neutral and do not lead the respondent to answer one way or another. Interviewers should also take care to keep their tone of voice neutral while still showing interest and support for the participants. There are a variety of resources available to guide in the development, implementation, and analysis of interview questions (Minichiello et al., 2008; Merriam & Tisdell, 2015; Tracy, 2019).

The previous methods involve asking participants to express, share, and reflect on their attitudes and actions. Sustainability literacy can also be assessed "real time." People often reveal their dispositions and personality traits through their behaviors; therefore, another assessment strategy that can be used is observations. Teachers can develop a tool – such as a rubric or a guide – to monitor behaviors they observe. In the literature, measurements of observation are often conducted in behavioral science and environmental psychology. In 2003, Osbaldiston and Sheldon conducted a meta-analysis to promote and assess responsible behaviors toward the environment. These studies often involve providing participants with a goal or incentive to take or change actions toward the environment. A review of the studies they analyzed will reveal a variety of instruments used to monitor changes in behavior. Tolppanen et al. (2019) conducted a case study to evaluate how a sustainability project affected students' attitudes toward environmentally responsible behavior. Their research involved students in doing a life cycle analysis of familiar projects. Their intention was to increase relevance of sustainability to their lives and encourage students to consider careers related to sustainability. Their case study involved observations, a questionnaire, and interviews. They found that while the learning experience increased students' attitudes toward responsible environmental actions, more strategies were needed for students to consider a sustainability-related career.

All of the previous strategies involve external researchers assessing others. Since dispositions and emotions are so personal, it is important to remember that self-assessments should be included in this section. Individuals can reflect upon and monitor their feelings through writing, such as in a journal. A variety of resources and guides for maintaining and assessing student journals are found in the literature (e.g., Bassot, 2020; Clarkeburn & Kettula, 2012; Thorpe, 2004). Similar to strategies used in action research, individuals can design their journal to investigate their practice and behaviors. They can set goals, make plans, record data, and analyze the results. Kemmis (2009) advocates for critical action research in support of education for sustainability. He explains that individuals can monitor their practice and seek out unsustainable behaviors. Through this process, educators:

… learn by doing; they collect data about their efforts; they consciously and self-consciously, critically and self-critically transform their ways of thinking, doing and relating in the world. They are exploring and reconstructing the practice architectures that construct their lives. (p. 472)

Through action research, writing can be used as a means of self-assessment. Generally, individuals maintain a journal that can be private and personal. In some cases, individuals wish to share some of their words in a blog or an essay. These can be analyzed by other researchers through a variety of means such as by narrative inquiry (Connelly & Clandinin, 1990; Kim, 2015). Hwang (2011) used narrative inquiry to analyze teachers' stories and to assess their curricular knowledge related to environmental education. He discusses the importance of reading and understanding the stories to insights into "sociocultural processes and school education, and ways in which culturally prevailing or dominant forms of narratives are operating, 'translated' into pedagogical modes, or even challenged" (p. 800). A key message here is the process of analyzing the written word. Through writing, individuals express their feelings, interpretations, reflections, and intentions. These attributes and others can be a focus of when inquiring into individual's narratives about sustainability.

17.2.3 The Gut

In addition to being passionate about and caring for the environment, it is important that individuals believe they have the capacity to take action to support sustainable development. Moreover, they need to believe that their actions will be worthwhile, effective, and fruitful (Bandura, 1997, 2006; Blackmore, 2016; Jackson, 2003; Reeve & Tseng, 2011). In everyday language, people talk about gut feelings, about a sense of moral duty, and about bravery and courage. There is also language related to identify self-esteem and self-confidence, and many of these terms have found their way to the literature (Bandura, 2006; Deakin Crick & Goldspink, 2014).

Historically, in environmental education several researchers explored the concept of locus of control related to citizen actions. Several studies have used and adapted an instrument developed by Smith-Sebasto and Fortner (1994) explored if participants agreed their efforts to protect the environment would have positive outcomes. In the sustainability literature, researchers are beginning to explore the role of agency and potential contribution to sustainable development (Giangrande et al., 2019; Waltner et al., 2019). Ernst et al. (2017) caution that given the self-reporting nature of these scales, it may be challenging to find positive relationship between intentions and actions. On the other hand, Fielding and Head (2012) were able to find positive relations between intentions and actions in their study. Their assessment locus of control included the following items that were correlated with other factors such as responsibility, concern, and knowledge:

- My individual actions can make a difference to the environment.
- I am only one person; I can't make a difference to the environment (reversed).
- I can influence decisions now; that will help protect the environment in the future (p. 175).

Burgos and Carnero (2020) stress the importance of promoting social responsibility in schools and note research about this attribute is missing in the literature. They adapted a software program used to by corporations to assess its employee's sense of social responsibility. The program, Measuring Attractiveness by a Categorical Evaluation Technique (MACBETH), involves participants in making decisions about situations based a set of comparisons. Burgos and Carnero (2020) used MACBETH to compare the social responsibility decisions of three cases. They found that institutions involved learners in community participation and action, including environmental activities such as school gardens and litter removal drives. They suggest that MACBETH can be used by institutions to improve the social responsibility of their learners, but acknowledge further investigations are needed to ensure its efficacy.

Other terms related to confidence and intention that have been used to discuss sustainability literacy are agency and self-efficacy. The definitions of these terms are frequently found in the works of Bandura (1989, 2006). Agency involves perceptions of one's ability to plan and to take action. As with assessments of locus of control, questionnaires about agency or self-efficacy often include items about individuals' perceptions or intentions to undertake environmentally responsible behaviors. Like the other self-reporting instruments discussed in this and other chapters in this book, the findings are limited by the honesty and transparency of the participants. Therefore, researchers often try to find other ways to assess and understand individuals' potential to take action. Similar to assessing dispositions, educators can observe the behaviors and analyze the writings of individuals to learn about their behaviors. One question that researchers should ask is how do individuals develop agency? A key way is through experience (Dunne & Edwards, 2010; Ernst et al., 2017). Therefore, to assess the "gut" of sustainability literacy, it is important to get students involved in planning, developing, and taking action. Throughout this process, educators can use formative assessment strategies to monitor students' learning experiences. These strategies, as discussed elsewhere in this chapter, involve observation tools, self-reflections, and rubrics. When students are taking action, they are physically involved in their learning. Therefore, assessment strategies involved in these types of learning are discussed in the next section.

17.2.4 Arms and Legs

The word "development" in Education for Sustainable Development connotes actual involvement in helping something improve and grow. An important part of sustainability literacy, as with environmental education, is for students to become physically involved in the process (Nation, 2008). They use their arms and legs to

canvas neighborhoods, pick up litter, and build park trails. Hungerford and Peyton (1980) labeled this type of environmental action "ecomanagement," explaining that it entails "any physical action taken by an individual or a group aimed directly at maintaining or improving existing ecosystems" (p. 149). As discussed in some chapters in this book, students can help improve the physical attributes of their community through learning activities called, among other things, project-based learning, performance-based learning, and service- or community-based learning.

Guidebooks on project-based learning (PBL) emphasize the importance of careful planning and implementation. Students need to have thoughtful, meaningful, and purposeful reasons for their actions. The activities should be linked to the curriculum and help students achieve subject area objectives. Therefore, another critical consideration for PBL is to ensure students are learning. Similar to other assessments described for sustainability literacy, strategies such as journals, rubrics, and observation tools can be used to monitor changes in students' thoughts and behaviors. There are a variety of performance-based assessments that teachers can use in PBL (e.g., Berman, 2008; Brennan et al., 2021; Gallavan, 2008; Stanely, 2019). Several researchers have conducted project-based learning with students and provided information on how they assessed student learning. A few of these studies are described below.

Bielefeldt et al. (2010) discuss the importance of service learning in environmental engineering. Their paper points out the benefits of project-based learning, including sustainability. In addition to sustainability, they identify outcomes of PBL such as critical thinking, self-efficacy, and increased knowledge (p. 537). Their paper provides several ways to assess these outcomes They suggest that ethnographic methods can be used to assess student journals. They note that qualitative data analysis using codes and themes can capture aspects of students' attitudes and dispositions. They caution that while insights into students' perceptions can be gained by analyzing a student reflection, assessment of multiple student entries will be needed to see changes over time. They note that in some cases students undertake a project in a community or culture different from their own. They recommended using assessment tools that measure changes in cultural sensitivity and awareness to learn if students' attitudes are changing. Their study includes a variety of other assessment tools that can be used to measure changes in leadership, critical thinking, and self-efficacy.

Another way to gain insights into students' learning during PBL is to ask them. Alacapınar (2008) describes how she interviewed participants in a community project to assess their learning. She conducted an experimental study comparing a group of fifth grade students who participated in PBL as part of their schooling to those who do not. She provides information about careful interviewing strategies and data analysis. She found that students' creativity increased through the project, and the interviews revealed student interest and willingness to work collaboratively.

As with other action projects, multiple measures will provide more comprehensive insights into students' learning during PBL. Doppelt (2003) used interviews in his study that involved Israeli high school students in PBL. He also describes the observation process he used to assess student learning. The students maintained a portfolio and responded to questionnaires. His study involved low-performing students, and he concluded that through the experience, students became more motivated to become involved and to become more responsible for their learning.

There are some recent studies of PBL in the sustainability literature (Khandakar et al., 2020; Kricsfalusy et al., 2018; Nation, 2008). Chang et al. (2018) describe an assessment tool they developed to measure the effectiveness of PBL undertaken by university students in a technology course. Their scales included measures of engagement (flow experience perception), self-efficacy, product evaluation, and learning motivation. They used t-tests and regression analysis to interpret the data. Their study showed that the four scales could be used to predict learning effectiveness.

There are other ways students become physically involved in their learning. Practical work in laboratory experiments can be assessed as part of their science classes. The means used to monitor students' lab safety and responsibility and following directions are relevant to sustainability literacy. Their lab reports need to meet the assignment criteria and be thoughtfully written. When students conduct field work and place-based education activities that take their learning outside the classroom and into the community, their work can be assessed through rubrics and traditional class exams.

Students' creativity can be developed and assessed through design classes. They can be challenged to improve the efficiency of common household appliances and living spaces. They can create recipes that use only local foods and envision a restaurant that partners with community farmers. When positively and constructively planned and implemented, contests and competitions are another way to measure student achievements. There are limitless opportunities for students to enter their work in contests to be judged. There are photography contests, mathematics quizzes, and debates where students can develop and display skills that are important for sustainability literacy. Mentzer and Becker (2009) measured student motivation during a competition where students design and race a full-sized electric vehicle. While they had mixed results as far as student achievement, they did find that students who struggled academically became more engaged in learning. There were also varying results in the study by Dadach (2013). Nevertheless, his work provides multiple measures that can be used to assess student learning when they actively engaged in a learning activity. In addition to traditional tests and measurements, they also suggest students can write an essay and provide a presentation. There were some students who did improve their grade point average; therefore, further research is needed to explore possible relationships between active learning and academic achievement. Studies do find that students are more engaged and motivated through hands-on projects; these motivations can lead to interest, caring, dispositions, and actions that are important for sustainability literacy.

17.3 Conclusion: Keeping the Whole Body in Mind

This chapter provided assessment ideas to monitor changes in students' knowledge, skills, and dispositions related to sustainability literacy during and after a variety of learning activities. The chapter began with a "whole body" approach to assessing students' sustainability literacy. A focal idea to this assessment involved students maintaining a portfolio. This portfolio can be organized in a variety of ways; one idea is to have a separate file for each of the goals of sustainable development. As educators develop learning experiences for students, they can keep these goals in mind. They can help students to address goals related to land, water, hunger, and equity. By remembering the body parts involved in learning, they can ensure students have comprehensive, holistic learning experiences that involve them in sustainable development. The assessment ideas provided for each of the body parts listed above can be incorporated into this portfolio. As students are responsible for maintaining their portfolio, they can ensure they keep a record of the evidence of their learning. As they review their portfolio, they can create a resume or brochure that highlights their key achievements. In this way, when they are asked what "certifies" them to be identified as sustainability literate, they can present their case, evidence that they have become involved in sustainable development in heart, mind, and soul.

References

Alacapınar, F. (2008). Effectiveness of project-based learning. *Eurasian Journal of Educational Research, 32*, 17–34.

Assaraf, O., & Orion, N. (2010). Four case studies six years later: Developing systems thinking skills in junior high school and sustaining them over time. *Journal of Research in Science Teaching, 47*(10), 1253–1280.

Ateşkan, A., & Lane, F. J. (2018). Assessing teachers' systems thinking skills during a professional development program. *Journal of Cleaner Production, 172,* 4338–4356.

Bandura, A. (1989). Human agency in social cognitive theory. *American Psychologist, 44*, 1175–1184.

Bandura, A. (1997). *Self-efficacy: The exercise of control.* Freeman.

Bandura, A. (2006). Toward a psychology of human agency. *Perspectives on Psychological Science, 1*(2), 164–180.

Bassot, B. (2020). *The reflective journal.* Bloomsbury Publishing.

Berman, S. (2008). *Performance-based learning: Aligning experiential tasks and assessment to increase learning.* Corwin Press.

Bielefeldt, A. R., Paterson, K. G., & Swan, C. W. (2010). Measuring the value added from service learning in project-based engineering education. *International Journal of Engineering Education, 26*(3), 535–546.

Black, P., Harrison, C., Lee, C., Marshall, B., & Wiliam, D. (2004). Working inside the black box: Assessment for learning in the classroom. *Phi Delta Kappan, 86*(1), 8–21.

Blackmore, C. (2016). Towards a pedagogical framework for global citizenship education. *International Journal of Development Education and Global Learning, 8*(1), 39–56.

Brennan, K., Blum-Smith, S., & Haduong, P. (2021). Four principles for assessing student-directed projects. *Phi Delta Kappan, 103*(4), 44–48.

Burgos, J., & Carnero, M. C. (2020). Assessment of social responsibility in education in secondary schools. *Sustainability, 12*(4849), 1–38.

Callea, A., De Rosa, D., Ferri, G., Lipari, F., & Costanzi, M. (2019). Are more intelligent people happier? Emotional intelligence as mediator between need for relatedness, happiness and flourishing. *Sustainability, 11*(4), 1022.

Chang, C. C., Kuo, C. G., & Chang, Y. H. (2018). An assessment tool predicts learning effectiveness for project-based learning in enhancing education of sustainability. *Sustainability, 10*(10), 3595.

Chao, Y. L., & Lam, S. P. (2011). Measuring responsible environmental behavior: Self-reported and other-reported measures and their differences in testing a behavioral model. *Environment and Behavior, 43*(1), 53–71.

Chawla, L. (1999). Life paths into effective environmental action. *The Journal of Environmental Education, 31*(1), 15–26.

Clarkeburn, H., & Kettula, K. (2012). Fairness and using reflective journals in assessment. *Teaching in Higher Education, 17*(4), 439–452.

Connelly, F. M., & Clandinin, D. J. (1990). Stories of experience and narrative inquiry. *Educational Researcher, 19*(5), 2–14.

Dadach, Z. E. (2013). Quantifying the effects of an active learning strategy on the motivation of students. *International Journal of Engineering Education, 29*(4), 1–10.

Deakin Crick, R., & Goldspink, C. (2014). Learner dispositions, self-theories and student engagement. *British Journal of Educational Studies, 62*(1), 19–35.

Di Fabio, A., & Saklofske, D. H. (2019). Positive relational management for sustainable development: Beyond personality traits. The contribution of emotional intelligence. *Sustainability, 11*(2), 330.

Doppelt, Y. (2003). Implementation and assessment of project-based learning in a flexible environment. *International Journal of Technology and Design Education, 13*(3), 255–272.

Dunlap, R. E. (2008). The new ecological paradigm scale: From marginality to worldwide use. *Journal of Environmental Education, 40*(1), 3–18.

Dunne, S., & Edwards, J. (2010). International schools as sites of social change. *Journal of Research in International Education, 9*(1), 24–39.

Eggert, S., Nitsch, A., Boone, W. J., Nückles, M., & Bögeholz, S. (2017). Supporting students' learning and socio-scientific reasoning about climate change—The effect of computer-based concept mapping scaffolds. *Research in Science Education, 47*(1), 137–159.

Ernst, J., & Monroe, M. (2004). The effects of environment-based education on students' critical thinking skills and disposition toward critical thinking. *Environmental Education Research, 10*(4), 507–522.

Ernst, J., Blood, N., & Beery, T. (2017). Environmental action and student environmental leaders: Exploring the influence of environmental attitudes, locus of control, and sense of personal responsibility. *Environmental Education Research, 23*(2), 149–175.

Estrada, M., Monferrer, D., Rodríguez, A., & Moliner, M. Á. (2021). Does emotional intelligence influence academic performance? The role of compassion and engagement in education for sustainable development. *Sustainability, 13*(1721), 1–18.

Fielding, K. S., & Head, B. W. (2012). Determinants of young Australians' environmental actions: The role of responsibility attributions, locus of control, knowledge and attitudes. *Environmental Education Research, 18*(2), 171–186.

Gallavan, K. (Ed.). (2008). *Developing performance-based assessments, grades 6–12.* Corwin Press.

Giangrande, N., White, R. M., East, M., Jackson, R., Clarke, T., Saloff Coste, M., & Penha-Lopes, G. (2019). A competency framework to assess and activate education for sustainable development: Addressing the UN sustainable development goals 4.7 challenge. *Sustainability, 11*(2832), 1–16.

Healey, R. L. (2012). The power of debate: Reflections on the potential of debates for engaging students in critical thinking about controversial geographical topics. *Journal of Geography in Higher Education, 36*(2), 239–257.

Heinrich, W. F., Habron, G. B., Johnson, H. L., & Goralnik, L. (2015). Critical thinking assessment across four sustainability-related experiential learning settings. *The Journal of Experimental Education, 38*(4), 373–393.

Hoppe, T., Renkl, A., Seidel, T., Rettig, S., & Riess, W. (2020). Exploring how teachers diagnose student conceptions about the cycle of matter. *Sustainability, 12*(4184), 1–15.

Hungerford, H. R., & Peyton, R. B. (1980). A paradigm for citizen responsibility: Environmental action. In A. B. Sacks, L. L. Burris-Bammel, C. B. Davis, & L. A. Iozzi (Eds.), *Current issues VI: The yearbook of environmental education and environmental studies* (pp. 173–192). Educational Resources Information Center and National Association for Environmental Education.

Hwang, S. (2011). Narrative inquiry for science education: Teachers' repertoire-making in the case of environmental curriculum. *International Journal of Science Education, 33*(6), 797–816.

Iman, J. N. (2017). Debate instruction in EFL classroom: Impacts on the critical thinking and speaking skill. *International Journal of Instruction, 10*(4), 87–108.

International Baccalaureate Organization. (n.d.-a). *Creativity, activity, service* [website]. https://ibo.org/programmes/diploma-programme/curriculum/creativity-activity-and-service/. Accessed 11 Apr 2021.

International Baccalaureate Organization. (n.d.-b). *Extended essay* [website]. https://ibo.org/programmes/diploma-programme/curriculum/extended-essay/. Accessed 11 Apr 2021.

Jackson, D. B. (2003). Education reform as if student agency mattered: Academic microcultures and student identity. *Phi Delta Kappan, 84*(8), 579–585.

Karaarslan-Semiz, G., & Teksöz, G. (2019). Sistemsel düşünme becerilerinin tanımlanması, ölçülmesi ve değerlendirilmesi üzerine bir çalışma: Kavram haritaları [A study on identifying, measuring and evaluating systems thinking Skills: Concept maps]. *Başkent University Journal of Education, 6*(1), 111–126.

Karaarslan-Semiz, G., & Teksöz, G. (2020). Developing the systems thinking skills of pre-service science teachers through an outdoor ESD course. *Journal of Adventure Education and Outdoor Learning, 20*(4), 337–356.

Kemmis, S. (2009). Action research as a practice-based practice. *Educational Action Research, 17*(3), 463–474.

Kennedy, R. (2007). In-class debates: Fertile ground for active learning and the cultivation of critical thinking and oral communication skills. *International Journal of Teaching & Learning in Higher Education, 19*(2), 183–190.

Khandakar, A., Chowdhury, M. E. H., Gonzales, A., Jr., Pedro, S., Touati, F., Emadi, N. A., & Ayari, M. A. (2020). Case study to analyze the impact of multi-course project-based learning approach on education for sustainable development. *Sustainability, 12*(2), 480.

Kim, J. H. (2015). *Understanding narrative inquiry: The crafting and analysis of stories as research*. Sage Publications.

Kricsfalusy, V., George, C., & Reed, M. G. (2018). Integrating problem-and project-based learning opportunities: Assessing outcomes of a field course in environment and sustainability. *Environmental Education Research, 24*(4), 593–610.

Lange, F., & Dewitte, S. (2019). Measuring pro-environmental behavior: Review and recommendations. *Journal of Environmental Psychology, 63*, 92–100.

Leander, K., & Boldt, G. (2013). Rereading "A pedagogy of multiliteracies" bodies, texts, and emergence. *Journal of Literacy Research, 45*(1), 22–46.

Maurer, M., Koulouris, P., & Bogner, F. X. (2020). Green awareness in action—How energy conservation action forces on environmental knowledge, values and behaviour in adolescents' school life. *Sustainability, 12*(955), 1–15.

McBeth, W., & Volk, T. L. (2009). The national environmental literacy project: A baseline study of middle grade students in the United States. *The Journal of Environmental Education, 41*(1), 55–67.

Mentzer, N., & Becker, K. (2009). Motivation while designing in engineering and technology education impacted by academic preparation. *Journal of STEM Teacher Education, 46*(3), 90–112.

Merriam, S. B., & Tisdell, E. J. (2015). *Qualitative research: A guide to design and implementation*. John Wiley & Sons.

Minichiello, V., Aroni, R., & Hays, T. (2008). *In-depth interviewing*. Pearson/Prentice Hall.

Ministry of National Education (MoNE). (2014). *Milli Eğitim Bakanlığı ortaöğretim kurumları yönetmeliği [Ministry of National Education Regulation on Secondary Education Institutions]*. http://mevzuat.meb.gov.tr/html/ortaogrkurumyon_0/ortaogrkurumyon_1.html. Accessed 14 Mar 2021.

Mobley, C., Vagias, W. M., & DeWard, S. L. (2010). Exploring additional determinants of environmentally responsible behavior: The influence of environmental literature and environmental attitudes. *Environment and Behavior, 42*(4), 420–447.

Narmaditya, B. S., & Omar, I. M. B. (2019). Debate-based learning and its impact on students' critical thinking skills. *Classroom Action Research Journal, 3*(1), 1–7.

Nation, M. L. (2008). Project-based learning for sustainable development. *Journal of Geography, 107*(3), 102–111.

Nolet, V. (2009). Preparing sustainability-literate teachers. *Teachers College Record, 111*(2), 409–442.

Osbaldiston, R., & Sheldon, K. M. (2003). Promoting internalized motivation for environmentally responsible behavior: A prospective study of environmental goals. *Journal of Environmental Psychology, 23*(4), 349–357.

Proulx, G. (2004). Integrating scientific method & critical thinking in classroom debates on environmental issues. *The American Biology Teacher, 66*(1), 26–33.

Ramsey, J., Hungerford, H., & Volk, T. (1989). A technique for analyzing environmental issues. *Journal of Environmental Education, 21*(1), 26–30.

Ramsey, J., Hungerford, H., & Volk, T. (1992). Environmental education in the K-12 curriculum: Finding a niche. *Journal of Environmental Education, 23*(2), 35–45.

Reeve, J., & Tseng, C. M. (2011). Agency as a fourth aspect of students' engagement during learning activities. *Contemporary Educational Psychology, 36*(4), 257–267.

Schneiderhan-Opel, J., & Bogner, F. X. (2019). Between environmental utilization and protection: Adolescent conceptions of biodiversity. *Sustainability, 11*(4517), 1–14.

Smith-Sebasto, N. J., & Fortner, R. W. (1994). The environmental action internal control index. *The Journal of Environmental Education, 25*(4), 23–29.

Stanely, T. (2019). *Using rubrics for performance-based assessment: A practical guide to evaluating student work*. Prufrock Press.

Thorpe, K. (2004). Reflective learning journals: From concept to practice. *Reflective Practice, 5*(3), 327–343.

Tolppanen, S., Jäppinen, I., Kärkkäinen, S., Salonen, A., & Keinonen, T. (2019). Relevance of life-cycle assessment in context-based science education: A case study in lower secondary school. *Sustainability, 11*(5877), 1–15.

Tracy, S. J. (2019). *Qualitative research methods: Collecting evidence, crafting analysis, communicating impact*. John Wiley & Sons.

Vaske, J. J., & Kobrin, K. C. (2001). Place attachment and environmentally responsible behavior. *The Journal of Environmental Education, 32*(4), 16–21.

Waltner, E. M., Rieß, W., & Mischo, C. (2019). Development and validation of an instrument for measuring student sustainability competencies. *Sustainability, 11*(1717), 1–20.

Jennie Farber Lane is an associate professor in the Graduate School of Education at Bilkent University. She received her PhD in curriculum and instruction from UW-Madison. Prior to coming to Bilkent, she was the director of the Wisconsin K-12 Energy Education Program (KEEP). Her other work experiences include co-authoring the *Project WET Curriculum and Activity Guide*, which is used throughout the world, teaching public school in New York City and Lewiston, and instructing pre-service teachers in Thailand and at the University of Wisconsin-Stevens Point. Her research areas include environmental, place-based, and sustainability education.

Armağan Ateşkan is an assistant professor in the Graduate School of Education at Bilkent University. She has an MA in biology teacher education from Bilkent University, and a PhD in computer education and instructional technologies from Middle East Technical University. Dr. Ateskan has been working as an instructor in the Graduate School of Education, Bilkent University, Türkiye, since 2002. In addition, she has worked at a high school in Türkiye as a biology teacher and at an international school in Belgium as an educational technologist. She directed several international and national projects related to IB DP, IB MYP, and environmental education.

Index

A

Action competence, 8, 10, 25, 26, 29–30, 35, 58, 59, 156–158, 162, 164
Air pollution, 68, 73, 75, 79–88, 90
Anthropocene, 57, 108, 167
Aspiration, 21, 124, 126, 134
Assessment, 5, 8–10, 66, 68, 69, 168, 181, 182, 184–189, 192, 195–198, 206, 209, 212, 214, 215, 222–227, 234, 235, 237–242, 248–256
Assessment of learning, 8, 69, 185, 186, 222, 249
Authentic learning, 55, 56, 64, 158

B

Biomimicry, 5, 7, 129–134

C

Climate change education, 41, 168, 171
Climate crisis and criticality, 76
Competence assessment, 196, 197
Competences, 5–10, 27, 29–32, 43, 55–59, 64–66, 70, 120, 139, 155, 157, 181–187, 191–198, 208, 222–224, 233–237, 240–242, 244
Cooperative learning, 98–102

D

Democracy in education, 5, 26, 29, 31

E

Education for sustainability, 6, 19, 39–51, 77, 96, 98, 108, 111–113, 252
Education for sustainable development (ESD), 3–10, 13–21, 25, 26, 28–35, 40, 42–45, 47, 51, 76–78, 84–86, 88, 90, 94–105, 108, 111, 120, 121, 126, 130, 133, 134, 153–157, 161–164, 167–171, 175, 176, 181–189, 192–195, 197–198, 206–209, 212–216, 221–225, 241, 242, 254
Education for sustainable development in primary schools, 168
Empathy with nature, 113, 114

E

Enquiry-based learning, 64
Environmental justice, 5, 7, 75–90
Environmental philosophies, 28

F

Functions of education, 31, 36

H

Holistic self-sufficiency, 139

I

Implementation, 7, 9, 20, 35, 94, 95, 105, 108, 138, 139, 153–164, 172, 175, 192, 207, 214–216, 222, 223, 241, 252, 254
Indicators, 104, 110, 146, 187, 205–216, 222

L

Lesson plan, 83, 95, 174
Literacy, 9, 43, 78, 95, 197, 222, 239, 240, 247–256

M

Measurement, 8, 9, 47, 49, 88, 104, 105, 185, 186, 196–198, 206–209, 212–215, 225, 250, 252, 255

N

Nature-embedded learning, 124

O

Outcome, 6–10, 15, 18–21, 30, 46, 61, 62, 70, 81, 95, 101, 121, 139, 155, 161, 162, 169–171, 175–176, 182–188, 192, 196, 197, 205–216, 221–242, 249–250, 253, 254
Outdoor education, 39–51, 114, 138–150, 188
Outdoor education environment, 50, 150
Outdoor learning, 7, 40, 41, 44–46, 48, 51, 70, 138–141, 145, 149–150, 168, 172

P
Polygon Dole, 138–141, 150
Portfolio, 247, 249, 255, 256
Primary and secondary schools, 4, 5, 9, 40, 41, 46,
    221–242, 248
Problem-based learning, 48, 97
Purpose of education, 31

R
A Rounder Sense of Purpose (RSP), 8, 183–185, 187, 195
Rubric, 64, 66, 69, 250–252, 254, 255

S
School leadership, 7, 155, 160
School organization, 160–164
Selective traditions, 27–29, 32–35
STEM, 64, 67, 90, 97, 156
Sustainability, 4–9, 17–21, 30, 31, 35, 36, 39–47, 50, 51,
    55–70, 75, 76, 86, 90, 95, 96, 98–101, 105, 112,
    115, 119–121, 124, 128, 129, 134, 138–150,
    153–159, 162–164, 168, 169, 171, 175, 183–188,
    191–198, 205–216, 221–242, 247–256
Sustainability competences, 5, 7–9, 55, 64, 187, 192,
    193, 195, 197, 198, 222, 223, 242

Sustainable development, 4, 6, 8, 18–21, 26, 30, 46, 58,
    64, 69, 76, 95–97, 111, 112, 120, 128, 129, 155,
    156, 161, 171, 172, 182–184, 192, 194, 195, 197,
    206, 209, 222, 248, 249, 253, 256
Sustainable development goals (SDGs), 4, 5, 7, 13, 20,
    21, 58, 59, 64, 74–77, 80, 82, 95, 97–101, 103,
    105, 119, 121, 124, 183, 191, 192, 206, 207, 215,
    216, 221, 222, 225, 227, 233–236, 239
Systems thinking, 7, 10, 39–51, 65, 97, 193, 197, 222,
    233, 235, 236, 240, 248, 250, 251

T
Teacher education, 9
Teaching traditions, 25–36, 154

W
Whole-school approach, 7, 8, 77, 153–164, 168,
    171–175

Y
Young people, 4, 7, 17, 20, 21, 30, 36, 40, 45, 46, 51, 58,
    62, 64, 69, 74, 78, 88, 90, 111, 112, 115, 125,
    126, 138, 168, 169, 182–184, 233

Milton Keynes UK
Ingram Content Group UK Ltd.
UKHW051906251023
431302UK00004B/17

9 783031 091148